SHIZUHIKO NISHISATO is a member of the Department of Measurement, Evaluation and Computer Applications at the Ontario Institute for Studies in Education (OISE) and a member of the Department of Educational Theory at the University of Toronto.

This volume presents a unified and up-to-date account of the theory and methods of applying one of the most useful and widely applicable techniques of data analysis, 'dual scaling.' It addresses issues of interest to a wide variety of researchers concerned with data that are categorical in nature or by design: in the life sciences, the social sciences, and statistics.

The eight chapters introduce the nature of categorical data and the concept of dual scaling and present the applications of dual scaling to different forms of categorical data: the contingency table, the response-frequency table, the response-pattern table for multiple-choice data, ranking and paired comparison data, multidimensional tables, partially ordered and successively ordered categories, and incomplete data. The book also includes appendices outlining a minimum package of matrix calculus and a small FORTRAN program.

Clear, concise, and comprehensive, *Analysis of Categorical Data* will be a useful textbook or handbook for students and researchers in a variety of fields.

MATHEMATICAL EXPOSITIONS

Volumes Published

1 *The Foundations of Geometry* G. DE B. ROBINSON
2 *Non-Euclidean Geometry* H.S.M. COXETER
3 *The Theory of Potential and Spherical Harmonics* W.J. STERNBERG and T.L. SMITH
4 *The Variational Principles of Mechanics* CORNELIUS LANCZOS
5 *Tensor Calculus* J.L. SYNGE and A.E. SCHILD (out of print)
6 *The Theory of Functions of a Real Variable* R.L. JEFFERY (out of print)
7 *General Topology* WACLAW SIERPINSKI
(translated by C. CECILIA KRIEGER) (out of print)
8 *Bernstein Polynomials* G.G. LORENTZ (out of print)
9 *Partial Differential Equations* G.F.D. DUFF
10 *Variational Methods for Eigenvalue Problems* S.H. GOULD
11 *Differential Geometry* ERWIN KREYSZIG (out of print)
12 *Representation Theory of the Symmetric Group* G. DE B. ROBINSON
13 *Geometry of Complex Numbers* HANS SCHWERDTFEGER
14 *Rings and Radicals* N.J. DIVINSKY
15 *Connectivity in Graphs* W.T. TUTTE
16 *Introduction to Differential Geometry and Riemannian Geometry* ERWIN KREYSZIG
17 *Mathematical Theory of Dislocations and Fracture* R.W. LARDNER
18 *n-gons* FRIEDRICH BACHMANN and ECKART SCHMIDT
(translated by CYRIL W.L. GARNER)
19 *Weighing Evidence in Language and Literature: A Statistical Approach*
BARRON BRAINERD
20 *Rudiments of Plane Affine Geometry* P. SCHERK and R. LINGENBERG
21 *The Collected Papers of Alfred Young, 1873–1940*
22 *From Physical Concept to Mathematical Structure. An Introduction to Theoretical Physics* LYNN E.H. TRAINOR and MARK B. WISE
23 *Lattice Path Combinatorics with Statistical Applications* T.V. NARAYANA
24 *Analysis of Categorical Data: Dual Scaling and Its Applications*
SHIZUHIKO NISHISATO

MATHEMATICAL EXPOSITIONS NO. 24

Analysis of categorical data: dual scaling and its applications

SHIZUHIKO NISHISATO

UNIVERSITY OF TORONTO PRESS
Toronto Buffalo London

Toronto Buffalo London
Printed in Canada

ISBN 0–8020–5489–7

Canadian Cataloguing in Publication Data

Nishisato, Shizuhiko, 1935–
Analysis of categorical data

(Mathematical expositions; 24 ISSN 0076–5333)
Bibliography: p.
Includes index.

ISBN 0–8020–5489–7

1. Multivariate analysis. 2. Matrices.
I. Title. II. Series.

QA278.N57 519.5′3 C80–094407–0

To LORRAINE and IRA

Contents

*Chapter 3 is optional. The reader may skip it first time through the text and return to it after looking at some of the applications discussed in the ensuing chapters.

Preface

The purpose of this book is, primarily, to present an up-to-date account of the theory and method of *dual scaling*, a 'neglected multivariate technique' (Hill 1974), in a simple and unified form. During some 47 years of its history, dual scaling has been given various other names by people in different disciplines, and has taken a number of directions in its development. Although this book does not cover all developments and variants of the technique, it attempts to explain the underlying theory from a single focal point. For the book I have adopted the new and relatively neutral name 'dual scaling' to avoid further confusion among readers in diverse areas of specialty. If successful, the book will demonstrate that dual scaling is one of the most useful, widely applicable techniques of data analysis.

Many numerical examples are included to illustrate derivations of particular formulas and applications of certain methods. In addition, a minimum package of matrix calculus is given in an appendix for those who may have no knowledge of the subject. For practical reasons, a small computer program, called DUAL1, is also included in an appendix. The inclusion of the program was motivated primarily by my observation of instances in which new methods were successfully put into practical use with the aid of effective computer programs (e.g., multidimensional scaling, cluster analysis); it may be relatively easy to recognize merits of a method, but to put it into routine use is a different matter.

The book was written for students, teachers, and researchers, with or without much mathematical training, in business, industry, and such disciplines as education, psychology, sociology, anthropology, botany, ecology, zoology, medicine, pharmacology, epidemiology, criminology, and statistics. In these areas, one often collects so-called categorical data, that is, observations recorded in terms of pre-specified categories such as 'excellent, good, poor,' 'always, often, sometimes, rarely, never,' 'yes, no,' 'male, female,' 'alive, moribund, dead,' and 'too effective, just effective, not effective.' Because of the ubiquity of categorical data, there has been a persistent interest in their quantification and analysis. Dual scaling is an approach to the problem, and offers a rigorous, effective, and versatile tool for elucidating the information contained in categorical data.

The author has been engaged in research on optimal scaling (another name for dual scaling) since 1969, primarily its generalization to accommodate different types of categorical data. The book represents a summary of the work during that period, and is mainly based on five technical reports on the Measurement and Evaluation of Categorical Data (Nishisato 1972b, 1973a, 1975, 1976, 1979b).

Many people have contributed to the publication of this book. I am greatly indebted to Dr R. Darrell Bock, who introduced me to a variety of scaling techniques, including the technique he christened 'optimal scaling,' while I was a graduate student at the University of North Carolina at Chapel Hill. He was also supervisor of my PHD research, which involved developing a log linear model. I am indebted to Dr Lyle V. Jones, the director of the Psychometric Laboratory; Dr Emir H. Shuford, my first advisor at Chapel Hill; and Dr Masanao Toda, my first teacher in mathematical psychology at Hokkaido University, Sapporo, Japan. At the Ontario Institute for Studies in Education and the University of Toronto, I am most indebted to Dr Roderick P. McDonald for many hours of stimulating discussion and Mr Richard G. Wolfe for his valuable comments on my work. I have also benefited from discussions and association with other colleagues, including Dr Ross E. Traub, Dr Leslie D. McLean, the late Dr John C. Ogilvie, the late Dr Robert W.B. Jackson, Dr Shmuel Avital, Dr Raghu P. Bhargava, Dr Clifford M. Christensen, Mr Prince S. Arri, Mr Colin Fraser, Mrs Muriel Fung, and Mr Richard Levine. I am also grateful for the help of many graduate students, in particular Yukihiko Torii, Yukio Inukai, Tomoichi Ishizuka, Maria Svoboda, Gwyneth M. Boodoo, Wan-Pui Poon, Hirotsugu Yamauchi, Wen-Jenn Sheu, and Thomas P. Howley. Special thanks are due to Dr Kuo-Sing Leong, former student and now director of Lee Kong Chian Computer Center, Nanyang University, Singapore, for his contributions to OPSCALE, a versatile computer program for dual scaling, which he, Mr R.G. Wolfe, and I developed. In 1978 the work was used as a text for a course in the Department of Educational Theory at the University of Toronto, and many modifications were made in response to criticisms from students Diana Oi-Heung Chan, Jamshid Etezadi, Wayne Johnstone, Hau Lei, Dalia Rachman, Wen-Jenn Sheu, and Pamela Young, as well as from Dr Meyer W. Starr of the University of Windsor, who was then on leave in Toronto.

I have also received helpful suggestions, encouragement, and constructive criticisms from many other people, to name a few, Dr Yoshio Takane, Dr Jan de Leeuw, Dr Forrest W. Young, Dr Michael J. Greenacre, Dr Joseph B. Kruskal, Dr J. Douglas Carroll, Dr Bert F. Green, Jr., Dr Michael Browne, Dr Gilbert Saporta, Dr Chikio Hayashi, and Professor Takayuki Saito. At the 1976 annual meeting of the Psychometric Society, a symposium on optimal scaling was held, at which Drs F.W. Young, J. de Leeuw, G. Saporta, and I were speakers and Dr J.B. Kruskal was the discussant. During the discussion period, Dr Joseph L. Zinnes posed a question about the adequacy of the name 'optimal scaling.' This eventually motivated my adopting the new name 'dual scaling' for this book. I am also greatly indebted to the two anonymous reviewers of the manuscript, whose suggestions and comments were most helpful in improving the manuscript.

I acknowledge, too, the assistance of Mrs Gloria Powell and Mrs Patricia O'Connor, who typed the drafts beautifully and with care. Some of the technical reports on which the book is mostly based were edited and typed by my wife Lorraine, and the bibliography of the final draft was carefully checked by my son Ira.

Finally, I wish to express my deep appreciation for the financial support provided by the Natural Sciences and Engineering Research Council Canada (Grant No. A4581), the Canada Council (Grants No. 73-0399 and 74-0300), the Ontario Ministry of Education (Grant No. 5012), and the Ontario Institute for Studies in Education. Publication of the book was partly supported by grants from the Natural Sciences and Engineering Research Council Canada (Scientific Publication Grant No. 618) and from the Publications Fund of University of Toronto Press.

S.N.
Toronto, Spring 1979

1 Introduction

1.1 CATEGORICAL DATA

In empirical research one often collects data in categorical form. The use of categories generally facilitates the work and helps one to retrieve information in a manageable way; in consequence, their use has become fairly common in many branches of the social and natural sciences. The categorical arrangement of data is sometimes arbitrarily imposed and sometimes arises naturally from the measurement processes. 'Socio-economic level' is an example of a variable which is intrinsically continuous but often treated as categorical, by splitting the continuous scale of annual income into three divisions: high, middle, and low. In contrast, 'sex' is a natural categorical variable which cannot be mapped onto a continuous scale. The distinction between the two types of data, however, does not alter the basic procedure of dual scaling.

In empirical studies more categorical data seem to be derived from intrinsically continuous variables than from intrinsically categorical variables, perhaps because investigators often find it easier to record data in terms of a finite number of categories than in terms of real numbers, or find that their subjects are too young or too naïve to make fine judgments and to respond in terms of a continuous scale. In the latter case, there is in fact a danger that if the difficult judgmental task is insisted upon the inability of the subjects might contaminate the data in a complex way. Expedience of data collection and consistency of response, aided by categorical representation, are often attained at the expense of information in the data, however. In other words, greater convenience in data collecting often results in less informative data. Regarding this conflict between convenience and information content, one tends to choose the former to be certain that one can collect data without too much difficulty. Such a course of action is often justified under the name of feasibility of the study. This choice of action seems to explain in part why categorical data of intrinsically continuous variables are ubiquitous in empirical research. A natural question then is how one should retrieve information from such categorical data and how much, and furthermore how one should extract quantitative information from data derived from both purely categorical and intrinsically continuous variables. Any step toward 'optimum' analysis of both types of categorical data deserves serious attention, and this consideration provides the basic motivation for developing dual scaling.

Let us consider some examples of categorical data. Typical data are obtained from responses to such questions as:

1. Have you had reactions to any of the following? (Please check) [ ] house dust, [ ] dandelion, [ ] ragweed, [ ] animal hair, [ ] egg, [ ] cigar-smoking.
2. Which of the criminals listed below do you think should (i) be given a long term of imprisonment, (ii) be heavily fined, (iii) receive psychiatric help, (iv) be rehabilitated? List as many as you think appropriate for each category, using the alphabetic codes: (a) shoplifter, (b) arsonist, (c) gambler, (d) burglar, (e) kidnapper, (f) impostor, (g) pickpocket, (h) forger.
3. How do you find meals at your cafeteria? [ ] very good, [ ] good, [ ] mediocre, [ ] not good, [ ] bad, [ ] no comment. Do you think your cafeteria's prices are [ ] too high, [ ] rather high, [ ] about average, [ ] rather low, [ ] very low, [ ] no comment?

Researchers often record their observations in terms of categories. For example:

4. The effects of five insecticides are investigated. The results are recorded in terms of the number of insects falling into each of the following categories: alive, moribund, dead.

In daily life, we also use categorical representation to summarize our observations. For instance:

5. Schoolteachers prepare reports of their students' academic standings in terms of grades A, B, C, D, E for the various subjects (english, mathematics, history, gymnastics, etc.).

In education and psychology, we often use multiple-choice tests, where subjects are asked to choose one of several alternatives (response options) for each question.

6. Where is the provincial capital of Ontario? [ ] Ottawa, [ ] Hamilton, [ ] Toronto, [ ] London.

The same format, i.e., multiple choice, is also used in many survey questionnaires.

7. Do you read magazine A? [ ] yes, [ ] no. Do you read magazine B? [ ] yes, [ ] no.

Once data are collected, they are usually summarized in tables. In this book we classify categorical representations of data into the following five groups: (a) contingency tables, (b) response-frequency tables; (c) response-pattern tables; (d) rank order tables, and (e) multidimensional tables. This classification is by no means exhaustive, but serves to identify several ways of formulating dual scaling. The types can be characterized as follows.

(a) *Contingency tables* Table 1.1 is an example of a contingency table, which presents response frequencies put in the *cross-tabulation* form of 'categories of one variable' by 'categories of another variable.' Contingency tables present data based on responses of a group of subjects to several questions, the cells representing *joint* occurrences of all the possible combinations of responses. The present example involves two questions, and may therefore be called 'two-item data,' in distinction from '*n*-item data' and 'multidimensional contingency tables' to be described later.

(b) *Response-frequency tables* Table 1.2 is an example of a response-frequency table, which is an object-by-category cross-tabulation. Response-frequency tables represent responses from *different groups* of subjects, and hence should be distinguished from contingency tables. More specifically, Table 1.2 presents data obtained from 250 subjects assigned randomly to five groups, the 50 subjects in each group being asked to classify *one* meal into one of the four response categories. This arrangement is quite different from that for a contingency table, and the difference can be explained as follows.

Suppose the data in Table 1.2 were obtained in the format of a contingency table. To obtain 250 responses, only 50 subjects are needed, for each of the 50 subjects would judge *five* meals (not one) in terms of the four response categories. Then the contingency table would be a $4\times4\times4\times4\times4$ table of joint responses. To construct a table of 1,024 ($=4^5$) cells is not practical, but what is more important is that it may be utterly meaningless to consider 1,024 cells for responses from only 50 subjects – there will be at least 974 empty

TABLE 1.1

Contingency table (categories × categories)

Do you read magazine B?	Do you read magazine A? Yes	No	Total
Yes	82	13	95
No	30	47	77
Total	112	60	172

TABLE 1.2

Response-frequency table (objects × categories)

Meals	Meals at our cafeteria taste (1)*	(2)	(3)	(4)	Total
A	5	4	31	10	50
B	0	3	15	32	50
C	23	25	2	0	50
D	2	36	11	1	50
E	4	19	21	6	50
Total	34	87	80	49	250

*(1) terrible, (2) mediocre, (3) good, (4) delicious.

cells. If the contingency table data are summarized in the form of Table 1.2, there is still a problem, however, in that it appears as though great loss of information has been incurred, namely the subjects' joint responses to the five meals.

From a practical point of view, response-frequency tables may be preferred to contingency tables. One should note, however, that response-frequency tables require a common set of response categories for all the objects, which is not required for contingency tables.

(c) *Response-pattern tables* Table 1.3 is an example of a response-pattern table, which indicates responses (1's) and non-responses (0's) in the 'subject-by-polychotomous-item option' table. Response-pattern tables are directly related to contingency tables. If one classifies all the response patterns in Table 1.3 into distinct patterns and finds frequencies for these patterns, one can convert Table 1.3 into a $2 \times 3 \times 3$ contingency table. The only difference between Table 1.3 and the new contingency table is that the subjects cannot be identified in the latter. In practice, response-pattern tables are often easier to deal with than contingency tables. For instance, the earlier example of a $4 \times 4 \times 4 \times 4 \times 4$ contingency table of responses from 50 subjects can be represented in a 50×20 response-pattern table (i.e., 50 subjects by five polychotomous items with four options per item). This table is relatively compact and contains the entire information in the data. To distinguish it from two-item data, the response-pattern table may be called '*n*-item data.' The major distinction between the contingency table and response-pattern table lies, however, in the way the data are arranged.

TABLE 1.3

Response-pattern table (subjects × polychotomous items)

	Item 1		Item 2			Item 3			
Subjects	(1)*	(2)	(3)	(4)	(5)	(6)	(7)	(8)	Total
1	1	0	0	1	0	1	0	0	3
2	1	0	1	0	0	1	0	0	3
3	0	1	0	0	1	0	1	0	3
4	1	0	0	0	1	0	0	1	3
⋮	⋮	⋮	⋮	⋮	⋮	⋮	⋮	⋮	⋮
45	0	1	0	1	0	0	1	0	3
Subtotal	30	15	11	28	6	25	18	2	135
Total	45		45			45			135

*Options: (1) yes, (2) no, (3) always, (4) sometimes, (5) never, (6) yes, (7) no, (8) no comment.

(d) *Rank order tables* A typical example of rank order data is obtained by asking N subjects to rank n objects according to the order of their preference. The data are often summarized in an $N \times n$ table (Table 1.4) with rank numbers as elements. If the data are represented in the response-pattern format, the size of the table becomes $N \times n^2$ (Table 1.5), which may not be practical. Another form of rank order data, which is very popular, is the 'object-by-rank' ($n \times n$) table (Table 1.6). Although this table is much more compact than Tables 1.4 and 1.5, it no longer contains information about individual differences in ranking, and is thus less informative.

One may wonder what the basic difference between Table 1.3 and Table 1.5 is. Though they look quite similar they differ in the kinds of constraints imposed on them. When ranking data are collected, it is common practice not to allow ties in ranking. Thus, if a subject assigns rank k to object a, he cannot assign rank k to any other object. In consequence, all the marginal frequencies of the object-by-rank table are the same (45 in Table 1.6). This constraint is not imposed on the response-pattern table in general. Even when one uses a common set of response categories (e.g., yes–no) for all the multiple-choice questions, the marginal frequencies of the item-by-category table constructed from the response-pattern table are in general not the same. Thus the condition imposed on ranking data is a tight constraint on subjects'

TABLE 1.4

Rank order table ($N \times n$) consisting of rank numbers $N=45$, $n=3$

	Objects		
Subjects	1	2	3
1	2	1	3
2	1	2	3
3	1	3	2
4	2	1	3
⋮	⋮	⋮	⋮
45	1	2	3

TABLE 1.5

Rank order table in the format of a response-pattern table ($N \times n^2$), $N=45$, $n=3$

	Object 1			Object 2			Object 3			
Subjects	1*	2	3	1	2	3	1	2	3	Total
1	0	1	0	1	0	0	0	0	1	3
2	1	0	0	0	1	0	0	0	1	3
3	1	0	0	0	0	1	0	1	0	3
4	0	1	0	1	0	0	0	0	1	3
⋮	⋮	⋮	⋮	⋮	⋮	⋮	⋮	⋮	⋮	⋮
45	1	0	0	0	1	0	0	0	1	3
Subtotal	33	11	1	10	30	5	2	4	39	135
Total		45			45			45		135

*Ranks.

TABLE 1.6

Rank order table of objects-by-ranks ($n \times n$), $n=3$

	Ranks			
Objects	1	2	3	Total
1	33	11	1	45
2	10	30	5	45
3	2	4	39	45
Total	45	45	45	

judgmental processes and seems to offer a reasonable basis for distinguishing the rank order table from the general response-pattern table.

When the number of stimuli, n, increases, the judgment involved in ranking may become rather demanding. One may then consider asking subjects to rank a subset of k stimuli ($k<n$) chosen from the entire set. The total number of such subsets is $n!/[k!(n-k)!]$. If n is an even number, the number of subsets becomes largest when $k=n/2$. Considering that the smaller the value of k the easier the judgment is for the subjects, the best strategy for large n seems to be to choose k less than $n/2$. The most popular choice seems to be $k=2$, which provides *paired comparison* data. When subsets of size k ($k<n$) are used, however, one may have to resort to some form of incomplete design, for, in addition to the large number of subsets, the same pairs, triplets, or quadruplets of stimuli will appear in many subsets and will be repeatedly compared under such a scheme. These duplications of judgments can be reduced without too much loss of information by effective use of incomplete designs. See, for example, Cochran and Cox (1957).

Let us consider paired comparison data (i.e., $k=2$) in some detail. The data can be represented in the format of a response-pattern table (Table 1.7). One of the two columns for each pair is completely redundant. Therefore, if the subsequent analysis employs a design matrix to specify the structure of each column, one of the two columns for each pair can be deleted without loss of information, which reduces the dimensions of the table from $N \times n(n-1)$ to $N \times n(n-1)/2$. It is interesting to note that Table 1.7 does not reflect the same type of constraint as indicated in Table 1.6, but that any number of objects can be assigned rank 1 or 2. These ranks, however, are

TABLE 1.7

Paired comparison data as a rank order table ($N \times n(n-1)$), $N=100$, $n=10$

	(X_1,X_2)		(X_1,X_3)			(X_9,X_{10})		
Subjects	1*	2	1	2	...	1	2	Total
1	1	0	0	1	...	1	0	45†
2	0	1	0	1	...	0	1	45
3	1	0	1	0	...	0	1	45
4	1	0	0	1	...	1	0	45
⋮	⋮	⋮	⋮	⋮	⋮	⋮	⋮	⋮
100	1	0	0	1	...	0	1	45
Subtotal	76	24	15	85	...	41	59	4,500
Total	100		100		...	100		4,500

*Ranks: 1 indicates judgment $X_j > X_k$ for the pair (X_j, X_k) and 2, $X_j < X_k$.
†This number is equal to $n(n-1)/2$. No position effect or order effect is assumed to exist, and hence (X_j, X_k) and (X_k, X_j) are regarded as the same.

meaningful only in reference to the objects in each pair. Paired comparison data presented in the response-pattern format thus differ from the same data presented as a general response-pattern table.

(e) *Multidimensional tables* Multidimensional categorical tables are generated by introducing two or more classification variables or by mixing formats (a) to (d) of data. Tables 1.8 and 1.9 are examples of a multidimensional contingency table (i.e., a table of joint responses to more than two categorical variables) and a multidimensional response-frequency table (i.e., a table of stimulus frequencies over two or more sets of response categories), respectively. The latter can also be represented as a $5 \times 4 \times 3$ mixture of formats (a) and (b) as in Table 1.10. This table can be further expanded by, for example, dividing the meals into two groups, oriental and western.

A multidimensional response-pattern table can be constructed from a battery of multiple-choice tests. For example, the battery may be divided into subbatteries such as verbal tests and quantitative tests, and each subbattery may be further divided into individual tests. The subjects themselves may be classified into subgroups, according to sex (boys, girls)-by-locality (urban, rural), for example, or according to particular biographical information.

TABLE 1.8

Multidimensional contingency table: 'Do you read magazines A, B, C, D?'

		Read A?			
		Yes		No	
		Read B?		Read B?	
Read C?	Read D?	Yes	No	Yes	No
Yes	Yes	28	4	12	3
	No	7	17	4	5
No	Yes	15	3	23	8
	No	1	6	4	32

TABLE 1.9

Multidimensional response-frequency table

	Meals at our cafeteria are:				The price is:			
Meals	(1)*	(2)	(3)	(4)	(5)†	(6)	(7)	Total
A	5	4	31	10	18	31	1	100
B	0	3	15	32	28	19	3	100
C	23	25	2	0	41	9	0	100
D	2	36	11	1	15	30	5	100
E	4	19	21	6	25	20	5	100

*(1) terrible, (2) mediocre, (3) good, (4) delicious.
†(5) very reasonable, (6) reasonable, (7) very high.

One can consider a multidimensional extension of the rank order table by introducing a few criteria (e.g., stimulus attributes) for ranking. For instance, meals may be ranked first according to the order in which they *look* tasty and second according to the order of their actual taste.

TABLE 1.10

Mixture of contingency and response-frequency tables

		Meals are:				
Meals	Prices	(1)*	(2)	(3)	(4)	Total
A	(5)†	5	3	5	5	
	(6)	0	1	26	4	50
	(7)	0	0	0	1	
B	(5)	0	2	1	25	
	(6)	0	1	13	5	50
	(7)	0	0	1	2	
⋮	⋮	⋮	⋮	⋮	⋮	⋮
E	(5)	3	6	15	1	
	(6)	1	13	4	2	50
	(7)	0	0	2	3	

*(1) terrible, (2) mediocre, (3) good, (4) delicious.
†(5) very reasonable, (6) reasonable, (7) very high.

Furthermore, one can merge several multidimensional tables, or mix tables of different formats, to generate a wide variety of multidimensional tables. In this book, these tables are all represented in the two-dimensional (row-by-column) form with subcategories within rows and columns. It may prove to be useful to classify a large number of possible multidimensional tables into a few groups, but this will have to wait for a future study.

In the social sciences, one often collects data which are referred to as *proximity* data, *similarity* data, or *dissimilarity* data. Proximity data are probably one of the most popular types. They might be considered as an additional category in the above classification scheme, but their categorical representation depends much on the ways the data are collected. For example, proximity data can be represented in such forms as ranking, paired comparisons of pairs, triadic comparisons, and response frequencies of pairs of stimuli in successive categories. Besides, it seems that a classification into the above five schemes can accommodate most of these different forms. Proximity data are therefore not specifically identified in the classification.

1.2
DUAL SCALING AND ITS HISTORY

Once data are collected and represented in a row-by-column table, the next stage is data analysis. The approach employed here involves quantification of rows and columns of the table so as to optimize a certain criterion. For example, given a two-way table of the incidences of three levels of anxiety and five categories of school achievement for a group of students, one can derive two real-valued variates from the two categorical variates in such a way that the correlation between the derived variates should be a maximum. The approach is not new, but has existed for at least 47 years. During that period, the technique has been discovered and rediscovered independently by several investigators, and has been advocated under a variety of names, such as 'the method of reciprocal averages,' 'additive scoring,' 'appropriate scoring,' 'canonical scoring,' 'Guttman weighting,' 'principal component analysis of qualitative data,' 'optimal scaling,' 'Hayashi's theory of quantification,' 'simultaneous linear regression,' 'analyse factorielle des correspondances,' 'correspondence factor analysis,' 'correspondence analysis,' and 'biplot.' Some of these names have obviously been derived from particular algorithms used by their proponents and are not appropriate for the purpose of this book, and others are not specific enough to characterize the technique. To avoid further possible confusion among users in different disciplines, I propose the name 'dual scaling,' which does not adhere to any specific algorithm but reflects a fundamental feature peculiar to the technique. The choice of this term will be clarified in Chapter 3 through a discussion of 'duality' in this scaling procedure.

The history of dual scaling can be traced in the articles of de Leeuw (1973), Greenacre (1978a, b) and Nishisato (1979a). Dual scaling has an enlightening precursor, called algebraic eigenvalue theory, which was pioneered in the 18th century. Euler, Cauchy, Jacobi, Cayley, and Sylvester, among others, contributed to the development of the theory. Its statistical applications were considered by Karl Pearson (1901b), who analysed dispersion matrices using some results of the theory. His problem was to find the best linear combination of variables in the least-squares sense, which led to solving the eigenequation of the dispersion matrix. Pearson found that the best linear combination was given by projections of the points on the principal axis. We now know that this is the basic operation of principal component analysis. It was not until Hotelling's study in 1933, however, that principal component analysis was clearly and succinctly formulated and that the eigenvalue theory became a useful statistical tool for those in applied areas. More specifically, Hotelling presented principal component analysis as a data reduction technique. In a similar vein, Hotelling (1936), Horst (1936), Edgerton and Kolbe

(1936), and Wilks (1938) developed techniques to derive 'optimum' linear combinations of 'continuous' variables by formulating eigenequations. These studies are important in the sense that they provided solid foundations for future statistical applications of the eigenvalue theory.

Eckart and Young (1936) presented another line of relevant work based on the eigenvalue theory, clarifying how a data matrix could be decomposed into the row structure and the column structure. This decomposition is often referred to as 'the Eckart-Young decomposition theorem,' 'basic structure' (Horst 1963), 'canonical form' (R.M. Johnson 1963), 'singular decomposition' (Good 1969), 'singular value decomposition' (Stewart 1973), and 'réduction tensorielle' (Benzécri et al. 1973b). Greenacre (1978a, b) refers to the study by Schmidt (1907) as possibly the first work on this topic.

The basic structure of a data matrix and its principal components are closely related, and it is known (e.g., Gower 1966) that the row and column structures of data matrix A can be obtained from the principal components of AA' and $A'A$, respectively. The determining relation will be used to define 'duality' for our procedure in Chapter 3.

Dual scaling, as the term is used here, is an extension of the same approach (i.e., the Eckart-Young decomposition through principal component analysis) to categorical data. Its origin is not easy to trace. Fisher (1940) and Guttman (1941) are often regarded as the method's inventors. The credit should, however, probably go to earlier investigators in the 1930s. One can find at least three relevant articles: Richardson and Kuder (1933), Horst (1935), and Hirschfeld (1935).

The study by Richardson and Kuder (1933) was cited by Horst (1935) as the first article on the method of reciprocal averages, which is now considered one of the techniques of dual scaling. The Richardson and Kuder article does not fully specify the procedure or name it, however, omissions which could have obscured their contribution. Richardson and Kuder discuss such basic ideas of dual scaling as reducing the within-item variation and increasing both item reliability and between-subject discriminability, but, although they describe the averaging process to derive weights for items, it is rather difficult to see how the weights and subjects' scores are successively corrected to optimize several criteria stated in their article. Horst (1935, p. 370) writes: 'The method which he [Richardson] suggested was based on the hypothesis that the scale value of each statement should be taken as a function of the average score of the men for whom the statement was checked and, further, that the score of each man should be taken as the average scale value of the statements checked for the man... This technique has been called for convenience the Method of Reciprocal Averages.'

In the same article, Horst considered that the statements (items) should be scaled so as to give the largest possible discrimination among the individuals measured. He stated that certain mathematical equations developed for his objective were found to bear a significant resemblance to the scaling equations underlying the method of reciprocal averages. His general approach was to assign one or zero to each item depending on whether it was checked or not, and to determine weights for items such that when the weighted scores are combined into a single score for each person the distribution of all the scores has the largest possible dispersion. In this instance, he treated dichotomous items as dummy variables taking the value 1 or 0, and considered a weighted linear combination of these variables. His method, then, is the same as dual scaling. He later extended the technique to continuous variables (Horst 1936), and thus departed from dual scaling.

A study which is quite different from those of Richardson and Kuder (1933) and Horst (1935) was published in 1935 by Hirschfeld. He posed the question: 'Given a discontinuous distribution p_{ij}, is it possible to introduce new values for the variates x_i and y_j such that *both* regressions are linear?' Assuming that one already knows a set of variates x_i and y_j, which together with the given bivariate distribution p_{ij} yield linear regressions, he derived a set of equations of what is now termed dual scaling. His treatment of the problem is mathematically clear and concise, and presents a contrast to the other two articles, in which no mathematical formulas were given. Hirschfeld's criterion of quantification, namely, linearity of both regressions of the derived variates, offers an interesting interpretation of dual scaling, and is undoubtedly one way to formulate the scaling procedure.

The three studies cited above suggest that the basic ideas of dual scaling were conceived and formulated in the 1930s. It is interesting to note, however, that not only these studies but also some later studies somehow failed to draw the attention of researchers to this general area, and that dual scaling was repeatedly rediscovered by several investigators during the next 47 years or so. Let us now look at some of the major studies from 1940 to date.

In 1940 Fisher considered weighting categorical variables in the context of discriminant analysis. Using Tocher's data for Caithness compiled by K. Maung of the Galton Laboratory, he asked for what eye colour scores the five hair colour classes are most distinct. He stated that the answer might be found in a variety of ways. As an example, he used a technique of starting with arbitrarily chosen scores for eye colours, determining from them average scores for hair colours, and using the latter to find new scores for eye colours. This process tends to a limit and yields scores for the two categorical variables, eye colour and hair colour. This is the core of the algorithm of the method of reciprocal averages, which Horst (1935) states is due to Richardson and Kuder (1933). Fisher noted that in the sample from which the two scores are derived each score has a linear regression on the other. As noted above, the linearity of regressions was the starting point in Hirschfeld's study (1935).

Interestingly enough, however, Fisher cited neither the article by Richardson and Kuder nor that of Hirschfeld, suggesting that he discovered the technique independently of these investigators.

Maung (1941) studied the measurement of correlation of a two-way table of qualitative characters and presented detailed formulations of three approaches to dual scaling, using 'a scoring system developed by Professor R.A. Fisher in the discriminant analysis' (Maung 1941, p. 189). His first approach is to determine scores for the rows and scores for the columns of the table so as to maximize the *canonical correlation*, defined for the two sets of scored responses. The second approach is to determine scores for the rows and scores for the columns so as to maximize the *product moment correlation*, defined over the pairs of scored responses. The third approach is to determine scores for the row (columns) so as to maximize the *ratio of the between-column (between-row) sum of squares to the total sum of squares* of scored responses. Maung stated that the three approaches would yield identical results.

In 1941 Guttman presented a clear and detailed formulation of dual scaling as applied to multiple-choice data. He employed the notion of *internal consistency* as the criterion for optimization. According to Guttman, weights for the response options should be determined so as to minimize the ratio of the within-subject sum of squares to the total sum of squares. Once weights are determined, scores for the subjects are given as proportional to the subjects' average weighted responses. Scores for the subjects should be determined so as to minimize the ratio of the within-option sum of squares to the total sum of squares. Once scores are determined, weights for the options are given as proportional to the options' average scores. These two procedures are equivalent to Maung's third approach and lead to identical results. Guttman also demonstrated that the same result can be obtained by determining simultaneously scores for the subjects and weights for the options so as to maximize the product-moment correlation, defined over the pairs of scores and weights. This is the same as Maung's second approach. Guttman provided a bibliographical note, in which he referred, as similar and relevant studies, to Thurstone and Chave (1929), Hotelling (1933), 'the method of reciprocal averages suggested by M.W. Richardson and sketched in Horst [1935],' Horst (1936), Edgerton and Kolbe (1936), and Wilks (1938). He stated that although his theory was motivated by these studies it was developed independently of them. Neither Hirschfeld's nor Fisher's study was cited by Guttman.

In 1946 Mosier presented a paper on the use of IBM accounting equipment in carrying through an iterative process, called reciprocal averaging. He listed the following properties as resulting from dual scaling: (1) the reliability of each item and the internal consistency of the weighted inventory are maximized; (2) the correlation between an item and the total score is maximized and the product-moment correlation coefficient becomes identical with the correlation ratio; (3) the relative variance of the distribution of scores (coefficient of variation) is maximized; (4) the relative variance of item scores within a single case is minimized; (5) the correlation between an item and the total score is proportional to the standard deviation of the option weights for that item; (6) questions which bear no relation to the total-score variable are automatically weighted so that they exert no effect on the scoring. These properties are important for multiple-choice testing and represent some of the basic features of dual scaling.

In 1946 Guttman extended his approach of dual scaling to quantification of paired comparison and rank-order data. Using his principle of internal consistency, he derived numerical values for objects being compared in such a way that the values of objects a given person judges higher than other objects should be as different as possible from the values of objects he judges lower than other objects. The study is unique in the sense that it indicated a different way of applying the technique of dual scaling.

In 1948 Fisher analysed reactions of 12 samples of human blood tested with 12 different sera. The data (reactions) were represented in a 12×12 table by five symbols: $-$, ?, w, (+) and +. His question was: 'Given a two-way table of non-numerical observations, what values, or scores, shall be assigned to them in order that the observations shall be as additive as possible?' After assigning 0 to $-$ and 1 to +, he defined the problem as that of determining three values, x, y, and z, for reactions ?, w, and (+), respectively, so as to maximize the between-row and the between-column sums of squares, relative to the total sum of squares. Fisher thus presented an example of the analysis of variance of non-numerical observations through dual scaling.

In 1950 Guttman reiterated his 1941 formulation in the context of his scalogram analysis, and provided an interesting mathematical exposition of the principal component properties of his 'perfect scale.' In the same year, Burt presented an application of principal component analysis to categorical data. With this study as the focal point, Guttman (1953) and Burt (1953) exchanged inspiring views on the analysis of categorical data. P.O. Johnson (1950) presented an example of the application of the Fisher-Maung formulation of discriminant analysis to determining appropriate scores for the letter grades in school. Bartlett (1951) presented another study on discriminant analysis of categorical variables.

In 1950 Hayashi launched a series of studies on quantification methods. Although his references include Guttman's studies in 1941, 1946, and 1950, he extended the approach much further than Guttman. Hayashi dealt with symmetric data matrices (e.g., similarity matrices, paired comparison matrices) and non-symmetric matrices (e.g., sociometry matrices, tournament data matrices) in 1952, multidimensional tables in 1954, and multidimensional quantification of paired comparison data in 1964 and 1967. He and his associates' have published a large number of other studies under the rubric 'Hayashi's theory of quantification.'

In 1952 Williams discussed significance tests for the Fisher-Maung formulation of dual scaling and the Yates formulation (1948) of contingency-table analysis in which quantitative values for one or both classifications are given. He showed that significance tests, developed for discriminant analysis for the interpretation of interactions, which are exact when the variates involved are normally distributed, may be applied as asymptotically exact tests to contingency tables.

In 1953 Lancaster presented a study of a reconciliation of the chi-square statistic by considering both its 'metrical and enumerative aspects,' citing the studies by Fisher (1940) and Maung (1941). Guttman (1955) provided further results on principal components of a perfect scale and discussed an additive metric.

In 1956 Bock presented an interesting application of dual scaling. He proposed a scheme for choosing a few individuals whose preferences for given objects are most representative of those of a larger group of individuals. In particular, his study considers quantifying the preferences of each individual so as to discriminate optimally among objects, testing whether or not a common continuum of preference may be assumed for the quantified preferences, constructing a linear estimator of values for the objects on this continuum, and selecting as judges the least number of individuals whose quantified preferences determine values for the objects with acceptable accuracy.

In 1958 Lancaster presented a thorough mathematical exposition of the structure of the two-way table, referring to such studies as Hirschfeld (1935), Hotelling (1936), Fisher (1940), Maung (1941), and Williams (1952). In the same year, Lord provided a proof that the scoring system derived by dual scaling maximizes the generalized Kuder-Richardson reliability coefficient, that is, Cronbach's alpha (Cronbach 1951). Torgerson (1958, pp. 338–45) provided a useful summary of Mosteller's formulation (Mosteller 1949) of dual scaling in his well-known book on scaling.

In 1960 Bock published a paper, 'Methods and Applications of Optimal Scaling,' based on his talk delivered at the 1957 annual meeting of the American Psychological Association. This paper has been widely cited and has helped publicize the method under the attractive name 'optimal scaling.' In 1960 also, Baker published a computer program of the method of reciprocal averages, which helped promote this scaling technique as a powerful tool of data analysis among psychologists and educational researchers. In 1961 Kendall and Stuart (pp. 568–78) presented canonical analysis of a two-way contingency table, along the lines of Maung (1941) and Lancaster (1958).

In 1962 Bradley, Katti, and Coons considered imposing the constraint of complete order on the categories so that the scaled weights of the categories might conform to their a priori order. They introduced a new iterative algorithm to dual scaling. Lingoes (1963, 1964, 1968) presented a formulation of dual scaling, called simultaneous linear regressions, and computer programs. This formulation is basically the same as Hirschfeld's (1935), but seems to have been developed independently, Lingoes referring instead to Guttman (1941).

In 1965 Kruskal investigated a way to eliminate the need for an interaction term in the analysis of factorial experiments. His method employs a family of monotone transformations and determines the 'best' transformation in terms of optimizing a squared-residuals criterion of fit to the assumed model. Kruskal noted the method's resemblance to the studies by Fisher (1948, Example 46.2) and Bradley, Katti, and Coons (1962), but extended the problem to multi-way factorial data and employed a well-known algorithm, called monotone regression analysis, which is widely used in non-metric multidimensional scaling (e.g., Shepard 1962a, b; Kruskal 1964a, b).

In 1965 Shiba presented his formulation of dual scaling for multi-category data. In 1966 McKeon published a monograph on 'relations between canonical correlation, factor analysis, discriminant analysis, and scaling theory.' The study clarifies the relation of dual scaling to other statistical techniques. In 1967 Whittaker presented a review of 'gradient analysis' and ordination methods, which were developed in plant ecology in the 1950s and 1960s. Although these methods are based on an idea similar to that of the method of reciprocal averages, they seem to have been developed independently of any approach to dual scaling.

In 1968 McDonald presented a general procedure for obtaining weighted linear combinations of variables, including multiple regression weights, canonical variate analysis, principal components, maximizing composite reliability, canonical factor analysis, and certain other well-known models. This is a systematization of the general weighting problem, in which dual scaling is regarded as one of the special cases.

In the 1960s several investigators presented dual-scaling-type approaches to paired comparison data, which were distinct from the traditional Thurstonian approach, the main contributors being Slater (1960a), Tucker (1960), Hayashi (1964, 1967), Carroll and Chang (1964, 1968), and de Leeuw (1968). (Recall that Guttman's pioneering study was published in 1946.) Further studies are reported by Bechtel, Tucker, and Chang (1971), Carroll (1972), de Leeuw (1973), and Nishisato (1976, 1978a, c).

In the 1960s Benzécri and his associates started a series of studies of data analysis, and developed under the name 'l'analyse des correspondances' (correspondence analysis) a technique which is intended for the same purpose as dual scaling. Among others, the papers by Benzécri (1969) and Escofier-Cordier (1969) are often cited by other investigators. The work of Benzécri and his associates culminated in two gigantic volumes on data analysis (Benzécri and others 1973a, b), one of which is totally devoted to *l'analyse des correspondances*. The method is popular in France.

It is interesting to note a substantive increase in publications on dual scaling in the 1970s. A few books on data analysis devote a number of pages and chapters to dual scaling (Hayashi, Higuchi, and Komazawa 1970; Takeuchi and Yanai 1972; Benzécri and others 1973b; Nishisato 1975; Lingoes 1977). A series of studies on dual scaling with graphical orientation was presented under the name 'biplot' by Gabriel (1971, 1972, 1978a, b) and Bradu and Gabriel (1978). Hill presented a concise formulation of the method of reciprocal averages in 1973 and an excellent summary of dual scaling under the name 'correspondence analysis' in 1974. Teil (1975) outlined Benzécri's formulation in English, using the name 'correspondence factor analysis.' Nishisato (1971a, 1972a, b, 1973a) extended dual scaling analysis of variance to the general multi-way design. Multidimensional qualitative data were considered for dual scaling by de Leeuw (1973), Iwatsubo (1971, 1974, 1975, 1978), Nishisato (1971a, 1972b, 1973a, 1976), and Yoshizawa (1975). Nishisato and Inukai (1972) presented a study on item weighting, as against option weighting, under the name 'partially optimal scaling.' Saito (1973) investigated the quantification of categorical data in terms of the generalized variance. The problem of partial order constraints on categories was considered by Nishisato (1973a) using a simple fast algorithm called the method of successive data modifications (SDM) and by Nishisato and Arri (1975) using a non-linear programming approach. Multiple discriminant analysis of qualitative data was investigated by Saporta (1975), Bouroche, Saporta, and Tenenhaus (1975), and Saporta (1976). A new direction in the dual scaling approach has been sought and extensively investigated by Young, de Leeuw, and Takane (Young, de Leeuw, and Takane 1976a, b; de Leeuw, Young, and Takane 1976; Takane, Young, and de Leeuw 1977; de Leeuw 1976; Young 1976; van Rijckevorsel and de Leeuw 1978). McDonald, Ishizuka, and Nishisato (1977; see also Ishizuka 1976) applied dual scaling to factor analysis, in which responses were optimally scaled in terms of the common factor model. Healy and Goldstein (1976) presented another formulation of dual scaling, covering a case where the use of generalized inverses is involved. McDonald, Torii, and Nishisato (1979; see also Torii 1977) reformulated the theorems of Rao and Mitra (1971) on proper eigenvalues and eigenvectors in the context of dual scaling of multidimensional tables, and considered several criteria for optimization. Their study clarified the necessary and sufficient conditions for the existence of proper eigenvalues and eigenvectors for a generalized eigenequation *involving a singular matrix* and in addition showed that the reparametrization method proposed by Nishisato (1976) would satisfy the conditions. Nishisato (1978a, b) discussed his general approach to dual scaling of multidimensional tables (Nishisato 1976) and extended it to the case in which the input data matrix has negative elements, a major step in the development of dual scaling.

During the past decade, several people have considered dual scaling in full detail and in wider perspective. In addition to the four books mentioned above, we can find several other publications. The article by Saito, Ogawa, and Nojima (1972) offers a concise description of dual scaling and related techniques. The monograph by de Leeuw (1973) is an excellent exposition of dual scaling, its extensions, and other related methods, and can be regarded as one of the most important references, together with the book by Benzécri and his associates (1973b). The work by Saporta (1975) is also a useful and interesting reference. The studies by Nishisato (1972a, 1973a, 1976, 1979b) offer dual scaling as a practical tool for data analysis and apply it to different types of categorical data. Greenacre (1978a; its English translation 1978b) provides an excellent compendium of dual scaling and related techniques and, more specifically, a concise summary of theoretical comparisons of biplot, principal component analysis, correspondence analysis, and multidimensional unfolding analysis. His work is of extreme importance from both theoretical and practical points of view, for it gives a thorough discussion of such concepts as distance, within-set distance, between-set distance, and interpretability of the graphical display of data, complete with numerous examples.

There are undoubtedly a number of other important publications on the topic of dual scaling, but the above review should give an indication of the general scope of the problems.

1.3 OVERVIEW

As reviewed above, dual scaling is a family of techniques which are intended mainly, but not exclusively, for categorical data. Its development has been rapid over the past few years, so it is difficult to cover every facet of its extensions in detail in one volume. In this book, the emphasis will be on the framework of the 'classical' approach with a modern flavour.

Chapter 2 provides a thorough description of the basic procedure of dual scaling; it is hoped that it will prove comprehensive and informative enough for most practical purposes. Chapter 3 discusses further mathematical aspects of dual scaling (those who are primarily interested in applications may skip this chapter and move on to Chapter 4 and the ensuing chapters). Chapters 4 to 7 present specific applications of the technique to different forms of categorical data: contingency and response-frequency tables, response-pattern tables, rank-order/paired comparison tables, and multidimensional tables. Each of these chapters starts with a thorough discussion of problems which are somewhat peculiar to the specific form of data, and then gives numerical examples to illustrate the procedure. Chapter 8, the last chapter, discusses miscellaneous problems, that is, dual scaling of ordered categories, the analysis of variance of categorical data, and incomplete data; the last section presents concluding remarks.

To simplify the formulation and to obtain generalizable equations, matrix notation is used throughout the book. Although the notation is introduced gradually with many examples, an understanding of basic matrix operations is required to follow some of the derivations of formulas. Such knowledge can be acquired from introductory books on matrix algebra (e.g., Horst 1963; Graybill 1969; Hammer 1971), but, for the convenience of some readers, a minimum package of matrix calculus is provided in Appendix A, which it is hoped will prove sufficient for readers to follow this book. Appendix B presents a small FORTRAN program, called DUAL1, which was prepared primarily to generate numerical examples for classroom teaching and workshops. It is hoped that DUAL1 will prove useful as a learning device and a means to put dual scaling methodology into practical use. For production runs, readers are referred to a large-scale FORTRAN program, OPSCALE, written by Leong, Nishisato, and Wolfe (1980). OPSCALE is also based on methods described in this book.

2
Dual scaling I

2.1
FORMULATION

In the historical review of dual scaling in Chapter 1 several ways of formulating the procedure were mentioned. As an introduction to the subject, one of those alternatives will now be described. Other approaches, which yield the same results, will be covered in Chapter 3.

The approach to be presented here was fully described by Guttman (1941) and Maung (1941). A precise basis for its procedure is given in Bock's statement: This approach 'is to assign numerical values to alternatives, or categories, so as to discriminate optimally among the objects...in some sense. Usually it is the least-squares sense, and the values are chosen so that the variance between objects after scaling is maximum with respect to that within objects' (Bock 1960, p. 1). To put this statement into mathematical formulas, let us consider a hypothetical numerical example. Suppose that three teachers were each rated by 10 randomly chosen students in terms of three categories – good, average, and poor – and the data were tabulated in a response-frequency table (Table 2.1). Suppose that, although each teacher was judged by 10 students, Teacher 3 received only nine judgments, suggesting that one of the 10 judgments was missing. Such a situation could arise in practice and mean several things: one student might have evaluated Teacher 3 in terms of a category not included in the instruction (e.g., 'no comment'), or that part of the data might have been lost. Anyway, 9 was used for Teacher 3 in place of 10 to derive a general, not simplified, formulation of dual scaling.

Of the three response categories, 'good' is the best, but there is no way off-hand to tell how much better it is than 'average' or 'poor,' or how much better 'average' is than 'poor.' Therefore we assign unknown weights x_1, x_2, and x_3 to categories 1 (good), 2 (average), and 3 (poor), respectively. Each response in category 1, for example, receives the weighted score x_1. Thus, in terms of the weighted scores, Table 2.1 can be presented as in Table 2.2. There are three unknown constants, x_1, x_2, and x_3, and we want to determine their values so as to optimize a certain criterion. We use Guttman's internal consistency (Guttman 1950) as the criterion. To apply this criterion, we

TABLE 2.1

Evaluation of teaching performance (response-frequency table)

Teachers	Good	Average	Poor	Total
1	1	3	6	10
2	3	5	2	10
3	6	3	0	9
Total	10	11	8	29

regard Table 2.2 as a one-way classification table, that is, a table of scores of the three teachers without response categories (Table 2.3), and use the framework of the analysis of variance.

According to the criterion of internal consistency, we want to determine the values of x_1, x_2, and x_3 so as to make the scores within teachers as similar as possible and the scores of different teachers as different as possible. In other words, we want to determine x_1, x_2, and x_3 so as to minimize the discrepancies between each teacher's scores and his average score and maximize the discrepancies between the three average scores of the individual teachers and the average of the three average scores. In statistics, the sum of squares of the discrepancies between each teacher's individual scores and his average score is referred to as the within-group sum of squares and is indicated by ss_w. The sum of squares of the discrepancies between the three average scores of individual teachers and the average of the three average

TABLE 2.2

Data expressed in terms of weighted scores

Teachers	Good	Average	Poor	Total
1	x_1	x_2, x_2, x_2	$x_3, x_3, x_3,$ x_3, x_3, x_3	$x_1+3x_2+6x_3$
2	x_1, x_1, x_1	x_2, x_2, x_2 x_2, x_2	x_3, x_3	$3x_1+5x_2+2x_3$
3	x_1, x_1, x_1 x_1, x_1, x_1	x_2, x_2, x_2		$6x_1+3x_2$
Total	$10x_1$	$11x_2$	$8x_3$	$10x_1+11x_2+8x_3$

TABLE 2.3

One-way classification table of weighted scores

Teachers	Scores	Total
1	$x_1, x_2, x_3, x_3, x_3,$ x_2, x_2, x_3, x_3, x_3	$x_1+3x_2+6x_3=y_1$
2	$x_1, x_1, x_2, x_2, x_3,$ x_1, x_2, x_2, x_2, x_3	$3x_1+5x_2+2x_3=y_2$
3	$x_1, x_1, x_1, x_2, x_2,$ x_1, x_1, x_1, x_2	$6x_1+3x_2=y_3$
Total		$10x_1+11x_2+8x_3$ $=y_1+y_2+y_3=y_t$

(0) $f_t=10+11+8=29.$
(1) c (correction term) $=y_t^2/f_t.$
(2) $\Sigma\Sigma y_{ji}^2=10x_1^2+11x_2^2+8x_3^2.$
(3) $\Sigma(y_j^2/f_j)=y_1^2/10+y_2^2/10+y_3^2/9.$
(4) $ss_t=(2)-(1).$
(5) $ss_b=(3)-(1).$
(6) $ss_w=(2)-(3).$

scores, weighted by the respective numbers of responses, is called the between-group sum of squares and is indicated by ss_b. The sum of squares of the discrepancies between each teacher's individual scores and the average of the three average scores is called the total sum of squares and is indicated by ss_t. The relation

$$ss_t = ss_b + ss_w \tag{2.1}$$

holds. In other words, the overall variation among scores can be decomposed into the variation across the teachers and the variation within the teachers.

According to the criterion of internal consistency, we want to determine the values of x_1, x_2, and x_3 so as to minimize ss_w and maximize ss_b. However, unless some constraints are imposed on x_1, x_2, and x_3, it is always possible to make ss_w as small as one likes and ss_b as large as one wants. Therefore we shall consider minimizing the ratio ss_w/ss_t and maximizing the ratio ss_b/ss_t with respect to x_1, x_2, and x_3. The ratio ss_b/ss_t is called the squared *correlation ratio* and is indicated by η^2, that is,

$$ss_b/ss_t = \eta^2. \tag{2.2}$$

If both sides of (2.1) are divided by ss_t, we obtain

$$1 = \eta^2 + ss_w/ss_t. \tag{2.3}$$

Thus

$$ss_w/ss_t = 1 - \eta^2. \tag{2.4}$$

Since ss_w, ss_b, and ss_t are all sums of squares of discrepancy terms and hence all positive, it follows that the squared correlation ratio η^2 (hereafter called simply the correlation ratio) is bounded between 0 and 1,

$$1 \geq \eta^2 \geq 0. \tag{2.5}$$

It also follows from (2.2) and (2.4) that minimizing ss_w/ss_t is the same as maximizing ss_b/ss_t. Thus either one of these two may be used to obtain the most internally consistent values x_1, x_2, and x_3.

In order to derive formulas for ss_t, ss_b, and ss_w, it is convenient to introduce a general notation for a one-way classification table such as Table 2.3:

y_{ji} = the ith element of the weighted scores given to teacher j,
y_t = the sum of the weighted scores,
f_t = the total number of weighted scores,
y_j = the total weighted score of teacher j,
f_j = the total number of weighted scores of teacher j,
$\bar{y}_j = y_j/f_j$ = the mean (average) weighted score of teacher j,
$\bar{y}_t = y_t/f_t$ = the grand mean of weighted scores,
$c = y_t^2/f_t$ = the correction term.

ss_t is the sum of squares of the discrepancies between individual scores and the grand mean, that is,

$$ss_t = \sum\sum(y_{ji} - \bar{y}_t)^2 \tag{2.6}$$

$$= \sum\sum\left[y_{ji}^2 - 2y_{ji}(y_t/f_t) + (y_t/f_t)^2\right]$$

$$= \sum\sum y_{ji}^2 - 2(y_t/f_t)\sum\sum y_{ji} + f_t(y_t/f_t)^2$$

$$= \sum\sum y_{ji}^2 - (y_t^2/f_t) = \sum\sum y_{ji}^2 - c. \tag{2.7}$$

(2.6) is a direct translation of the definition of ss_t, and (2.7) is another expression for ss_t which is often convenient for computations. ss_b is the sum of squares of the discrepancies between the group means and the grand mean, weighted by the respective numbers of responses,

$$ss_b = \sum f_j(\bar{y}_j - \bar{y}_t)^2. \tag{2.8}$$

A formula often used for computation is

$$ss_b = \sum(y_j^2/f_j) - (y_t^2/f_t) = \sum(y_j^2/f_j) - c. \tag{2.9}$$

ss_w is the sum of squares of the discrepancies between the individual scores and the group means,

$$ss_w = \sum\sum(y_{ji} - \bar{y}_j)^2. \tag{2.10}$$

An alternative formula for computation is

$$ss_w = \sum \sum y_{ji}^2 - \sum (y_j^2/f_j). \tag{2.11}$$

It is obvious from (2.7), (2.9), and (2.11) that $ss_t = ss_b + ss_w$. Since our objective is to determine the weights for the categories that maximize the ratio ss_b/ss_t, we shall consider only the expressions for ss_b and ss_t.

In our example, we obtain from Table 2.1

$$f_1 = 10, \quad f_2 = 10, \quad f_3 = 9, \quad f_t = 29,$$

and from Table 2.2

$$y_1 = x_1 + 3x_2 + 6x_3,$$
$$y_2 = 3x_1 + 5x_2 + 2x_3,$$
$$y_3 = 6x_1 + 3x_2 + 0x_3 = 6x_1 + 3x_2,$$
$$y_t = 10x_1 + 11x_2 + 8x_3.$$

y_{ji} are given in Table 2.3. Therefore ss_b is given by (2.9) as

$$ss_b = \frac{(x_1+3x_2+6x_3)^2}{10} + \frac{(3x_1+5x_2+2x_3)^2}{10} + \frac{(6x_1+3x_2)^2}{9} - \frac{(10x_1+11x_2+8x_3)^2}{29}. \tag{2.12}$$

Similarly, using (2.7), we obtain

$$\begin{aligned} ss_t = {} & (x_1^2+x_2^2+x_2^2+x_2^2+x_3^2+x_3^2+x_3^2+x_3^2+x_3^2+x_3^2 \\ & +x_1^2+x_1^2+x_1^2+x_2^2+x_2^2+x_2^2+x_2^2+x_2^2+x_3^2+x_3^2 \\ & +x_1^2+x_1^2+x_1^2+x_1^2+x_1^2+x_1^2+x_2^2+x_2^2+x_2^2) \\ & -(10x_1+11x_2+8x_3)^2/29 \\ = {} & (10x_1^2+11x_2^2+8x_3^2)-(10x_1+11x_2+8x_2)^2/29. \end{aligned} \tag{2.13}$$

These expressions are also indicated in Table 2.3. We want to maximize the correlation ratio η^2, that is, ss_b/ss_t, in terms of x_1, x_2, and x_3. Since the correlation ratio is invariant under a shift of the origin of original measurements, we set the sum of weighted scores equal to zero (hence set correction term c equal to zero) to simplify the expression of the ratios,

$$y_t = 10x_1 + 11x_2 + 8x_3 = 0. \tag{2.14}$$

Then

$$ss_b = \frac{(x_1+3x_2+6x_3)^2}{10} + \frac{(3x_1+5x_2+2x_3)^2}{10} + \frac{(6x_1+3x_2)^2}{9} \tag{2.15}$$

and

$$ss_t = 10x_1^2 + 11x_2^2 + 8x_3^2. \tag{2.16}$$

The ratio ss_b/ss_t of these last expressions can be algebraically maximized with respect to the weights, and is easier to deal with than the ratio formed by (2.12) and (2.13).

At this stage, it seems useful to derive a more general expression for η^2 so that the same formulation may be applied to a variety of categorical data. Such a general expression can be obtained in terms of matrix notation. Let us illustrate by using the example given in Table 2.1. We define $F=(f_{ji})$ as the data matrix. In our example,

$$F = \begin{bmatrix} 1 & 3 & 6 \\ 3 & 5 & 2 \\ 6 & 3 & 0 \end{bmatrix}.$$

$\mathbf{f}$ is the 3×1 $(m\times 1)$ vector of the column totals of F,

$$\mathbf{f} = \begin{bmatrix} 10 \\ 11 \\ 8 \end{bmatrix}.$$

D is the 3×3 $(m\times m)$ diagonal matrix of the column totals of F,

$$D = \begin{bmatrix} 10 & 0 & 0 \\ 0 & 11 & 0 \\ 0 & 0 & 8 \end{bmatrix}.$$

D_n is the 3×3 $(n\times n)$ diagonal matrix of the row totals of F,

$$D_n = \begin{bmatrix} 10 & 0 & 0 \\ 0 & 10 & 0 \\ 0 & 0 & 9 \end{bmatrix}.$$

$\mathbf{x}$ is a 3×1 $(m\times 1)$ vector of weights (unknown) for the categories,

$$\mathbf{x} = \begin{bmatrix} x_1 \\ x_2 \\ x_3 \end{bmatrix}.$$

In terms of these matrices and vectors, we can write formula (2.14) in the form (see Appendix A for matrix calculus)

$$\begin{bmatrix} 10 & 11 & 8 \end{bmatrix} \begin{bmatrix} x_1 \\ x_2 \\ x_3 \end{bmatrix} = \mathbf{f}'\mathbf{x} = 0 \tag{2.17}$$

and formula (2.16) obtained by the condition (2.14) or (2.17), in the form

$$ss_t = \begin{bmatrix} x_1 & x_2 & x_3 \end{bmatrix} \begin{bmatrix} 10 & 0 & 0 \\ 0 & 11 & 0 \\ 0 & 0 & 8 \end{bmatrix} \begin{bmatrix} x_1 \\ x_2 \\ x_3 \end{bmatrix} = \mathbf{x}'D\mathbf{x}. \tag{2.18}$$

ss_b of (2.15) is relatively difficult to express in matrix notation. First, note that

$$F\mathbf{x} = \begin{bmatrix} 1 & 3 & 6 \\ 3 & 5 & 2 \\ 6 & 3 & 0 \end{bmatrix} \begin{bmatrix} x_1 \\ x_2 \\ x_3 \end{bmatrix} = \begin{bmatrix} x_1+3x_2+6x_3 \\ 3x_1+5x_2+2x_3 \\ 6x_1+3x_2 \end{bmatrix}.$$

Hence

$$\mathbf{x}'F'F\mathbf{x} = (x_1+3x_2+6x_3)^2 + (3x_1+5x_2+2x_3)^2 + (6x_1+3x_2)^2.$$

Noting that each term must be divided by the corresponding element of D_n, we finally obtain

$$ss_b = \mathbf{x}'F' \begin{bmatrix} \frac{1}{10} & 0 & 0 \\ 0 & \frac{1}{10} & 0 \\ 0 & 0 & \frac{1}{9} \end{bmatrix} F\mathbf{x} = \mathbf{x}'F'D_n{}^{-1}F\mathbf{x}. \tag{2.19}$$

Therefore the expression for η^2, simplified by (2.17), becomes

$$\eta^2 = \mathbf{x}'F'D_n{}^{-1}F\mathbf{x}/\mathbf{x}'D\mathbf{x}. \tag{2.20}$$

It may be useful to derive also the general form of η^2, that is, the expression without condition (2.17). Noting that the sum of the weighted scores can be expressed as $\mathbf{f}'\mathbf{x}$, we find the expression for the so-called correction term

$$(10x_1+11x_2+8x_3)^2/29 = \mathbf{x}'\mathbf{f}\mathbf{f}'\mathbf{x}/f_t, \tag{2.21}$$

where f_t is the total number of responses, that is, 29 in our example. Therefore the general forms of ss_b and ss_t, that is, (2.12) and (2.13) respectively, are

$$ss_b = \mathbf{x}'F'D_n{}^{-1}F\mathbf{x} - \mathbf{x}'\mathbf{f}\mathbf{f}'\mathbf{x}/f_t = \mathbf{x}'[F'D_n{}^{-1}F - \mathbf{f}\mathbf{f}'/f_t]\mathbf{x} \tag{2.22}$$

and

$$ss_t = \mathbf{x}'D\mathbf{x} - \mathbf{x}'\mathbf{f}\mathbf{f}'\mathbf{x}/f_t = \mathbf{x}'[D - \mathbf{f}\mathbf{f}'/f_t]\mathbf{x}. \tag{2.23}$$

Hence the general form of η^2 is given by

$$\eta^2 = \frac{ss_b}{ss_t} = \frac{\mathbf{x}'[F'D_n{}^{-1}F - \mathbf{f}\mathbf{f}'/f_t]\mathbf{x}}{\mathbf{x}'[D - \mathbf{f}\mathbf{f}'/f_t]\mathbf{x}}. \tag{2.24}$$

Maximization of η^2 can be carried out by a standard procedure. There are several problems which are peculiar to dual scaling of categorical data, however, so it may be instructive to discuss the procedure step by step. The next section is devoted to this optimization problem.

2.2 OPTIMIZATION

η^2 is a ratio of two quadratic forms. In order to determine the optimum vector $\mathbf{x}$, one can maximize η^2 directly, or maximize ss_b under the constraint that ss_t is fixed. Since the two approaches yield the same set of formulas, we shall use the latter, which is somewhat the simpler.

Let us consider the simplified formula for η^2, that is, (2.20), and the constraint that

$$\mathbf{x}'D\mathbf{x} = f_t \text{ (constant).} \tag{2.25}$$

Our problem then is

$$\text{to maximize} \quad \mathbf{x}'F'D_n^{-1}F\mathbf{x} \quad \text{subject to} \quad \mathbf{x}'D\mathbf{x} = f_t. \tag{2.26}$$

The subsequent optimization procedure can be somewhat simplified by transforming $\mathbf{x}$ to $\mathbf{w}$, where

$$\mathbf{w} = D^{1/2}\mathbf{x}, \quad \text{that is,} \quad \mathbf{x} = D^{-1/2}\mathbf{w}. \tag{2.27}$$

In terms of $\mathbf{w}$, (2.26) can be expressed as

$$\text{to maximize} \quad \mathbf{w}'D^{-1/2}F'D_n^{-1}FD^{-1/2}\mathbf{w} \quad \text{subject to} \quad \mathbf{w}'\mathbf{w} = f_t. \tag{2.28}$$

Using the Lagrange method of multipliers, we can reduce the problem to that of solving the following set of homogeneous equations of the first partial derivatives:

$$\partial Q(\mathbf{w})/\partial\mathbf{w} = \mathbf{0} \quad \text{and} \quad \partial Q(\mathbf{w})/\partial\lambda = 0, \tag{2.29}$$

where

$$Q(\mathbf{w}) = \mathbf{w}'D^{-1/2}F'D_n^{-1}FD^{-1/2}\mathbf{w} - \lambda(\mathbf{w}'\mathbf{w} - f_t), \tag{2.30}$$

λ is an unknown constant, called a Lagrange multiplier, and $\mathbf{0}$ is the null vector. The first equation of (2.29) is

$$\partial Q(\mathbf{w})/\partial\mathbf{w} = 2D^{-1/2}F'D_n^{-1}FD^{-1/2}\mathbf{w} - 2\lambda I\mathbf{w} = \mathbf{0} \tag{2.31}$$

and the second

$$\partial Q(\mathbf{w})/\partial\lambda = \mathbf{w}'\mathbf{w} - f_t = 0. \tag{2.32}$$

From (2.31), we obtain

$$D^{-1/2}F'D_n^{-1}FD^{-1/2}\mathbf{w} = \lambda\mathbf{w}. \tag{2.33}$$

Premultiplying both sides of (2.33) by $\mathbf{w}'$, we obtain

$$\mathbf{w}'D^{-1/2}F'D_n^{-1}FD^{-1/2}\mathbf{w} = \lambda\mathbf{w}'\mathbf{w}. \tag{2.34}$$

Hence

$$\lambda = \frac{\mathbf{w}'D^{-1/2}F'D_n^{-1}FD^{-1/2}\mathbf{w}}{\mathbf{w}'\mathbf{w}} = \frac{\mathbf{x}'F'D_n^{-1}F\mathbf{x}}{\mathbf{x}'D\mathbf{x}}. \tag{2.35}$$

From (2.20) and (2.35), we obtain

$$\lambda = \eta^2. \tag{2.36}$$

Thus the Lagrange multiplier of the present problem is the same as the correlation ratio. The equations to be solved are therefore

$$\left(D^{-1/2}F'D_n^{-1}FD^{-1/2} - \eta^2 I\right)\mathbf{w} = \mathbf{0}, \tag{2.37}$$

$$\mathbf{w}'\mathbf{w} = f_t. \tag{2.38}$$

Note that the weight vector $\mathbf{x}$ can be obtained from $\mathbf{w}$ by (2.27). Note also that the transformation of $\mathbf{x}$ to $\mathbf{w}$ by (2.27) resulted in the standard form of eigenequation (2.37) of a *symmetric* matrix. Without this transformation, the result of the optimization would lead to the expression $(F'D_n^{-1}F - \eta^2 D)\mathbf{x} = 0$, or $(D^{-1}F'D_n^{-1}F - \eta^2 I)\mathbf{x} = 0$. The latter is a standard eigenequation of a *non-symmetric* matrix, and this non-symmetry would later prove rather inconvenient when trying to extract more than one eigenvalue.

The same set of equations can be obtained by maximizing the general form of ss_b, that is, (2.22), under two constraints: (i) $ss_t = f_t$ and (ii) $\mathbf{x}'\mathbf{f}\mathbf{f}'\mathbf{x} = 0$ (i.e., the sum of the weighted scores $\mathbf{f}'\mathbf{x}$ is zero). We then have two Lagrange multipliers, λ_1 and λ_2, and the Lagrangian function is given by

$$Q(\mathbf{x}) = ss_b - \lambda_1(ss_t - f_t) - \lambda_2 \mathbf{x}'\mathbf{f}\mathbf{f}'\mathbf{x}. \tag{2.39}$$

The homogeneous equations to be solved are

$$\partial Q(\mathbf{x})/\partial \mathbf{x} = 2\left[F'D_n^{-1}F - \mathbf{f}\mathbf{f}'/f_t\right]\mathbf{x} - 2\lambda_1\left[D - \mathbf{f}\mathbf{f}'/f_t\right]\mathbf{x} - 2\lambda_2 \mathbf{f}\mathbf{f}'\mathbf{x} = \mathbf{0}, \tag{2.40}$$

$$\partial Q(\mathbf{x})/\partial \lambda_1 = \mathbf{x}'\left[D - \mathbf{f}\mathbf{f}'/f_t\right]\mathbf{x} - f_t = 0, \tag{2.41}$$

$$\partial Q(\mathbf{x})/\partial \lambda_2 = \mathbf{x}'\mathbf{f}\mathbf{f}'\mathbf{x} = 0. \tag{2.42}$$

Some algebraic manipulation of these three equations and (2.27) yields the set of equations (2.37) and (2.38), as expected.

In our example, we first obtain

$$F'D_n^{-1}F = \begin{bmatrix} 1 & 3 & 6 \\ 3 & 5 & 3 \\ 6 & 2 & 0 \end{bmatrix} \begin{bmatrix} \frac{1}{10} & 0 & 0 \\ 0 & \frac{1}{10} & 0 \\ 0 & 0 & \frac{1}{9} \end{bmatrix} \begin{bmatrix} 1 & 3 & 6 \\ 3 & 5 & 2 \\ 6 & 3 & 0 \end{bmatrix}$$

$$= \begin{bmatrix} 5.0 & 3.8 & 1.2 \\ 3.8 & 4.4 & 2.8 \\ 1.2 & 2.8 & 4.0 \end{bmatrix}.$$

Thus

$$D^{-1/2}F'D_n^{-1}FD^{-1/2}$$

$$=\begin{bmatrix} \frac{1}{\sqrt{10}} & 0 & 0 \\ 0 & \frac{1}{\sqrt{11}} & 0 \\ 0 & 0 & \frac{1}{\sqrt{8}} \end{bmatrix}\begin{bmatrix} 5.0 & 3.8 & 1.2 \\ 3.8 & 4.4 & 2.8 \\ 1.2 & 2.8 & 4.0 \end{bmatrix}\begin{bmatrix} \frac{1}{\sqrt{10}} & 0 & 0 \\ 0 & \frac{1}{\sqrt{11}} & 0 \\ 0 & 0 & \frac{1}{\sqrt{8}} \end{bmatrix}$$

$$=\begin{bmatrix} 0.5000 & 0.3623 & 0.1342 \\ 0.3623 & 0.4000 & 0.2985 \\ 0.1342 & 0.2985 & 0.5000 \end{bmatrix}.$$

For this example, (2.37) can be written as

$$\left[\begin{bmatrix} 0.5000 & 0.3623 & 0.1342 \\ 0.3623 & 0.4000 & 0.2985 \\ 0.1342 & 0.2985 & 0.5000 \end{bmatrix} - \eta^2\begin{bmatrix} 1 & 0 & 0 \\ 0 & 1 & 0 \\ 0 & 0 & 1 \end{bmatrix}\right]\begin{bmatrix} w_1 \\ w_2 \\ w_3 \end{bmatrix} = \begin{bmatrix} 0 \\ 0 \\ 0 \end{bmatrix}.$$

or

$$(0.5000-\eta^2)w_1 + 0.3623w_2 + 0.1342w_3 = 0,$$

$$0.3623w_1 + (0.4000-\eta^2)w_2 + 0.2985w_3 = 0,$$

$$0.1324w_1 + 0.2985w_2 + (0.5000-\eta^2)w_3 = 0.$$

The ordinary procedure is to solve (2.37) first, and then rescale **w** so that (2.38) is satisfied. The objective is to find **x** that maximizes η^2. Therefore, once **w** is obtained, **w** has to be converted to **x** by (2.27).

The distinction between **x** and **w** may be somewhat confusing, and a few words of explanation might help. As stated in the foregoing discussion, it is **x**, not **w**, that maximizes η^2. Recall that **w** was determined so as to maximize

$$\frac{\mathbf{w}'D^{-1/2}F'D_n^{-1}FD^{-1/2}\mathbf{w}}{\mathbf{w}'\mathbf{w}} = \frac{\mathbf{x}'F'D_n^{-1}F\mathbf{x}}{\mathbf{x}'D\mathbf{x}} = \eta^2(\mathbf{x}), \text{ say,} \tag{2.43}$$

and not the quantity

$$\mathbf{w}'F'D_n^{-1}F\mathbf{w}/\mathbf{w}'D\mathbf{w} = \eta^2(\mathbf{w}), \text{ say.}$$

Therefore the **w** that maximizes $\eta^2(\mathbf{x})$ is in general not the optimal weight vector, but the so-called principal component vector of the standardized F, that is, $D_n^{-1/2}FD^{-1/2}$. However, **w** becomes proportional to the optimal weight vector **x** when the diagonal elements of D and D_n are constant, that is, $D=c_1I$ and $D_n=c_2I$, where c_1 and c_2 are non-zero constants and I is the identity matrix.

Relevant to the above discussion is the following observation. When one wants to maximize ss_b with a constraint on $\mathbf{x}$, commonly one sets the constraint $\mathbf{x}'\mathbf{x}=1$ (or any other constant). This practice is equivalent to maximizing the following ratio with respect to $\mathbf{x}$:

$$\mathbf{x}'F'D_n^{-1}F\mathbf{x}/\mathbf{x}'\mathbf{x} = \psi, \text{ say.} \tag{2.44}$$

A crucial point, relevant to our problem, lies in the fact that maximization of ψ does not necessarily mean maximization of η^2. It is true that maximization of η^2 also involves maximization of the relative contribution of ss_b, but an important point is that ψ of (2.44) does not utilize the information contained in the distribution of the column marginals of F, which is a part of η^2. In other words, ψ is in general different from η^2. Once the vector $\mathbf{x}$ that maximizes η^2 is obtained, one can, of course, rescale $\mathbf{x}$ so that $\mathbf{x}'\mathbf{x}=1$. This rescaling, called normalization, does not change η^2, but is quite different from maximizing ψ. To summarize, let us indicate by $\eta^2(\mathbf{x}'D\mathbf{x})$ the correlation ratio obtained by maximizing ss_b, given $\mathbf{x}'D\mathbf{x}=1$ (or any other constant), and by $\eta^2(\mathbf{x}'\mathbf{x})$ the correlation ratio obtained from $\mathbf{x}$ that maximizes ψ. Then the relation

$$\eta^2(\mathbf{x}'D\mathbf{x}) \geq \eta^2(\mathbf{x}'\mathbf{x}) \tag{2.45}$$

holds. When $\mathbf{x}'\mathbf{x}$ is fixed, and used as the constraint for maximization of ss_b, it maximizes ψ and not η^2. Thus, using a similar notation, one can also state that

$$\psi(\mathbf{x}'\mathbf{x}) \geq \psi(\mathbf{x}'D\mathbf{x}), \tag{2.46}$$

where $\psi(\mathbf{x}'\mathbf{x})$ is the maximum of (2.44) and $\psi(\mathbf{x}'D\mathbf{x})$ is the value of ψ obtained from $\mathbf{x}$ that maximizes η^2. The above discussion simply points out that $\eta^2(\mathbf{x}'D\mathbf{x})$ in (2.45) and $\psi(\mathbf{x}'\mathbf{x})$ in (2.46) are upper limits of η^2 and ψ, respectively, for all possible vectors $\mathbf{x}$, except $\mathbf{x}=\mathbf{1}$ for η^2.

2.2.1 *Trivial solutions*

Before proceeding further, we should note that (2.37) has two *trivial* solutions, that is, solutions which are unrelated to the given data. The first trivial solution is the familiar one

$$\text{Trivial solution I: } \mathbf{w} = \mathbf{0}. \tag{2.47}$$

It is obvious that (2.47) is a solution since the right-hand side of (2.37) is $\mathbf{0}$. The second trivial solution is

$$\text{Trivial solution II: } \mathbf{w} = D^{1/2}\mathbf{1} \quad (\text{i.e., } \mathbf{x}=\mathbf{1}). \tag{2.48}$$

$\mathbf{1}$ is the vector of 1's. This vector $\mathbf{w}$ satisfies (2.37), whatever the data matrix may be, and provides that $\eta^2=1$. To illustrate, let us substitute $D^{1/2}\mathbf{1}$ for $\mathbf{w}$ in the left-hand side of (2.37),

$$\begin{aligned}(D^{-1/2}F'D_n^{-1}FD^{-1/2}-\eta^2I)D^{1/2}\mathbf{1} &= D^{-1/2}F'D_n^{-1}F\mathbf{1}-\eta^2D^{1/2}\mathbf{1}\\ &= D^{-1/2}F'D_n^{-1}D_n\mathbf{1}-\eta^2D^{1/2}\mathbf{1} = D^{-1/2}F'\mathbf{1}-\eta^2D^{1/2}\mathbf{1}\\ &= D^{-1/2}D\mathbf{1}-\eta^2D^{1/2}\mathbf{1} = (1-\eta^2)D^{1/2}\mathbf{1}.\end{aligned} \tag{2.49}$$

The right-hand side of (2.37) is $\mathbf{0}$. Therefore, when $\mathbf{w}=D^{1/2}\mathbf{1}$, we obtain $\eta^2=1$, no matter what F we may have.

A few interesting observations can be made about the second trivial solution. First, when $\eta^2=1$ it follows from (2.1) and (2.2) that $ss_w=0$. This condition is too restrictive to hold for real data. Second, the solution $\mathbf{w}=D^{1/2}\mathbf{1}$ does not satisfy the condition $\mathbf{f}'\mathbf{x}=0$, for

$$\mathbf{f}'\mathbf{x} = \mathbf{f}'D^{-1/2}\mathbf{w} = \mathbf{f}'D^{-1/2}D^{1/2}\mathbf{1} = \mathbf{f}'\mathbf{1} = f_t\ (\neq 0).$$

Third, if the weight vector $\mathbf{x}$ is equal to $\mathbf{1}$ all the categories receive the same weight, which amounts to placing all the responses into a single category, resulting in no variations within groups (i.e., $ss_w=0$). Thus it is understandable why the vector $\mathbf{x}=\mathbf{1}$ gives the absolute maximum of η^2 and why it does so, irrespective of the data.

The solution (2.48) is not only trivial but also an artefact arising from our formulation of the problem. It is an outcome of the condition indicated by (2.17). One may then argue why (2.17) should be employed. The reason is simply a matter of computational convenience. Under (2.17), the general form (2.24) is reduced to (2.20) and then yields (2.37), which is readily amenable to a standard procedure of computation. If (2.17) is not employed, we note that the matrix of the quadratic form of ss_t is always singular, yielding an equation of the type

$$(A-\lambda B)\mathbf{u} = \mathbf{0}, \tag{2.50}$$

where A and B are both singular. One cannot easily obtain the standard equation of an eigenvalue λ and the corresponding eigenvector $\mathbf{u}$ of matrix R, say, namely,

$$(R-\lambda I)\mathbf{u} = \mathbf{0}. \tag{2.51}$$

Although some satisfactory approaches to problems presented in the form (2.50) (e.g., McDonald, Torii, and Nishisato 1979) will be discussed later, our current problem can be handled rather easily under the simplifying condition (2.17), which generates the second trivial solution. What needs to be done then is to eliminate the two trivial solutions from the computation.

The first trivial solution can be eliminated by the standard method which involves solving the determinantal equation of (2.37), and this will be discussed later. For the second trivial solution, we can use a procedure that has been widely used in multivariate correlational analysis. Since this procedure is also applicable to the extraction of orthogonal scales, a topic to be discussed later, it will be described in detail.

Consider the equation (2.51) and let R be a symmetric matrix. Then we can eliminate the contribution of any solution, say $(\lambda_k, \mathbf{u}_k)$, from the equation. To do so, we only need to calculate the residual matrix R_k from the equation

$$R_k = R - \frac{R\mathbf{u}_k\mathbf{u}_k'R}{\mathbf{u}_k'R\mathbf{u}_k}. \tag{2.52}$$

The rationale of this procedure will be made clear when we present Lagrange's theorem in the next paragraph. Now, since $\mathbf{u}_k$ is a solution vector, we can substitute $\lambda_k\mathbf{u}_k$ for $R\mathbf{u}_k$. In other words, $R\mathbf{u}_k = \lambda_k\mathbf{u}_k$. Therefore (2.52) can be rewritten as

$$R_k = R - \frac{\lambda_k\mathbf{u}_k\mathbf{u}_k'\lambda_k}{\lambda_k\mathbf{u}_k'\mathbf{u}_k} = R - \lambda_k\frac{\mathbf{u}_k\mathbf{u}_k'}{\mathbf{u}_k'\mathbf{u}_k}. \tag{2.53}$$

If we normalize $\mathbf{u}_k$ into $\mathbf{u}_k{}^*$ by applying

$$\mathbf{u}_k{}^* = \mathbf{u}_k/\sqrt{\mathbf{u}_k'\mathbf{u}_k}\ , \tag{2.54}$$

then we obtain another expression for R_k,

$$R_k = R - \lambda_k\mathbf{u}_k{}^*\mathbf{u}_k{}^{*\prime}. \tag{2.55}$$

Normalization here means scaling the unit of $\mathbf{u}_k$ so that $\mathbf{u}_k{}^{*\prime}\mathbf{u}_k{}^* = 1$. The equation

$$(R_k - \lambda I)\mathbf{u} = \mathbf{0} \tag{2.56}$$

is now free from the effect of the solution $(\lambda_k, \mathbf{u}_k)$. In other words, we have eliminated the solution $(\lambda_k, \mathbf{u}_k)$ from equation (2.51).

Another important aspect of the above elimination process is what Lagrange's theorem states, namely, if the rank of R is r, the residual matrix R_k, defined by (2.52), is exactly of rank $r-1$. Lagrange's theorem has been presented by Rao (1965, p. 55) in a more general form:

LAGRANGE'S THEOREM *Let S be any square matrix of order m and rank $r>0$, and $\mathbf{x}$ and $\mathbf{y}$ be column vectors such that $\mathbf{x}'S\mathbf{y} \neq 0$. Then the residual matrix*

$$S_1 = S - \frac{S\mathbf{y}\mathbf{x}'S}{\mathbf{x}'S\mathbf{y}} \tag{2.57}$$

is exactly of rank $r-1$.

In our case, S is symmetric and $\mathbf{x} = \mathbf{y}$ is an eigenvector. Rao presents another, more general form of the theorem on the same page:

THEOREM *If S is $n \times m$ of rank $r > 0$, and A and B are of order $s \times n$ and $s \times m$ respectively, where $s \leq r$ and ASB' is non-singular, the residual matrix*

$$S_1 = S - SB'(ASB')^{-1}AS \tag{2.58}$$

is exactly of rank $r-s$ and S_1 is Gramian (i.e., symmetric and positive semi-definite) *if S is Gramian.*

This theorem will be later applied to one of the variants of dual scaling, the so-called method of reciprocal averages.

Let us now apply the above procedure to eliminate the second trivial solution, that is, $\eta^2=1$ and $\mathbf{w}=D^{1/2}\mathbf{1}$, from (2.37). Using (2.53) and indicating the residual matrix by C_1, we obtain

$$\begin{aligned} C_1 &= D^{-1/2}F'D_n^{-1}FD^{-1/2} - \eta^2\frac{\mathbf{w}\mathbf{w}'}{\mathbf{w}'\mathbf{w}} \\ &= D^{-1/2}F'D_n^{-1}FD^{-1/2} - \frac{D^{1/2}\mathbf{1}\mathbf{1}'D^{1/2}}{\mathbf{1}'D\mathbf{1}} \\ &= D^{-1/2}F'D_n^{-1}FD^{-1/2} - \frac{1}{f_t}D^{1/2}\mathbf{1}\mathbf{1}'D^{1/2} \end{aligned} \tag{2.59}$$

since $\mathbf{1}'D\mathbf{1}=f_t$. Once we define C_1 by (2.59), the equation

$$C_1\mathbf{w} = \eta^2\mathbf{w}, \quad \text{or} \quad (C_1-\eta^2 I)\mathbf{w} = \mathbf{0}, \tag{2.60}$$

is free from the effect of the trivial solution, that is, $\eta^2=1$ and $\mathbf{w}=D^{1/2}\mathbf{1}$.

In our example,

$$D^{1/2}\mathbf{1}\mathbf{1}'D^{1/2}/f_t$$

$$= \frac{1}{29}\begin{bmatrix} \sqrt{10} & 0 & 0 \\ 0 & \sqrt{11} & 0 \\ 0 & 0 & \sqrt{8} \end{bmatrix}\begin{bmatrix} 1 \\ 1 \\ 1 \end{bmatrix}\begin{bmatrix} 1 & 1 & 1 \end{bmatrix}\begin{bmatrix} \sqrt{10} & 0 & 0 \\ 0 & \sqrt{11} & 0 \\ 0 & 0 & 0\sqrt{8} \end{bmatrix}$$

$$= \begin{bmatrix} 0.3448 & 0.3617 & 0.3084 \\ 0.3617 & 0.3793 & 0.3235 \\ 0.3084 & 0.3235 & 0.2759 \end{bmatrix}.$$

Therefore, using this matrix and the matrix $D^{-1/2}F'D_n^{-1}FD^{-1/2}$, calculated earlier, we obtain the residual matrix

$$C_1 = \begin{bmatrix} 0.1552 & 0.0006 & -0.1742 \\ 0.006 & 0.0207 & -0.0250 \\ -0.1742 & -0.0250 & 0.2241 \end{bmatrix}.$$

(2.60) is still subject to the first trivial solution, $\mathbf{w}=\mathbf{0}$. In order to avoid this, we follow the standard procedure. We note that η^2 in (2.60) must satisfy the determinantal equation

$$|C_1 - \eta^2 I| = 0. \tag{2.61}$$

When C_1 is $m\times m$, (2.61) is a polynomial equation of the mth degree in $-\eta^2$, or, more specifically,

$$|C_1 - \eta^2 I| = (-\eta^2)^m + \sigma_1(\eta^2)^{m-1} + \sigma_2(-\eta^2)^{m-2} + \cdots + \sigma_{m-1}(-\eta^2) + \sigma_m$$
$$= \sum_{k=0}^{m} (-\eta^2)^{m-k} \sigma_k, \tag{2.62}$$

where $\sigma_0 = 1$, σ_1 is the sum of the diagonal elements of C_1, that is, $\text{tr}(C_1)$, σ_k is the sum of all the k-rowed principal minor determinants, and hence $\sigma_m = |C_1|$. In our example,

$$\sigma_1 = 0.1552 + 0.0207 + 0.2241,$$

$$\sigma_2 = \begin{vmatrix} 0.1552 & 0.0006 \\ 0.0006 & 0.0207 \end{vmatrix} + \begin{vmatrix} 0.1552 & -0.1742 \\ -0.1742 & 0.2241 \end{vmatrix} + \begin{vmatrix} 0.0207 & -0.0250 \\ -0.0250 & 0.2241 \end{vmatrix},$$

$$\sigma_3 = |C_1| = 0$$

(according to Lagrange's theorem, the rank of C_1 is 2, and hence C_1 is singular). Therefore

$$|C_1 - \eta^2 I| = (-\eta^2)^3 + \sigma_1(-\eta^2)^2 + \sigma_2(-\eta^2)$$
$$= -\eta^2\left[(\eta^2)^2 - \sigma_1(\eta^2) + \sigma_2\right] = 0.$$

Hence there are three values of η^2 that satisfy the determinantal equation. One of them is obviously zero.

Note that our task started with maximizing η^2 with respect to $\mathbf{x}$. Of the m possible values of η^2 that satisfy (2.61), we are thus interested only in the largest value. Once this is identified, we substitute the value in (2.60) and solve the equation for $\mathbf{w}$. The weight vector $\mathbf{x}$ can then be calculated from $\mathbf{w}$ by (2.27).

Generally speaking, to solve a polynomial equation of high order is a difficult and laborious task. However, we are not interested in extracting all the eigenvalues, but just the largest one. For this purpose, we present an iterative method.

2.2.2 *Iterative procedure*

There are many numerical methods for extracting only the largest eigenvalue and the corresponding eigenvector from an equation (e.g., Hotelling 1936; Frazer, Duncan, and Collar 1938; Faddeev and Faddeeva 1963; Fox 1964). These methods employ essentially the same technique of generating a convergent sequence of vectors and computing the largest eigenvalue by an iterative process. These methods use the relation that the *eigenvalue of the pth power of a matrix is the pth power of the eigenvalue of the matrix*, that is,

$$R^p \mathbf{u} = \lambda^p \mathbf{u}. \tag{2.63}$$

The derivation of a general procedure is based on the following rationale. Let us first assume that all eigenvalues are distinct and different from zero. Then there are m independent vectors $\mathbf{u}_i$ associated with the m eigenvalues, where m is the order of R. Therefore any arbitrary $m \times 1$ vector $\mathbf{b}_0$, say, can be expressed as a linear combination of $\mathbf{u}_i$,

$$\mathbf{b}_0 = \sum_{i=1}^{m} c_i \mathbf{u}_i, \tag{2.64}$$

where c_i are constants. Since $R\mathbf{u} = \lambda\mathbf{u}$, we obtain the relation

$$R\mathbf{b}_0 = \sum_{i=1}^{m} \lambda_i c_i \mathbf{u}_i. \tag{2.65}$$

Consider now the sequence of vectors

$$\mathbf{b}_0,\quad R\mathbf{b}_0 = \mathbf{b}_1,\quad R\mathbf{b}_1 = R^2\mathbf{b}_0 = \mathbf{b}_2,\quad R\mathbf{b}_2 = R^3\mathbf{b}_0 = \mathbf{b}_3,\quad \ldots,$$
$$R\mathbf{b}_{p-1} = R^p\mathbf{b}_0 = \mathbf{b}_p. \tag{2.66}$$

Then we can obtain from (2.63) and (2.65) the following expression for $\mathbf{b}_p$:

$$\mathbf{b}_p = R^p\mathbf{b}_0 = \sum_{i=1}^{m} \lambda_i^p c_i \mathbf{u}_i. \tag{2.67}$$

Let us assume that the eigenvalues are ordered, with $\lambda_1 > \lambda_2 > \cdots > \lambda_m$, and rewrite (2.67) as

$$\mathbf{b}_p = \lambda_1^p \left[c_1\mathbf{u}_1 + \sum_{j=2}^{m} \left(\frac{\lambda_j}{\lambda_1} \right)^p c_j \mathbf{u}_j \right]. \tag{2.68}$$

Hence

$$\lim_{p \to \infty} \mathbf{b}_p = \lambda_1^p c_1 \mathbf{u}_1. \tag{2.69}$$

Therefore, if p is sufficiently large, the largest eigenvalue can be approximated by, for example,

$$\lambda_1 \cong \frac{\mathbf{b}_p'\mathbf{b}_p}{\mathbf{b}_p'\mathbf{b}_{p-1}} = \frac{\mathbf{b}_0' R^{2p} \mathbf{b}_0}{\mathbf{b}_0' R^{2p-1} \mathbf{b}_0}. \tag{2.70}$$

Once λ_1 is obtained, the corresponding eigenvector $c_1\mathbf{u}_1$ can easily be obtained from $\mathbf{b}_p$ (note that if $\mathbf{u}_1$ is a solution vector, then $c_1\mathbf{u}_1$ is also a solution vector). In this book, we use the following iterative method, which is based on the same principle:

Step I Multiply C_1 by an *arbitrary vector* $\mathbf{b}_0$, and indicate the product by $\mathbf{b}_1$, that is, $C_1\mathbf{b}_0 = \mathbf{b}_1$.

Step II Divide the elements of $\mathbf{b}_1$ by the greatest absolute value of the elements in $\mathbf{b}_1$, say $|k_1|$, and indicate the resulting vector by $\mathbf{b}_1^*$, that is, $\mathbf{b}_1/|k_1| = \mathbf{b}_1^*$.

Step III Compute $C_1\mathbf{b}_1^* = \mathbf{b}_2$, and standardize $\mathbf{b}_2$ in the same way as in Step II, that is, $\mathbf{b}_2/|k_2| = \mathbf{b}_2^*$, where $|k_2|$ is the greatest absolute value of the elements of $\mathbf{b}_2$.

Step IV Repeat the process

$$C_1\mathbf{b}_j^* = \mathbf{b}_{j+1},\ \mathbf{b}_{j+1}/|k_{j+1}| = \mathbf{b}_{j+1}^*, \tag{2.71}$$

until $\mathbf{b}_j^*$ becomes identical or almost identical with $\mathbf{b}_{j+1}^*$.

When the difference between $\mathbf{b}_j^*$ and $\mathbf{b}_{j+1}^*$ becomes negligible, $\mathbf{b}_{j+1}^*$ (or $\mathbf{b}_{j+1}$) provides an eigenvector, associated with the largest eigenvalue of C_1, which is given by $|k_{j+1}|$. Thus, this iterative method solves simultaneously two equations (2.60) and (2.61) with respect to the largest eigenvalue. $|k_{j+1}|$ corresponds to the maximum η^2 and $\mathbf{b}_{j+1}^*$ is a vector collinear with $\mathbf{w}$. If one wishes faster convergence of this algorithm through modifications of the iterative formula, see, for example, Faddeev and Faddeeva (1963) and Ramsay (1975). The vector $\mathbf{w}$ that satisfies (2.38) can be obtained from $\mathbf{b}_{j+1}^*$ by

$$\mathbf{w} = \left(\frac{f_t}{\mathbf{b}_{j+1}^{*\prime}\mathbf{b}_{j+1}^*}\right)^{1/2}\mathbf{b}_{j+1}^*, \tag{2.72}$$

for then

$$\mathbf{w}'\mathbf{w} = \frac{f_t}{\mathbf{b}_{j+1}^{*\prime}\mathbf{b}_{j+1}^*}\mathbf{b}_{j+1}^{*\prime}\mathbf{b}_{j+1}^* = f_t. \tag{2.73}$$

Once $\mathbf{w}$ is obtained, the weight vector $\mathbf{x}$ that maximizes η^2 is obtained from (2.27), that is,

$$\mathbf{x} = D^{-1/2}\mathbf{w}. \tag{2.74}$$

Since $\mathbf{x}$ maximizes η^2, it is called the *optimal* weight vector (Bock 1960).

Let us apply this iterative method to our example. Since C_1 is 3×3, the initial vector $\mathbf{b}_0$ has to be 3×1. Although $\mathbf{b}_0$ is an arbitrary vector, there are occasions when a particular choice of $\mathbf{b}_0$ causes a serious problem. An obvious example is a choice of $\mathbf{b}_0$ such that $C_1\mathbf{b}_0 = \mathbf{b}_1 = \mathbf{0}$. Then the iterative process has to be immediately terminated because it is already trapped at the location of the first trivial solution. Such a situation can easily arise if we choose $\mathbf{b}_0$ equal to $\mathbf{1}$ because the sum of each row of C_1 is very close to zero, owing to the elimination of the second trivial solution. The problem can be avoided by choosing a $\mathbf{b}_0$ which is different from the vector of constant elements. Another example is a choice of $\mathbf{b}_0$ as a vector orthogonal to the solution vector. The convergence of $\mathbf{b}_1^*$ to the solution vector may then not be possible. This situation occurs rarely, but it can arise when a small matrix is subjected to the iterative method. There may be other instances where a choice of the initial vector $\mathbf{b}_0$ makes the iterative scheme unworkable. But, generally speaking, such possibilities seldom materialize in practice, for data matrix F is a complex array of numbers, and the likelihood that it will possess such highly regular patterns that the scheme will be unworkable is very slight. We would, however, advise that when the order of magnitudes of the weights can be intuitively judged, the elements of the initial vector be set up to reflect that order. Following such a practice can lead to a quick convergence of the sequence to the solution. In our example, we choose the following vector as $\mathbf{b}_0$:

$$\mathbf{b}_0 = \begin{bmatrix} 1 \\ 0 \\ -1 \end{bmatrix}.$$

Then

$$C_1\mathbf{b}_0 = \begin{bmatrix} 0.1552 & 0.0006 & -0.1742 \\ 0.0006 & 0.0207 & -0.0250 \\ -0.1742 & -0.0250 & 0.2241 \end{bmatrix} \begin{bmatrix} 1 \\ 0 \\ -1 \end{bmatrix} = \begin{bmatrix} 0.3294 \\ 0.0256 \\ -0.3983 \end{bmatrix} = \mathbf{b}_1.$$

The largest absolute value in $\mathbf{b}_1$ is 0.3983. Therefore $|k_1| = 0.3983$, and

$$\mathbf{b}_1{}^* = \frac{\mathbf{b}_1}{0.3983} = \begin{bmatrix} 0.8270 \\ 0.0643 \\ -1.0000 \end{bmatrix}.$$

Similarly,

$$C_1\mathbf{b}_1{}^* = \begin{bmatrix} 0.3026 \\ 0.0268 \\ -0.3698 \end{bmatrix} = \mathbf{b}_2, \qquad |k_2| = 0.3698,$$

$$\mathbf{b}_2{}^* = \frac{\mathbf{b}_2}{0.3698} = \begin{bmatrix} 0.8183 \\ 0.0725 \\ -1.0000 \end{bmatrix},$$

$$C_1\mathbf{b}_2{}^* = \begin{bmatrix} 0.3012 \\ 0.0270 \\ -0.3685 \end{bmatrix} = \mathbf{b}_3, \qquad |k_3| = 0.3685,$$

$$\mathbf{b}_3{}^* = \frac{\mathbf{b}_3}{0.3685} = \begin{bmatrix} 0.8174 \\ 0.0733 \\ -1.0000 \end{bmatrix},$$

$$C_1\mathbf{b}_3{}^* = \begin{bmatrix} 0.3011 \\ 0.0270 \\ -0.3683 \end{bmatrix} = \mathbf{b}_4, \qquad |k_4| = 0.3683,$$

$$\mathbf{b}_4{}^* = \frac{\mathbf{b}_4}{0.3683} = \begin{bmatrix} 0.8175 \\ 0.0733 \\ -1.0000 \end{bmatrix},$$

$$C_1\mathbf{b}_4{}^* = \begin{bmatrix} 0.3011 \\ 0.0270 \\ -0.3683 \end{bmatrix} = \mathbf{b}_5, \qquad |k_5| = 0.3683,$$

$$\mathbf{b}_5{}^* = \frac{\mathbf{b}_5}{0.3683} = \begin{bmatrix} 0.8175 \\ 0.0733 \\ -1.0000 \end{bmatrix}.$$

Since $\mathbf{b}_4{}^* = \mathbf{b}_5{}^*$ and $|k_5| = 0.3683$, we obtain the solution

$$\eta^2 = 0.3683 \quad \text{and} \quad \mathbf{b}_5{}^* = \begin{bmatrix} 0.8175 \\ 0.0733 \\ -1.0000 \end{bmatrix}.$$

In this example, $f_t = 29$. Therefore $\mathbf{w}$ can be obtained from $\mathbf{b}_5{}^*$ by (2.72):

$$\mathbf{w} = \left(\frac{29}{\mathbf{b}_5{}^{*\prime}\mathbf{b}_5{}^*}\right)^{1/2} \mathbf{b}_5{}^* = \begin{bmatrix} 3.4029 \\ 0.3051 \\ -4.1626 \end{bmatrix}.$$

The optimal weight vector is given by

$$\mathbf{x} = D^{-1/2}\mathbf{w} = \begin{bmatrix} \frac{1}{\sqrt{10}} & 0 & 0 \\ 0 & \frac{1}{\sqrt{11}} & 0 \\ 0 & 0 & \frac{1}{\sqrt{8}} \end{bmatrix} \begin{bmatrix} 3.4029 \\ 0.3051 \\ -4.1626 \end{bmatrix} = \begin{bmatrix} 1.0761 \\ 0.0920 \\ -1.4717 \end{bmatrix}.$$

The elements of $\mathbf{x}$, that is, 1.0761, 0.0920, and -1.4717, are optimal weights for categories good, average, and poor, respectively. These are the values that maximize η^2.

In practice, the ultimate interest often lies in finding scores, y_j, for the rows ('teachers' in our example) of the data matrix. It is known (e.g., Fisher 1940; Guttman 1941) that scores y_j are *proportional* to the means of the responses weighted by x_i. Furthermore, the constant of proportionality is known to be equal to $1/\eta$ (Nishisato 1976), where η is the positive square-root of η^2 (η will be used in this sense hereafter). Therefore the optimal vector of y_j is

$$\mathbf{y} = D_n{}^{-1}F\mathbf{x}/\eta. \tag{2.75}$$

If we indicate by $\mathbf{g}$ the vector of row totals of F, we obtain

$$\mathbf{g}'\mathbf{y} = \mathbf{g}'D_n{}^{-1}F\mathbf{x}/\eta = \mathbf{1}'F\mathbf{x}/\eta = \mathbf{f}'\mathbf{x}/\eta = 0, \tag{2.76}$$

since $\mathbf{f}'\mathbf{x} = 0$ by (2.17). In our example, the optimal score vector is

$$\begin{bmatrix} y_1 \\ y_2 \\ y_3 \end{bmatrix} = \frac{1}{\sqrt{0.3682}} \begin{bmatrix} \frac{1}{10} & 0 & 0 \\ 0 & \frac{1}{10} & 0 \\ 0 & 0 & \frac{1}{9} \end{bmatrix} \begin{bmatrix} 1 & 3 & 6 \\ 3 & 5 & 2 \\ 6 & 3 & 0 \end{bmatrix} \begin{bmatrix} 1.0761 \\ 0.0920 \\ -1.4717 \end{bmatrix} = \begin{bmatrix} -1.2322 \\ 0.1227 \\ 1.2326 \end{bmatrix}.$$

These are the most discriminative scores for the three teachers, Teacher 3 being the best and Teacher 1 the worst of the three.

2.3 STATISTICAL TESTS

The solution $[\eta^2, \mathbf{x}, \mathbf{y}]$ is optimum in the sense that it maximizes the criterion, that is, η^2. Once the optimum solution is obtained, we would like to know whether the scores y_j can discriminate significantly among the three teachers, for example, and also whether the solution $[\eta^2, \mathbf{x}, \mathbf{y}]$ explains the original data in an exhaustive manner. These are two separate questions since a set of highly discriminative scores does not guarantee an exhaustive account of the total variance.

To answer the above questions, let us briefly review some statistical tests for similar problems, developed on the basis of normal theory assumptions. Let us indicate by T, B, and W, respectively, the total sum of products matrix of p populations on q variables, the sum of products matrix for between-populations, and the sum of products matrix for within-populations. Consider the eigenvalues λ, μ, and ϕ of the equations

$$|W-\lambda T| = 0, \quad |B-\mu T| = 0, \quad |B-\phi W| = 0. \tag{2.77}$$

These determinantal equations are obtained respectively from the minimization or maximization of the following ratios in terms of $\mathbf{x}$:

$$\left(\frac{\mathbf{x}'W\mathbf{x}}{\mathbf{x}'T\mathbf{x}}\right), \quad \left(\frac{\mathbf{x}'B\mathbf{x}}{\mathbf{x}'T\mathbf{x}}\right), \quad \left(\frac{\mathbf{x}'B\mathbf{x}}{\mathbf{x}'W\mathbf{x}}\right).$$

The eigenvalues are the ratios of the quadratic forms expressed in terms of the solution vector, that is,

$$\lambda = \frac{\mathbf{x}'W\mathbf{x}}{\mathbf{x}'T\mathbf{x}}, \qquad \mu = \frac{\mathbf{x}'B\mathbf{x}}{\mathbf{x}'T\mathbf{x}}, \qquad \phi = \frac{\mathbf{x}'B\mathbf{x}}{\mathbf{x}'W\mathbf{x}}.$$

Since $T = B + W$, simple algebraic manipulations lead to the relation

$$\lambda = 1 - \mu = 1/(1+\phi). \tag{2.78}$$

Let N_i be the number of subjects in group i, and

$$N = \sum_{i=1}^{p} N_i.$$

Define the ratio of the determinants of W and T by

$$|W|/|T| = \Lambda, \text{ say,} \tag{2.79}$$

where Λ is called Wilks' lambda. Λ varies between zero and unity. If the means of the p groups are equal, Λ becomes unity. Thus, in order to test the equality of p group means, one can consider the statistic

$$\chi^2 = -N \log_e \Lambda, \tag{2.80}$$

which is approximately a chi-square (χ^2) with $q(p-1)$ degrees of freedom (see, e.g., Kendall 1957, p. 132). A refinement of (2.80) has been proposed by

Bartlett (1947), namely

$$\chi^2 = -\left[N-1-\tfrac{1}{2}(p+q)\right]\log_e\Lambda, \tag{2.81}$$
$$\mathrm{df} = q(p-1).$$

Rao (1952) proposed another statistic,

$$F = \frac{1-\Lambda^{1/s}}{\Lambda^{1/s}}\,\frac{ms-2\lambda}{q(p-1)}, \tag{2.82}$$
$$\mathrm{df}_1 = q(p-1), \qquad \mathrm{df}_2 = ms-2\lambda,$$

where

$$m = N-1-\tfrac{1}{2}(p+q), \qquad s = \sqrt{\frac{q^2(1-p)^2-4}{(1-p)^2+q^2-5}},$$
$$\lambda = \left[q(p-1)-2\right]/4.$$

Suppose that there are t eigenvalues associated with (2.77). Recalling that the determinant of a symmetric matrix is equal to the product of the eigenvalues, we obtain from (2.78) and (2.79)

$$\Lambda = \prod_{i=1}^{t}\lambda_i = \prod_{i=1}^{t}(1-\mu_i) = \prod_{i=1}^{t}\frac{1}{1+\phi_i}. \tag{2.83}$$

Thus Λ is a function of many eigenvalues. Therefore, if we express (2.80), (2.81), and (2.82) in terms of λ_i, μ_i, or ϕ_i using (2.83), it is obvious that these statistics would not tell us in which dimensions significant differences exist. Recall that in dual scaling we are primarily interested in testing the discriminability of the dimension associated with the largest correlation ratio η^2. We therefore need a test statistic which deals with individual eigenvalues, λ_i, μ_i, or ϕ_i, rather than a function of many eigenvalues such as Λ.

Of the three eigenvalues λ, μ, and ϕ in (2.77), μ is the one that corresponds to η^2 in dual scaling. Let t eigenvalues μ_j be arranged according to the order of their values $\mu_1>\mu_2>\cdots>\mu_t$. Rao (1952, p. 373) states that $q(p-1)$ degrees of freedom are distributed among the various eigenvalues μ_1, μ_2, $\mu_3,\ldots$:

$$q(p-1) = (p+q-2)+(p+q-4)+(p+q-6)+\cdots \tag{2.84}$$

Then a χ^2 may be calculated for each eigenvalue by the formula

$$\chi^2 = -\left[N-1-\tfrac{1}{2}(p+q)\right]\log_e(1-\mu_j), \tag{2.85}$$
$$\mathrm{df} = p+q-2j, \quad \text{where } j = 1,2,3,\ldots,t.$$

This statistic has been widely used in discriminant analysis.

There is an alternative criterion to that of Λ for judging the overall group differences, namely that

$$\chi^2 = N(\mu_1+\mu_2+\cdots+\mu_t) \tag{2.86}$$

is a χ^2 approximation with $q(p-1)$ degrees of freedom. Using this test statistic, one may wish to introduce the statistic $N\mu_j$ as a χ^2 approximation with $p+q-2j$ degrees of freedom for testing the significance of μ_j. According to Lancaster (1969), some researchers such as Kendall and Stuart (1961) and Williams (1952) have attempted to avoid the difficulties of the distribution theory by assuming that $N\mu_j$ is distributed as χ^2, but Lancaster (1963) showed that this assumption is not well founded. $N\mu_1$ is stochastically greater than χ^2 of the corresponding degrees of freedom and $N\mu_t$ is stochastically smaller than the corresponding value of χ^2.

In dual scaling, data are generally categorical and are different from those for which the test statistics discussed above were derived. In consequence, some attempts have been made to develop different statistics for dual scaling (de Leeuw 1973, 1976). In the present study, we have chosen Bock's adaptation (1960) of Bartlett's χ^2 approximation for dual scaling for several reasons. First, dual scaling derives continuous variates as linear combinations of categorical variates. Although η^2 is an eigenvalue obtained from categorical data matrices, it is a ratio of two quadratic forms of the derived variates. Second, as Kendall and Stuart (1961, p. 569) state, the theoretical implication of dual scaling is clear: 'if we seek separate scoring systems for the two categorized variables such as to maximize their correlation, we are basically trying to produce a bivariate normal distribution by operation upon the margins of the table.' The evaluation of η^2 then may be considered in the context of the foregoing discussion of Λ and χ^2. Third, (2.85) is relatively simple to calculate. In our notation, Bartlett's approximation, adapted by Bock for dual scaling, is given by

$$\chi^2 = -\left[f_t - 1 - \tfrac{1}{2}(n+m-1)\right]\log_e\left(1-\eta_j^2\right), \tag{2.87}$$

$$\text{df} = n+m-1-2j, \qquad j = 1,2,\ldots,t,$$

where f_t is the total number of responses in the table, n the number of rows, m the number of columns, and $\eta_1^2 > \eta_2^2 > \cdots > \eta_t^2$.

One may also consider an alternative suggested by Fisher (1948; see also Williams 1952 and Bock 1956). This is an approximate test for equality of group means, based on the analysis of variance, and is given by

$$F = \frac{\eta_1^2/\text{df}_b}{(1-\eta_1^2)/\text{df}_w}, \tag{2.88}$$

$$\text{df}_1 = \text{df}_b, \qquad \text{df}_2 = \text{df}_w,$$

where $\text{df}_b = n+m-2$ and $\text{df}_w = f_t - n - m + 1$. The degrees of freedom df_b are adjusted for the constants fitted by adding $m-1$ to the degrees of freedom between groups. In consequence,

$$\text{df}_w = \text{df}_t - \text{df}_b = (f_t - 1) - (n+m-2) = f_t - n - m + 1.$$

In the light of (2.86) and Lancaster's remark (1963), however, it appears difficult to justify the application of the statistic (2.88) to other eigenvalues $\eta_2^2, \eta_3^2, \ldots, \eta_t^2$. For this reason, the present book will not use this alternative for analysis of several dimensions.

Let us now consider applications of Bartlett's approximation to dual scaling. When the optimum solution $[\eta_1^2, \mathbf{x}, \mathbf{y}]$, say, is obtained, the significance of discriminability of the scoring scheme $[\mathbf{x}, \mathbf{y}]$ can be tested by (2.87), where $j=1$. If this χ^2 is significant, it indicates not only that the scoring scheme is useful in discriminating among subjects, but also that there may exist a second scoring scheme associated with η_2^2 which is significantly discriminative. In this case, dual scaling will continue following the procedure to be discussed later. If, however, η_1^2 is not significant, the corresponding scoring scheme $[\mathbf{x}, \mathbf{y}]$ is still the best one can obtain, and it is perhaps not worth looking at the second scoring scheme associated with η_2^2. Regardless of the significance of the χ^2, one would be interested in knowing what percentage of the total variance of the data matrix is accounted for by the optimum scoring scheme $[\mathbf{x}, \mathbf{y}]$. It is known that the sum of the diagonal elements of C_1, called the trace of C_1 and indicated by $\mathrm{tr}(C_1)$, is equal to the sum of the eigenvalues of C_1,

$$\mathrm{tr}(C_1) = \sum_{j=1}^{t} \eta_j^2. \tag{2.89}$$

Therefore the percentage of the total variance of the data matrix accounted for by the optimum scoring scheme is

$$\frac{\eta_1^2}{\mathrm{tr}(C_1)} \times 100\% = \delta_1\%, \text{ say,} \tag{2.90}$$

δ_1 tells us how much information contained in the data matrix can be explained by the two sets of numbers assigned to the marginals, that is, $[\mathbf{x}, \mathbf{y}]$. Similarly, the percentage of the variance accounted for by the first k solutions can be calculated by $(\eta_1^2 + \eta_2^2 + \cdots + \eta_k^2) \times 100\% / \mathrm{tr}(C_1) = \delta_k$, say.

In a more exhaustive analysis, one may wish to obtain a more inferentially oriented statistic than δ. One example of such a statistic is suggested by Bartlett (1947; see also Rao 1965). Its adaptation for dual scaling may be

$$\chi^2 = -\left[f_t - 1 - \tfrac{1}{2}(n+m-1) \right] \sum_{j=2}^{t} \log_e\left(1 - \eta_j^2\right), \tag{2.91}$$

$$\mathrm{df} = (n-2)(m-2).$$

This is an approximate test for significance of the total variance unaccounted for by the optimum solution. If this χ^2 is significant, dual scaling may continue to extract the second major dimension, which is orthogonal to the optimum dimension (see Section 2.4). In practice, the investigator may have to decide whether or not to extract more dimensions in conjunction with other factors such as the interpretability of dimensions other than the optimum one and the usefulness of extracting more than one dimension. There may be very few occasions when the investigator will use more than one scoring system, the optimum one.

Let us now apply the χ^2 test to our example. The largest correlation ratio, η_1^2, is 0.3683, $n=3$ (teachers), $m=3$ (categories), and $f_t=29$. Therefore Bartlett's χ^2 approximation is

$$\chi^2 = -\left[29-1-\tfrac{1}{2}(3+3-1)\right]\log_e(1-0.3683)$$

$$= -25.5\log_e 0.6317 = 11.71,$$

$$\text{df} = 3+3-1-2 = 3.$$

From the χ^2 table in any statistics book, we find that 11.71 for three degrees of freedom is statistically significant at the 0.01 level, and we therefore reject the null hypothesis of no discrimination. In other words, the optimum scoring system discriminates among the teachers significantly, or one may simply say that the scores y_j can be used to distinguish among the teachers. The trace of C_1 is

$$\text{tr}(C_1) = 0.1552 + 0.0207 + 0.2241 = 0.4000.$$

Therefore the percentage of the total variance accounted for by the optimum solution is

$$\delta_1 = \frac{0.3683\times 100\%}{0.4000} = 92.075\%.$$

Thus only 7.925% of the total variance remains to be explained. In this case, the investigator is likely to be satisfied with using only the optimum solution. However, one may wonder whether a substantial amount of information remains unaccounted for. To answer this question, we can use (2.91). Note that $r(C_1)=2$ by the Lagrange theorem. Therefore $\eta_1^2+\eta_2^2=\text{tr}(C_1)$, and we obtain $\eta_2^2=0.4000-0.3683=0.0317$. Since there are only two eigenvalues, (2.87) and (2.91) provide identical tests, the former being a test of significance of η_2^2 and the latter a test of significance of the residual after eliminating η_1^2, which is η_2^2. The number of degrees of freedom for (2.87), $j=2$, is $3+3-1-4=1$, and for (2.91), $(3-2)(3-2)=1$. The χ^2 value is

$$\chi^2 = -\left[29-1-\tfrac{1}{2}(3+3-1)\right]\log_e(1-0.0317) = 0.83.$$

This value is not significant for one degree of freedom. The contribution of the second scoring scheme can be assessed by δ as

$$\delta_2-\delta_1 = \frac{0.0317\times 100\%}{0.4000} = 7.925\%.$$

2.4 MULTIDIMENSIONAL ANALYSIS

When the maximized correlation ratio, η_1^2, is statistically significant, it may be of some interest to see whether or not there exists another significant weighting system which is independent of the first. In pursuing this problem, it is convenient to draw a distinction between the optimum weight vector **x** and eigenvector **w** of C_1. As discussed in Section 2.2, **x** is the non-trivial vector which maximizes the ratio of ss_b to ss_t, that is, (2.20). One can also say that **x** is the non-trivial right eigenvector of the non-symmetric matrix $D^{-1}F'D_n^{-1}F$ which maximizes η^2. It is important to note that when a square matrix, say S, is not symmetric there exist two eigenvectors (right and left), say **u** and **v**, associated with an eigenvalue, λ, namely

$$S\mathbf{u} = \lambda\mathbf{u} \quad \text{and} \quad \mathbf{v}'S = \lambda\mathbf{v}'. \tag{2.92}$$

Therefore, if one wants to eliminate the contribution of a solution $(\lambda, \mathbf{u})$, one must also calculate **v** and apply the general form of Lagrange's theorem (2.57). To avoid this extra step of computation, one may deal with a symmetric matrix such as $D^{-1/2}F'D_n^{-1}FD^{-1/2}$; then eigenvector **w**, for example, may be converted to **x** as we did in Section 2.2. This procedure creates a problem when one wants to extract several weighting systems, however. Two vectors $\mathbf{x}_j$ and $\mathbf{x}_k$, associated with two eigenvalues η_j^2 and η_k^2, are in general not orthogonal, but are orthogonal in the metric of D, that is, $\mathbf{x}_j'D\mathbf{x}_k=0, j\neq k$. In contrast, eigenvectors $\mathbf{w}_j$ and $\mathbf{w}_k, j\neq k$, are always orthogonal to each other, that is, $\mathbf{w}_j'\mathbf{w}_k=0$. From a practical point of view, therefore, it may be convenient to adopt the following definitions:

1. Let $\eta_1^2, \eta_2^2, \eta_3^2, \ldots$ be the eigenvalues of C_1 and $\mathbf{w}_1, \mathbf{w}_2, \mathbf{w}_2, \ldots$ be the corresponding eigenvectors obtained under the specified unit of scaling. Then these eigenvectors are referred to as *orthogonal vectors*. Note that C_1 can be expressed as

$$C_1 = \sum_{j=1}^{t} \eta_j^2 \frac{\mathbf{w}_j\mathbf{w}_j'}{\mathbf{w}_j'\mathbf{w}_j}, \tag{2.93}$$

where $t = r(C_1)$.

2. Define

$$\mathbf{x}_j = D^{-1/2}\mathbf{w}_j, \qquad j = 1, 2, \ldots, t. \tag{2.94}$$

$\mathbf{x}_1$ maximizes η^2, and is therefore called the *optimal weight vector* (Bock 1960). $\mathbf{x}_2, \mathbf{x}_3, \ldots$ are vectors associated with local maxima of η^2 and hence are referred to as *locally optimal weight vectors*.

In practice, it is obvious that one should use $\mathbf{x}_1$ and the corresponding score vector, $\mathbf{y}_1$ say, if one is interested only in one scoring system. There are occasions in which data are multidimensional, however. In other words, the analysis may reveal that δ_1 is relatively small and that the χ^2 test for the remaining eigenvalues is significant. This situation arises quite often, such as when (a) the survey consists of heterogeneous questions, (b) subjects use different criteria of judgment, (c) questions are ambiguous, (d) individual differences of the subjects are substantial, and (e) questions are too difficult to answer.

When dual scaling reveals multidimensionality of data, one may question which set of weights, $(\mathbf{w}_1, \mathbf{w}_2, \ldots)$ or $(\mathbf{x}, \mathbf{x}_2, \ldots)$, one should use. Some may prefer $(\mathbf{w}_1, \mathbf{w}_2, \ldots)$ to $(\mathbf{x}_1, \mathbf{x}_2, \ldots)$ because vectors $\mathbf{w}_j$ are orthogonal; others may indicate the reversed order of preference because vectors $\mathbf{x}_j$ optimize the criterion η^2. It is important to note, however, that the two sets of vectors have different meanings, and it is not simply a matter of preference. When one is interested in *weights* (relative importance) of categories, one should look at $\mathbf{x}_j$, not $\mathbf{w}_j$. For example, recall the trivial solution (2.48), which was interpreted as amounting to 'placing all the responses into a single category.' This interpretation came from $\mathbf{x}=\mathbf{1}$ (all the categories receiving the same weights), and not from $\mathbf{w}=D^{1/2}\mathbf{1}$. If one is interested in relative importance of categories good, average, and poor, one should look at $\mathbf{x}_j$ and use $\mathbf{x}_j$ to weight the categories of F; one should not weight the categories by $\mathbf{w}_j$, for the data will then be doubly weighted by $D^{1/2}$. Since vectors $\mathbf{w}_j$ absorb the information about the marginal frequencies, $\mathbf{w}_j$ can be used to reproduce matrix C_1 in such an additive manner as (2.93), whereas vectors $\mathbf{x}_j$ are independent of the marginal frequencies and indicate only the relative amount of response consistency. $\mathbf{x}_j$ and $\mathbf{w}_j$ become equal in information content only when data matrix F has constant marginals within each set of rows and columns.

The extraction of the second and the remaining solutions is relatively simple. We first eliminate the largest eigenvalue and the corresponding vector, $(\eta_1^2, \mathbf{w}_1)$ say, from C_1, and then find the largest eigenvalue and the corresponding vector $(\eta_2^2, \mathbf{w}_2)$ of the residual matrix, C_2 say, under the constraint that $\mathbf{w}_2'\mathbf{w}_2 = f_t$. Similarly, we eliminate $(\eta_2^2, \mathbf{w}_2)$ from C_2 to obtain the residual matrix, C_3, and then find the largest eigenvalue and the corresponding eigenvector of C_3, $(\eta_3^2, \mathbf{w}_3)$, given that $\mathbf{w}_3'\mathbf{w}_3 = f_t$. This process may be continued until we obtain a statistically non-significant value of η^2. For each vector $\mathbf{w}_j$, formula (2.94) can be used to calculate the locally optimal weight vector $\mathbf{x}_j$.

In the above procedure, we are solving essentially the same set of equations

$$(C - \eta^2 I)\mathbf{w} = \mathbf{0}, \qquad \mathbf{w}'\mathbf{w} = f_t, \tag{2.95}$$

using the iterative method discussed in Section 2.2.2. The only change we have to make is to modify C from C_1 to C_2, then to C_3, and so on, successively by the formula

$$C_{j+1} = C_j - \eta_j^2 \frac{\mathbf{w}_j\mathbf{w}_j'}{\mathbf{w}_j'\mathbf{w}_j}, \qquad j = 1, 2, \ldots, t-1. \tag{2.96}$$

$\mathbf{x}_j$ and $\mathbf{w}_j$ are vectors for the columns of the data matrix. Similarly, we define the corresponding vectors for the rows. First, the vector for the rows which corresponds to $\mathbf{x}_1$ is called the *optimal score vector* and is indicated by $\mathbf{y}_1$. This vector can be calculated by (2.75). As discussed earlier, η_1^2, $\mathbf{x}_1$, and $\mathbf{y}_1$ constitute the optimal solution set. Corresponding to locally optimal weight vectors $\mathbf{x}_j$, $j=2,3,\ldots,t$, are *locally optimal score vectors* $\mathbf{y}_j$, defined by

$$\mathbf{y}_j = D_n^{-1}F\mathbf{x}_j/\eta_j. \tag{2.97}$$

The vectors for the rows which correspond to orthogonal vectors $\mathbf{w}_j$ are indicated by $\mathbf{u}_j$ and are defined by

$$\mathbf{u}_j = D_n^{1/2}\mathbf{y}_j \tag{2.98}$$

$$= D_n^{-1/2}F\mathbf{x}_j/\eta_j \tag{2.99}$$

$$= D_n^{-1/2}FD^{-1/2}\mathbf{w}_j/\eta_j. \tag{2.100}$$

$\mathbf{u}_j$ are called *orthogonal vectors*. The orthogonality of $\mathbf{u}_j$ can be shown as follows:

$$\mathbf{u}_j'\mathbf{u}_k = \frac{1}{\eta_j}\mathbf{w}_j'D^{-1/2}F'D_n^{-1/2}\frac{1}{\eta_k}D_n^{-1/2}FD^{-1/2}\mathbf{w}_k$$

$$= \frac{1}{\eta_j\eta_k}\mathbf{w}_j'D^{-1/2}F'D_n^{-1}FD^{-1/2}\mathbf{w}_k = \frac{1}{\eta_j\eta_k}\mathbf{w}_j'C_0\mathbf{w}_k, \tag{2.101}$$

where

$$C_0 = D^{-1/2}F'D_n^{-1}FD^{-1/2}. \tag{2.102}$$

If $j=k$,

$$\mathbf{w}_j'C_0\mathbf{w}_k/\eta_j\eta_k = \mathbf{w}_j'C_0\mathbf{w}_j/\eta_j^2 = \mathbf{w}_j'\eta_j^2\mathbf{w}_j/\eta_j^2 \qquad \text{since } C_0\mathbf{w}_j = \eta_j^2\mathbf{w}_j$$

$$= \mathbf{w}_j'\mathbf{w}_j = f_t.$$

If $j\neq k$, $\mathbf{w}_j$ and $\mathbf{w}_k$ are orthogonal to one another (i.e., $\mathbf{w}_j'\mathbf{w}_k=0$) and therefore

$$\mathbf{w}_j'C_0\mathbf{w}_k/\eta_j\eta_k = \eta_a^2\mathbf{w}_j'\mathbf{w}_k/\eta_j\eta_k = 0, \qquad a = j,k.$$

Thus

$$\mathbf{u}_j'\mathbf{u}_k = \begin{cases} f_t & \text{if } j=k, \\ 0 & \text{if } j\neq k. \end{cases} \tag{2.103}$$

It is also obvious that the following relations hold:

$$\mathbf{w}_j'\mathbf{w}_k = \begin{cases} f_t & \text{if } j=k, \\ 0 & \text{if } j\neq k; \end{cases} \tag{2.104}$$

$$\mathbf{x}_j'D\mathbf{x}_k = \begin{cases} f_t & \text{if } j=k, \\ 0 & \text{if } j\neq k; \end{cases} \tag{2.105}$$

$$\mathbf{y}_j'D_n\mathbf{y}_k = \begin{cases} f_t & \text{if } j=k, \\ 0 & \text{if } j\neq k. \end{cases} \tag{2.106}$$

Formulas (2.103) and (2.104) indicate that the norms of $\mathbf{w}_j$ and $\mathbf{u}_j$ are adjusted equal and constant for all the dimensions. Similarly, (2.105) and (2.106) indicate that the sum of squares of responses weighted by $\mathbf{x}_j$ and that of responses scored by $\mathbf{y}_j$ are adjusted equal and constant for all the dimensions. As we have seen earlier, the relative contribution of dimension j, $\delta_j - \delta_{j-1}$, is proportional to the magnitude of η_j^2. Thus the scaling units used in (2.103) to (2.106) have the effect of inflating the values of those vectors associated with relatively 'unimportant' dimensions, that is those of very small values of η^2. This unit of scaling can be quite misleading, and may make inter-dimensional comparisons of score and weight vectors very difficult. For multidimensional analysis it is therefore recommended that $\mathbf{x}_j$, $\mathbf{y}_j$, $\mathbf{w}_j$, and $\mathbf{u}_j$ be rescaled to $\mathbf{x}_j^*$, $\mathbf{y}_j^*$, $\mathbf{w}_j^*$, and $\mathbf{u}_j^*$, respectively, by applying the relations

$$\mathbf{x}_j^* = \eta_j \mathbf{x}_j, \tag{2.107}$$

$$\mathbf{y}_j^* = \eta_j \mathbf{y}_j, \tag{2.108}$$

$$\mathbf{w}_j^* = \eta_j \mathbf{w}_j, \tag{2.109}$$

$$\mathbf{u}_j^* = \eta_j \mathbf{u}_j, \tag{2.110}$$

where $j = 1, 2, \ldots, t$. Then we obtain

$$\mathbf{w}_j^{*\prime} \mathbf{w}_j^* = \mathbf{u}_j^{*\prime} \mathbf{u}_j^* = \eta_j^2 f_t, \tag{2.111}$$

$$\mathbf{x}_j^{*\prime} D \mathbf{x}_j^* = \mathbf{y}_j^{*\prime} D_n \mathbf{y}_j^* = \eta_j^2 f_t. \tag{2.112}$$

The multidimensional decomposition of C_1 can be expressed as

$$C_1 = \sum_{j=1}^{t} \eta_j^2 \frac{\mathbf{w}_j \mathbf{w}_j'}{\mathbf{w}_j' \mathbf{w}_j} = \sum_{j=1}^{t} \frac{1}{f_t} \mathbf{w}_j^* \mathbf{w}_j^{*\prime}. \tag{2.113}$$

Note that the contribution of η_j^2 to dimension j is now absorbed into $\mathbf{w}_j^*$.

Let us now carry out multidimensional analysis of our example. We have so far obtained η_1^2, $\mathbf{x}_1$, $\mathbf{y}_1$, and $\mathbf{w}_1$. The orthogonal score vector $\mathbf{u}_1$ can be calculated by (2.98) as

$$\mathbf{u}_1 = D_n^{1/2} \mathbf{y}_1 = \begin{bmatrix} \sqrt{10} & 0 & 0 \\ 0 & \sqrt{10} & 0 \\ 0 & 0 & \sqrt{9} \end{bmatrix} \begin{bmatrix} -1.2322 \\ 0.1228 \\ 1.2327 \end{bmatrix} = \begin{bmatrix} -3.8966 \\ 0.3880 \\ 3.6978 \end{bmatrix}.$$

The second solution can be obtained in the following way. First, obtain residual matrix C_2 by eliminating the first solution, $(\eta_1^2, \mathbf{w}_1)$,

$$C_2 = C_1 - \eta_1^2 \frac{\mathbf{w}_1\mathbf{w}_1'}{\mathbf{w}_1'\mathbf{w}_1}$$

$$= C_1 - 0.3683 \frac{\begin{bmatrix} 3.4029 \\ 0.3051 \\ -4.1626 \end{bmatrix} \begin{bmatrix} 3.4029 & 0.3051 & -4.1626 \end{bmatrix}}{\begin{bmatrix} 3.4029 & 0.3051 & -4.1626 \end{bmatrix} \begin{bmatrix} 3.4029 \\ 0.3051 \\ -4.1626 \end{bmatrix}}$$

$$= \begin{bmatrix} 0.1552 & 0.0006 & -0.1742 \\ 0.0006 & 0.0207 & -0.0250 \\ -0.1742 & -0.0250 & 0.2241 \end{bmatrix} - \begin{bmatrix} 0.1471 & 0.0132 & -0.1799 \\ 0.0132 & 0.0012 & -0.0161 \\ -0.1799 & -0.0161 & 0.2201 \end{bmatrix}$$

$$= \begin{bmatrix} 0.0081 & -0.0126 & 0.0057 \\ -0.0126 & 0.0195 & -0.0089 \\ 0.0057 & -0.0089 & 0.0040 \end{bmatrix}.$$

We now subject this matrix to the iterative method to extract the largest eigenvalue and the corresponding eigenvector **w** such that $\mathbf{w}'\mathbf{w}=f_t$. Recall that the elements of the first solution vector $\mathbf{w}_1$ were $w_1>w_2>w_3$. Thus this order will not be maintained by the elements of the second solution vector $\mathbf{w}_2$, which is orthogonal to $\mathbf{w}_1$. Using this knowledge, we may choose an arbitrary trial vector for the iterative scheme. As an example, let us use the following vector as $\mathbf{b}_0$, the elements of which have a completely reversed order to those of $\mathbf{w}_1$:

$$\mathbf{b}_0 = \begin{bmatrix} -1 \\ 0 \\ 1 \end{bmatrix}.$$

The iterative procedure starts with multiplication of C_2 by $\mathbf{b}_0$,

$$C_2\mathbf{b}_0 = \begin{bmatrix} 0.0081 & -0.0126 & 0.0057 \\ -0.0126 & 0.0195 & -0.0089 \\ 0.0057 & -0.0089 & 0.0040 \end{bmatrix} \begin{bmatrix} -1 \\ 0 \\ 1 \end{bmatrix} = \begin{bmatrix} -0.0024 \\ 0.0037 \\ -0.0017 \end{bmatrix} = \mathbf{b}_1.$$

The largest absolute value in $\mathbf{b}_1$ is 0.0037. Therefore

$$\mathbf{b}_1^* = \frac{\mathbf{b}_1}{0.0037} = \begin{bmatrix} -0.6486 \\ 1.0000 \\ -0.4595 \end{bmatrix}.$$

The next step gives

$$C_2\mathbf{b}_1^* = \begin{bmatrix} -0.0205 \\ 0.0318 \\ -0.0144 \end{bmatrix} = \mathbf{b}_2, \qquad \mathbf{b}_2^* = \frac{\mathbf{b}_2}{0.0318} = \begin{bmatrix} -0.6447 \\ 1.0000 \\ -0.4528 \end{bmatrix}.$$

Since the discrepancy between $\mathbf{b}_1^*$ and $\mathbf{b}_2^*$ is still large, the iterative process continues:

$$C_2\mathbf{b}_2^* = \begin{bmatrix} -0.0204 \\ 0.0317 \\ -0.0144 \end{bmatrix} = \mathbf{b}_3, \qquad \mathbf{b}_3^* = \frac{\mathbf{b}_3}{0.0317} = \begin{bmatrix} -0.6435 \\ 1.0000 \\ -0.4543 \end{bmatrix},$$

$$C_2\mathbf{b}_3^* = \begin{bmatrix} -0.0204 \\ 0.0317 \\ -0.0144 \end{bmatrix} = \mathbf{b}_4 = \mathbf{b}_3.$$

Thus $\mathbf{b}_3^* = \mathbf{b}_4^*$ and the iteration stops at this point with the solution

$$\eta_2^2 = 0.0317, \quad \mathbf{b}_3^* = \mathbf{b}_4^* = \begin{bmatrix} -0.6435 \\ 1.0000 \\ -0.4543 \end{bmatrix}.$$

Eigenvector $\mathbf{w}_2$ that satisfies (2.95) can be calculated:

$$\mathbf{w}_2 = \left(\frac{f_t}{\mathbf{b}_4^{*\prime}\mathbf{b}_4^*}\right)^{1/2} \mathbf{b}_4^* = \begin{bmatrix} -2.7222 \\ 4.2304 \\ -1.9219 \end{bmatrix}.$$

The locally optimal weight vector $\mathbf{x}_2$ can be calculated by (2.94):

$$\mathbf{x}_2 = D^{-1/2}\mathbf{w}_2 = \begin{bmatrix} \frac{1}{\sqrt{10}} & 0 & 0 \\ 0 & \frac{1}{\sqrt{11}} & 0 \\ 0 & 0 & \frac{1}{\sqrt{8}} \end{bmatrix} \begin{bmatrix} -2.7222 \\ 4.2304 \\ -1.9219 \end{bmatrix} = \begin{bmatrix} -0.8608 \\ 1.2755 \\ -0.6795 \end{bmatrix}.$$

Score vectors corresponding to $\mathbf{x}_2$ and $\mathbf{w}_2$ can be calculated as

$$\mathbf{y}_2 = \frac{1}{\eta_2} D_n^{-1} F\mathbf{x}_2 = \begin{bmatrix} -0.6242 \\ 1.3683 \\ -0.8352 \end{bmatrix}, \qquad \mathbf{u}_2 = D_n^{1/2}\mathbf{y}_2 = \begin{bmatrix} -1.9739 \\ 4.3269 \\ -2.5056 \end{bmatrix},$$

respectively. If we rescale all the results in terms of the relative size of each dimension, we obtain

$$\mathbf{x}_1^* = \eta_1\mathbf{x}_1 = \sqrt{0.3683}\begin{bmatrix} 1.0761 \\ 0.0920 \\ -1.4717 \end{bmatrix} = \begin{bmatrix} 0.6531 \\ 0.0558 \\ -0.8931 \end{bmatrix},$$

$$\mathbf{x}_2^* = \eta_2\mathbf{x}_2 = \sqrt{0.0317}\begin{bmatrix} -0.8608 \\ 1.2755 \\ -0.6795 \end{bmatrix} = \begin{bmatrix} -0.1533 \\ 0.2271 \\ -0.1210 \end{bmatrix}.$$

Similarly,

$$\mathbf{y}_1^* = \begin{bmatrix} -0.7478 \\ 0.0745 \\ 0.7481 \end{bmatrix}, \quad \mathbf{w}_1^* = \begin{bmatrix} 2.0651 \\ 0.1851 \\ -2.5262 \end{bmatrix}, \quad \mathbf{u}_1^* = \begin{bmatrix} -2.3648 \\ 0.2357 \\ 2.2433 \end{bmatrix},$$

$$\mathbf{y}_2^* = \begin{bmatrix} -0.1111 \\ 0.2436 \\ -0.1487 \end{bmatrix}, \quad \mathbf{w}_2^* = \begin{bmatrix} -0.4847 \\ 0.7532 \\ -0.3422 \end{bmatrix}, \quad \mathbf{u}_2^* = \begin{bmatrix} -0.3514 \\ 0.7704 \\ -0.4461 \end{bmatrix}.$$

In terms of these rescaled vectors, one can easily see the relative importance of the first dimension over the second dimension.

As has already been discussed in Section 2.3, the second solution is not significant, that is to say, locally optimal weight $\mathbf{x}_2$ does not provide significantly discriminative scores $\mathbf{y}_2$. The analysis thus ends at this point. In addition, we note that the two solutions exhaust the information in the data since $\eta_1^2 + \eta_2^2 = \mathrm{tr}(C_1)$ and $r(C_1) = 2$. Major results of analysis can be summarized as in Table 2.4.

This small example illustrates basic aspects of dual scaling. When the size of the data matrix becomes larger than 3×3, one may obtain more than one significant solution. Thus, in general, χ^2 analyses of η_j^2 and the residual after

TABLE 2.4

Performance evaluation of teachers

Teachers	Good	Average	Poor	Total	Optimal score ($\mathbf{y}_1$)
1	1	3	6	10	−1.2322
2	3	5	2	10	0.1228
3	6	3	0	9	1.2327
Total	10	11	8	29	
Optimal weight ($\mathbf{x}_1$)	1.0761	0.0920	−1.4717		$\delta_1 = 92.075$

Summary statistics

Dimension	η^2	δ	χ^2	df	p
1	0.3683	92.075	11.71	3	<0.05
2	0.0317	100.000	0.83	1	(not significant)
Total	0.4000			4	

TABLE 2.5

χ^2 analysis of η^2 $(\eta_1^2 > \eta_2^2 > \cdots > \eta_j^2 \cdots)$*

Dimension	χ^2	df
1	$-\nu \log_e(1-\eta_1^2)$	$n+m-3$
2	$-\nu \log_e(1-\eta_2^2)$	$n+m-5$
3	$-\nu \log_e(1-\eta_3^2)$	$n+m-7$
⋮	⋮	⋮
j	$-\nu \log_e(1-\eta_j^2)$	$n+m-(2j+1)$
⋮	⋮	⋮
Total	$-\nu \Sigma_k \log_e(1-\eta_k^2)$	$(n-1)(m-1)$

*$\nu = [f_t - 1 - \frac{1}{2}(n+m-1)]$.

TABLE 2.6

χ^2 analysis of remaining eigenvalues $(\eta_1^2 > \eta_2^2 > \cdots > \eta_j^2 > \cdots)$*

Remaining eigenvalues	χ^2	df
$\eta_1^2, \eta_2^2, \ldots, \eta_j^2, \ldots, \eta_t^2$	$-\nu \log_e \Lambda = -\nu \sum_{k=1}^{t} \log_e(1-\eta_k^2)$	$(n-1)(m-1)$
$\eta_2^2, \ldots, \eta_j^2, \ldots, \eta_t^2$	$-\nu \sum_{k=2}^{t} \log_e(1-\eta_k^2)$	$(n-2)(m-2)$
⋮	⋮	⋮
$\eta_j^2, \ldots, \eta_t^2$	$-\nu \sum_{k=j}^{t} \log_e(1-\eta_k^2)$	$(n-j)(m-j)$

*$\nu = [f_t - 1 - \frac{1}{2}(n+m-1)]$.

eliminating $\eta_1^2, \eta_2^2, \ldots, \eta_j^2$ can be summarized as in Tables 2.5 and 2.6. Note that the general formula for testing the residual after eliminating $\eta_1^2, \eta_2^2, \ldots, \eta_{j-1}^2$ in Table 2.6 can also be expressed as

$$\chi^2 = -\left[f_t - 1 - \tfrac{1}{2}(n+m-1)\right] \log_e\left(\frac{\Lambda}{(1-\eta_1^2)(1-\eta_2^2)\cdots(1-\eta_{j-1}^2)}\right),$$

$$\text{df} = (n-j)(m-j). \tag{2.114}$$

In this chapter we have presented some basic aspects of dual scaling; the information provided here is probably sufficient for most practical purposes. If the reader wishes, he can skip Chapter 3 and turn to the applications of dual scaling to different forms of categorical data described in the ensuing chapters. Chapter 3 provides additional information about dual scaling, which may help the investigator understand characteristics of the technique that may be hidden behind the mathematics discussed in the present chapter.

3

Dual scaling II

3.1

DUALITY

In addition to the basic aspects of dual scaling discussed in Chapter 2, there are other aspects which are equally important in characterizing the technique. In this chapter we first describe one of its key features, called duality, then present four of a number of alternative formulations of the procedure, and finally provide a summary of its characteristics and mathematical structure.

Dual scaling has the property called duality. This property can be characterized by symmetric relations between derived weight vectors for the columns and rows of the data matrix. Some of the relations are:

1. **y** is a linear function of **x**; **x** is a linear function of **y**. Similarly, **u** is a linear function of **w**; **w** is a linear function of **u**.
2. If **x** minimizes the within-row sum of squares and maximizes the between-row sum of squares, then **y** minimizes the within-column sum of squares and maximizes the between-column sum of squares.
3. The distribution of the eigenvalues of the data matrix weighted by **x** is the same as for the data matrix weighted by **y**.

The following discussion of duality in our scaling procedure will, we hope, serve to provide a coherent and synthesized account of dual scaling.

Suppose we wish to determine **y** first, rather than **x**. The procedure then is to determine **y** so as to maximize the ratio of the between-category (between-column) sum of squares to the total sum of squares. The correlation ratio can then be expressed as

$$\eta^2 = \frac{\mathbf{y}'[FD^{-1}F' - \mathbf{g}\mathbf{g}'/f_t]\mathbf{y}}{\mathbf{y}'[D_n - \mathbf{g}\mathbf{g}'/f_t]\mathbf{y}}, \tag{3.1}$$

where **g** is the vector of row totals of data matrix F. Let us simplify (3.1) by setting $\mathbf{g}'\mathbf{y}$ equal to zero, as indicated by (2.76), and transforming **y** to **u** by

$$\mathbf{u} = D_n^{1/2}\mathbf{y}, \quad \text{or} \quad \mathbf{y} = D_n^{-1/2}\mathbf{u}. \tag{3.2}$$

Then (3.1) can be reduced to

$$\eta^2 = \mathbf{u}'D_n^{-1/2}FD^{-1}F'D_n^{-1/2}\mathbf{u}/\mathbf{u}'\mathbf{u}. \tag{3.3}$$

Following the same procedure as before, we eliminate the trivial solution due to the marginal constraint (i.e., $\mathbf{g}'\mathbf{y}=0$) by calculating residual matrix C_1^*, where

$$C_1^* = D_n^{-1/2}FD^{-1}F'D_n^{-1/2} - D_n^{1/2}\mathbf{11}'D_n^{1/2}/f_t. \tag{3.4}$$

The equations to be solved are

$$(C_1^* - \eta^2 I)\mathbf{u} = \mathbf{0}, \tag{3.5}$$

$$\mathbf{u}'\mathbf{u} = f_t. \tag{3.6}$$

Our objective is to determine $\mathbf{u}$ that maximizes η^2. Thus we can solve (3.5) by the iterative method discussed in Section 2.2.2. Once the solution vector $\mathbf{b}^*$ is obtained, $\mathbf{u}$ is calculated by

$$\mathbf{u} = \left(\frac{f_t}{\mathbf{b}^{*\prime}\mathbf{b}^*}\right)^{1/2}\mathbf{b}^*, \tag{3.7}$$

and $\mathbf{y}$ is obtained from $\mathbf{u}$ by (3.2).

When optimal score vector $\mathbf{y}$ is obtained, the optimal weight vector for the categories, $\mathbf{x}$, can be calculated as the mean vector of responses weighted by $\mathbf{y}$ times the constant of proportionality, $1/\eta$:

$$\mathbf{x} = D^{-1}F'\mathbf{y}/\eta. \tag{3.8}$$

We would now like to show that η^2, $\mathbf{x}$, and $\mathbf{y}$, obtained in this section, are identical with those obtained in Section 2.2. First, set

$$D_n^{-1/2}FD^{-1/2} = B. \tag{3.9}$$

Then (3.3) can be expressed as

$$\eta^2 = \mathbf{u}'BB'\mathbf{u}/\mathbf{u}'\mathbf{u}. \tag{3.10}$$

In previous sections η^2 was first expressed in terms of $\mathbf{x}$, and then $\mathbf{x}$ was transformed to $\mathbf{w}$ so that

$$\eta^2 = \mathbf{w}'D^{-1/2}F'D_n^{-1}FD^{-1/2}\mathbf{w}/\mathbf{w}'\mathbf{w} \tag{3.11}$$

$$= \mathbf{w}'B'B\mathbf{w}/\mathbf{w}'\mathbf{w}. \tag{3.12}$$

It is well known in matrix algebra that the two matrices BB' and $B'B$ have the same set of eigenvalues. Thus maximization of η^2, defined by (3.10), and that of (3.12) provide the same set of correlation ratios.

Suppose now that we obtain $\mathbf{x}$ first and then $\mathbf{y}$ by (2.75). Since (2.75) can also be rewritten as

$$D_n^{-1/2}F\mathbf{x} = \eta D_n^{1/2}\mathbf{y}, \tag{3.13}$$

we obtain

$$\eta^2 = \frac{\mathbf{x}'F'D_n^{-1}F\mathbf{x}}{\mathbf{x}'D\mathbf{x}} = \eta^2\frac{\mathbf{y}'D_n\mathbf{y}}{\mathbf{x}'D\mathbf{x}}. \tag{3.14}$$

Since $\mathbf{x}'D\mathbf{x}$ is set equal to f_t, it follows from (3.14) that $\mathbf{y}'D_n\mathbf{y}=f_t$. Now, suppose we obtain **y** first and then **x** by (3.8). Then, in exactly the same manner as above, we obtain

$$\eta^2 = \frac{\mathbf{y}'FD^{-1}F'\mathbf{y}}{\mathbf{y}'D_n\mathbf{y}} = \eta^2\frac{\mathbf{x}'D\mathbf{x}}{\mathbf{y}'D_n\mathbf{y}}. \tag{3.15}$$

Since $\mathbf{y}'D_n\mathbf{y}$, that is, $\mathbf{u}'\mathbf{u}$, is set equal to f_t, it follows from (3.15) that $\mathbf{x}'D\mathbf{x}=f_t$.

As shown above, the two analyses employ the same scale unit, $\mathbf{x}'D\mathbf{x}=\mathbf{y}'D_n\mathbf{y}=f_t$, and provide the same distribution of η^2. These two points lead to the conclusion that the two analyses provide the same set of solutions $(\eta_j^2, \mathbf{x}_j, \mathbf{y}_j)$. As we note, one formulation can be obtained from the other by interchanging the words 'rows' and 'columns.'

For orthogonal solutions $(\eta_j^2, \mathbf{w}_j, \mathbf{u}_j)$, duality relations are even more obvious since **w** and **u** are eigenvectors associated with $B'B$ and BB', respectively. If one obtains $\mathbf{w}_j$ first, $\mathbf{u}_j$ can be calculated from

$$\mathbf{u}_j = D_n^{-1/2}FD^{-1/2}\mathbf{w}_j/\eta_j = B\mathbf{w}_j/\eta_j \tag{3.16}$$

$$= D_n^{-1/2}F\mathbf{x}_j/\eta_j \tag{3.17}$$

$$= D_n^{1/2}\mathbf{y}_j. \tag{3.18}$$

If, however, one obtains $\mathbf{u}_j$ first, $\mathbf{w}_j$ can be calculated from

$$\mathbf{w}_j = D^{-1/2}F'D_n^{-1/2}\mathbf{u}_j/\eta_j = B'\mathbf{u}_j/\eta_j \tag{3.19}$$

$$= D^{-1/2}F'\mathbf{y}_j/\eta_j \tag{3.20}$$

$$= D^{1/2}\mathbf{x}_j. \tag{3.21}$$

Both ways result in the same set of solutions $(\eta_j^2, \mathbf{w}_j, \mathbf{u}_j)$.

Let us now indicate by $\mathbf{w}_0$ and $\mathbf{u}_0$ the eigenvectors of $B'B$ and BB', respectively, associated with the second type of trivial solution, that is, $(\eta^2=1, \mathbf{w}_0 = D^{1/2}\mathbf{1}, \mathbf{u}_0 = D_n^{1/2}\mathbf{1})$. Define the $(t+1)\times m$ matrix W and the $(t+1)\times n$ matrix U, which consist of the $t+1$ eigenvectors $\mathbf{w}_j$ and the $t+1$ eigenvectors $\mathbf{u}_j$ in the rows, respectively, namely,

$$W = \begin{bmatrix} \mathbf{w}_0' \\ \mathbf{w}_1' \\ \vdots \\ \mathbf{w}_t' \end{bmatrix}, \tag{3.22}$$

$$U = \begin{bmatrix} \mathbf{u}_0' \\ \mathbf{u}_1' \\ \vdots \\ \mathbf{u}_t' \end{bmatrix}. \tag{3.23}$$

The elements of these matrices are scaled in such a way that

$$WW' = UU' = f_t I. \tag{3.24}$$

Define also the matrices

$$X = WD^{-1/2}, \tag{3.25}$$

$$Y = UD_n^{-1/2}, \tag{3.26}$$

$$\Lambda^2 = \begin{bmatrix} 1 & 0 & 0 & \cdots & 0 \\ 0 & \eta_1^2 & 0 & \cdots & 0 \\ 0 & 0 & \eta_2^2 & \cdots & 0 \\ \vdots & \vdots & \vdots & \ddots & \vdots \\ 0 & 0 & 0 & \cdots & \eta_t^2 \end{bmatrix}. \tag{3.27}$$

Then, to summarize the above discussion of duality, we can present the following theorem:

THEOREM *Given an* $n \times m$ *frequency table,* F*, with frequency* f_{ij} *in the ith row and the jth column, let* B *be defined by* (3.9). *Then*

(a) *The* $t+1$ *non-zero eigenvalues of* $B'B$*, that is,* $1, \eta_1^2, \eta_2^2, \ldots, \eta_t^2$*, are the same as the* $t+1$ *non-zero eigenvalues of* BB'.

(b) *There exists a relation between* W*,* B*, and* U *such that* U *consists of the rows of* WB' *scaled to the constant norm, namely,*

$$U = \Lambda^{-1}WB'. \tag{3.28}$$

(c) *Similarly,*

$$W = \Lambda^{-1}UB. \tag{3.29}$$

(d) *In consequence, there exist the relations*

$$UBW' = f_t\Lambda, \tag{3.30}$$

$$YFX' = f_t\Lambda. \tag{3.31}$$

Studies by Lancaster (1958), Dempster (1969), and Greenacre (1978a, b), for example, provide useful references for this theorem. (3.30) can be obtained by substitutions of (3.24) and (3.28) or (3.29). (3.31) can be proved equal to (3.30) by substitutions,

$$YFX' = UD_n^{-1/2}FD^{-1/2}W' = UBW'. \tag{3.32}$$

The computation of UBW' or YFX' may provide a useful means to check the accuracy of the results. In our example of the previous chapter,

$$B = D_n^{-1/2} F D^{-1/2} = \begin{bmatrix} 0.1000 & 0.2860 & 0.6708 \\ 0.3000 & 0.4767 & 0.2236 \\ 0.6325 & 0.3015 & 0.0000 \end{bmatrix},$$

$$W = \begin{bmatrix} \mathbf{1}'D^{1/2} \\ \mathbf{w}_1' \\ \mathbf{w}_2' \end{bmatrix} = \begin{bmatrix} \sqrt{10} & \sqrt{11} & \sqrt{8} \\ 3.4029 & 0.3051 & -4.1626 \\ -2.7222 & 4.2304 & -1.9219 \end{bmatrix},$$

$$U = \begin{bmatrix} \mathbf{1}'D_n^{1/2} \\ \mathbf{u}_1' \\ \mathbf{u}_2' \end{bmatrix} = \begin{bmatrix} \sqrt{10} & \sqrt{10} & \sqrt{9} \\ -3.8966 & 0.3880 & 3.6978 \\ -1.9739 & 4.3269 & -2.5056 \end{bmatrix},$$

$$\eta_1^2 = 0.3683, \qquad \eta_1 = 0.6069,$$

$$\eta_2^2 = 0.0317, \qquad \eta_2 = 0.1780,$$

$$f_t = 29.$$

Therefore

$$UBW' = \begin{bmatrix} 28.9984 & 0.0002 & -0.0149 \\ -0.0011 & 17.6047 & 0.0182 \\ -0.0763 & 0.0638 & 5.1477 \end{bmatrix}$$

$$= 29\begin{bmatrix} 0.9999 & 0.0000 & -0.0005 \\ 0.0000 & 0.6071 & 0.0006 \\ -0.0026 & 0.0022 & 0.1774 \end{bmatrix},$$

and

$$f_t\Lambda = 29\begin{bmatrix} 1 & 0 & 0 \\ 0 & 0.6069 & 0 \\ 0 & 0 & 0.1780 \end{bmatrix}.$$

For most practical purposes, the discrepancies between UBW' and $f_t\Lambda$ may be regarded as negligible.

The duality of our scaling procedure is not only a mathematically elegant property but also has several implications of theoretical and practical interest. First, it implies that dual scaling minimizes *simultaneously* the within-row and the within-column sums of squares. In other words, dual scaling determines the vector for the rows and the vector for the columns of F so as to minimize *the row-by-column interaction* of scaled responses. This aspect is reflected in the name 'additive scoring' used by Fisher (1948) for dual scaling.

Secondly, it provides a means to reduce computer time. If the data matrix is $n \times m$ and $n > m$, one should subject the $m \times m$ matrix $B'B$ to the iterative procedure for eigenvalues and eigenvectors. If, however, $n < m$, one should subject the $n \times n$ matrix BB' to the iterative procedure. This flexibility helps reduce computer time since the iterative process is the most time-consuming part of computation. The duality (symmetry) is also reflected in the statistical tests discussed earlier. In the normal theory of statistics, Bartlett's χ^2 for testing the equality of p means on q variables has $q(p-1)$ degrees of freedom, as indicated in (2.81), but the corresponding statistic for dual scaling of the $n \times m$ table has $(n-1)(m-1)$ degrees of freedom, which are symmetric with respect to n and m. Similarly, the significance of residual eigenvalues, $\eta_r^2, \eta_{r+1}^2, \ldots, \eta_t^2$, can be tested by Bartlett's χ^2 with degrees of freedom $(p-r)(q-r-1)$ under normal theory and $(n-r)(m-r)$ under dual scaling. Again, the degrees of freedom under dual scaling are symmetric with respect to n and m. This symmetry also implies that the statistic can be used irrespective of whether $n>m$, $n=m$, or $n<m$.

Thirdly, duality implies that two equations (2.75) and (3.8), that is,

$$\mathbf{y} = D_n^{-1}F\mathbf{x}/\eta, \tag{2.75}$$

$$\mathbf{x} = D^{-1}F'\mathbf{y}/\eta, \tag{3.8}$$

are simultaneously satisfied when $\mathbf{x}$ and $\mathbf{y}$ are solution vectors. These two equations provide a basis for the algorithm employed in the so-called *method of reciprocal averages*. As the scaling unit is arbitrary, one can choose the constant of proportionality, $1/\eta$, as the unit. This choice has the effect of replacing $1/\eta$ by unity. In consequence, $\mathbf{y}$ and $\mathbf{x}$ are now simply the mean response vectors weighted by $\mathbf{x}$ and $\mathbf{y}$, respectively:

$$\{\mathbf{y} = D_n^{-1}F\mathbf{x}, \quad \mathbf{x} = D^{-1}F'\mathbf{y}\}. \tag{3.33}$$

The algorithm of reciprocal averaging starts with an arbitrarily chosen trial vector for $\mathbf{x}$ to calculate the initial vector $\mathbf{y}$ by the first formula of (3.33). It is convenient to standardize $\mathbf{y}$, that is, to divide $\mathbf{y}$ by the largest absolute value of its elements. The initial vector of standardized $\mathbf{y}$ is then used to calculate $\mathbf{x}$ by the second formula of (3.33). This $\mathbf{x}$ is then standardized and used to calculate a new $\mathbf{y}$, which is then standardized and used to calculate a new $\mathbf{x}$. The process continues until the successive vectors of sets, $[\mathbf{x}_{(p)}, \mathbf{y}_{(p)}]$ and $[\mathbf{x}_{(p+1)}, \mathbf{y}_{(p+1)}]$, say, become identical or almost identical. We shall show in the next section that this algorithm leads to the optimum solution.

3.2
OTHER APPROACHES TO DUAL SCALING

The approach we have been considering may be called the 'one-way analysis-of-variance approach.' In terms of the one-way classification table of responses weighted by x_i, the correlation ratio was introduced as the objective function for optimization. The results obtained from this formulation were proved to be identical, by virtue of duality, with those obtained by maximizing the correlation ratio defined for the one-way classification table of responses weighted by y_k. There are several other approaches which also lead to the same results.

3.2.1 *The method of reciprocal averages*

As mentioned in Chapter 1, Richardson and Kuder discussed the idea of reciprocal averaging in their 1933 paper, and Horst (1935) called it the method of reciprocal averages. The same technique was also suggested by Fisher (1940) as one of a variety of ways to solve our scaling problem. Mosier (1946) and Baker (1960) contributed to the popularization of the technique. Hill (1973, 1974) presented this method as a statistical tool for ecological research, and stated that the method had been developed under the title 'analyse factorielle des correspondances' by Benzécri and his co-workers in France (Benzécri 1969; Escofier-Cordier 1969). A technique which is similar to the method of reciprocal averages was discussed by Whittaker (1967) under the name 'gradient analysis.'

The name 'reciprocal averages' is obvious from (3.33) since the scores for the rows are averages of the column weights and reciprocally the column weights are averages of the row scores. Let us now prove that the reciprocal averaging process provides the optimum solution. To maintain simplicity in the proof, let us leave out the standardization or rescaling step of successive vectors from our discussion. Then successive vectors of $\mathbf{y}$ and $\mathbf{x}$ calculated by (3.33) are:

$$
\begin{aligned}
\mathbf{y}_{(1)} &= D_n^{-1}F\mathbf{x}_{(0)}, & \mathbf{x}_{(1)} &= D^{-1}F'\mathbf{y}_{(1)},\\
\mathbf{y}_{(2)} &= D_n^{-1}F\mathbf{x}_{(1)}, & \mathbf{x}_{(2)} &= D^{-1}F'\mathbf{y}_{(2)},\\
&\ \vdots & &\ \vdots\\
\mathbf{y}_{(p)} &= D_n^{-1}F\mathbf{x}_{(p-1)}, & \mathbf{x}_{(p)} &= D^{-1}F'\mathbf{y}_{(p)}.
\end{aligned}
$$

Note that

$$
\begin{aligned}
\mathbf{y}_{(p)} &= D_n^{-1}F\mathbf{x}_{(p-1)} = D_n^{-1}FD^{-1}F'\mathbf{y}_{(p-1)}\\
&= D_n^{-1}FD^{-1}F'D_n^{-1}F\mathbf{x}_{(p-2)} = \cdots = D_n^{-1}FD^{-1}F'D_n^{-1}F\cdots D_n^{-1}F\mathbf{x}_{(0)}\\
&= D_n^{-1/2}\left(D_n^{-1/2}FD^{-1/2}\right)\left(D^{-1/2}F'D_n^{-1/2}\right)\cdots\\
&\quad \times\left(D^{-1/2}F'D_n^{-1/2}\right)D_n^{-1/2}F\mathbf{x}_{(0)}\\
&= D_n^{-1/2}(BB')^{p-1}D_n^{-1/2}F\mathbf{x}_{(0)}.
\end{aligned}
\tag{3.34}
$$

Similarly,

$$
\mathbf{x}_{(p)} = D^{-1/2}(B'B)^{p-1}D^{-1/2}F'\mathbf{y}_{(1)}.
\tag{3.35}
$$

$\mathbf{x}_{(0)}$ is the arbitrary initial vector. Recall that $\mathbf{u}_j$ and $\mathbf{w}_j$ are eigenvectors of BB' and $B'B$, respectively, and that $\mathbf{y}_j$ and $\mathbf{x}_j$ are given by $D_n^{-1/2}\mathbf{y}_j$ and $D^{-1/2}\mathbf{w}_j$, respectively. Thus, if we set

$$D_n^{-1/2}F\mathbf{x}_{(0)} = \mathbf{d}_0, \qquad D^{-1/2}F'\mathbf{y}_{(1)} = D^{-1/2}F'D_n^{-1}F\mathbf{x}_{(0)} = \mathbf{d}_0^*, \tag{3.36}$$

and consider $\mathbf{d}_0$ and $\mathbf{d}_0^*$ to be initial trial vectors for the iterative process, we obtain from (3.33), (3.34), (2.68), and (2.69) the result

$$\begin{aligned}\lim_{p\to\infty} \mathbf{y}_{(p)} &= \lim_{p\to\infty}\left[D_n^{-1/2}(BB')^{p-1}\mathbf{d}_0\right] = \lim_{p\to\infty}\left(D_n^{-1/2}\sum_{j=0}^{t} c_j\left(\eta_j^2\right)^{p-1}\mathbf{u}_j\right)\\ &= c^*D_n^{-1/2}\mathbf{u}_0 = c^*\mathbf{y}_0,\end{aligned} \tag{3.37}$$

where c_j are constants, not all of which are zero, $c^* = c_0(\eta_0^2)^{p-1}$, η_j^2 are eigenvalues of BB', $\eta_0^2 > \eta_1^2 > \eta_2^2 > \cdots > \eta_t^2$, $\mathbf{u}_j$ are eigenvectors associated with η_j^2, and $r(BB') = t+1$. The above derivations are based on (2.63) to (2.69). Similarly, we obtain

$$\lim_{p\to\infty} \mathbf{x}_{(p)} = b^*D^{-1/2}\mathbf{w}_0 = b^*\mathbf{x}_0, \tag{3.38}$$

where $b^* = b_0(\eta_0^2)^{p-1}$, b_0 is a non-zero constant, and $\mathbf{w}_0$ is an eigenvector of $B'B$ associated with η_0^2.

Thus we can conclude that as p increases $\mathbf{y}_{(p)}$ and $\mathbf{x}_{(p)}$ tend to become collinear with $\mathbf{y}_0$ and $\mathbf{x}_0$, respectively, which are associated with the largest eigenvalue. This conclusion proves the optimality of the reciprocal averaging process. It is easy to see from (3.34) and (3.35) that the reciprocal averaging process is basically the same as the iterative method discussed in section 2.2.2.

In reciprocal averaging, one deals with data matrix F instead of $B'B$ or BB', and obtains $\mathbf{x}$ and $\mathbf{y}$ directly. In order to extend this method to multidimensional analysis, one can follow the general procedure discussed in Section 2.4. The calculation of the residual matrix now requires formula (2.57). To eliminate the effect of the first (trivial) solution, $(\mathbf{x}_0, \mathbf{y}_0)$, one can calculate the residual matrix, F_1 say, from

$$F_1 = F - \frac{F\mathbf{x}_0\mathbf{y}_0'F}{\mathbf{y}_0'F\mathbf{x}_0}, \tag{3.39}$$

where F is the data matrix, from which $\mathbf{x}_0$ and $\mathbf{y}_0$ were extracted. F_1 is now subjected to the reciprocal averaging process to obtain $(\mathbf{x}_1, \mathbf{y}_1)$. Then the contribution of the second solution is eliminated from F_1, resulting in residual matrix F_2. In general,

$$F_{j+1} = F_j - \frac{F_j\mathbf{x}_j\mathbf{y}_j'F_j}{\mathbf{y}_j'F_j\mathbf{x}_j}. \tag{3.40}$$

One can extract successively $(\mathbf{x}_0, \mathbf{y}_0)$, $(\mathbf{x}_1, \mathbf{y}_1)$, and so on. Note that

$$F = \sum_{j=0}^{t} \frac{F_j\mathbf{x}_j\mathbf{y}_j'F_j}{\mathbf{y}_j'F_j\mathbf{x}_j}, \tag{3.41}$$

where $r(F)=t+1$. This procedure of multidimensional analysis, however, cannot always be recommended computationally, for it is not necessarily faster than the procedure discussed previously, and it also involves the $n\times m$ matrix even when n is much larger than m or vice versa.

3.2.2 *Bivariate correlation approach*

Suppose that **x** and **y** are weights assigned, respectively, to the columns and the rows of data matrix F. Then the response falling in the ith row and the jth column of F can be replaced by the pair (y_i, x_j). There are in total f_t pairs of numbers. According to our criterion of internal consistency, we wish to determine the two numbers in each pair, y_i and x_j, assigned to the same response, to be as similar to each other as possible. In other words, we wish to determine **x** and **y** so as to maximize the product-moment correlation defined over the f_t pairs of (x_i, y_j). This approach was presented by Guttman (1941) and Maung (1941).

Since the correlation coefficient is invariant under a change in origin, we employ conditions (2.17) and (2.76), that is, $\mathbf{f}'\mathbf{x}=\mathbf{g}'\mathbf{y}=0$. Then the product-moment correlation, ρ, can be expressed as

$$\rho = \mathbf{y}'F\mathbf{x}/\sqrt{\mathbf{x}'D\mathbf{x}\mathbf{y}'D_n\mathbf{y}}\ . \tag{3.42}$$

Our problem is to determine **x** and **y** so as to maximize ρ. This can also be stated as the problem of maximizing $\mathbf{y}'F\mathbf{x}$ subject to the condition that $\mathbf{x}'D\mathbf{x}=\mathbf{y}'D_n\mathbf{y}=f_t$ (or any other constant). Using the latter, we define the Lagrangian function

$$Q = \mathbf{y}'F\mathbf{x} - \tfrac{1}{2}\lambda(\mathbf{x}'D\mathbf{x}-f_t) - \tfrac{1}{2}\mu(\mathbf{y}'D_n\mathbf{y}-f_t), \tag{3.43}$$

where $\frac{1}{2}\lambda$ and $\frac{1}{2}\mu$ are Lagrange multipliers. Differentiating (3.43) with respect to **x**, **y**, λ, and μ, and setting the derivatives equal to zero, we obtain

$$F'\mathbf{y} - \lambda D\mathbf{x} = \mathbf{0}, \tag{3.44}$$

$$F\mathbf{x} - \mu D_n\mathbf{y} = \mathbf{0}, \tag{3.45}$$

$$\mathbf{x}'D\mathbf{x} = f_t, \tag{3.46}$$

$$\mathbf{y}'D_n\mathbf{y} = f_t. \tag{3.47}$$

Premultiplying (3.44) by $\mathbf{x}'$ and (3.45) by $\mathbf{y}'$, we obtain

$$\lambda = \mathbf{x}'F'\mathbf{y}/\mathbf{x}'D\mathbf{x}, \tag{3.48}$$

$$\mu = \mathbf{y}'F\mathbf{x}/\mathbf{y}'D_n\mathbf{y}. \tag{3.49}$$

Since the two denominators are equal by (3.46) and (3.47), it follows from (3.42), (3.48), and (3.49) that

$$\lambda = \mu = \rho. \tag{3.50}$$

Furthermore, we obtain from (3.50), (3.44), and (3.45)

$$\mathbf{x} = D^{-1}F'\mathbf{y}/\rho, \tag{3.51}$$

$$\mathbf{y} = D_n^{-1}F\mathbf{x}/\rho. \tag{3.52}$$

Note the resemblance of these formulas to those derived earlier, (2.75) and (3.8). Their identities can be established if ρ^2 is proved to be equal to η^2, which can be done easily. We only need to substitute $\mathbf{y}$ of (3.52) for (3.48), that is,

$$\lambda = \rho = \frac{1}{\rho}\frac{\mathbf{x}'F'D_n^{-1}F\mathbf{x}}{\mathbf{x}'D\mathbf{x}} = \frac{1}{\rho}\eta^2. \tag{3.53}$$

Or, substituting $\mathbf{x}$ of (3.51) for (3.49), we obtain

$$\mu = \rho = \frac{1}{\rho}\frac{\mathbf{y}'FD^{-1}F'\mathbf{y}}{\mathbf{y}'D_n\mathbf{y}} = \frac{1}{\rho}\eta^2. \tag{3.54}$$

Hence

$$\rho^2 = \eta^2, \tag{3.55}$$

and (3.51) and (3.52) are identical with (3.8) and (2.75), respectively.

It is obvious that maximization of ρ results in the same solution vectors, $\mathbf{x}$ and $\mathbf{y}$, as maximization of η^2. Furthermore, equations (3.51) and (3.52) clearly indicate the identity of the method of reciprocal averages with the other approaches discussed so far.

3.2.3 *Canonical correlation approach*

The concept of canonical correlation (Hotelling 1936) can be applied to a two-way table. This is one of the approaches taken by Maung (1941). His presentation of the topic, however, might be too brief to comprehend in detail. We shall therefore reformulate the approach, using our numerical example.

The canonical correlation is an index of relationship between two sets of variables. More specifically, it is the maximum correlation between a linear combination of the first set and that of the second set of variables. In our example, the two sets of variables are three response categories (columns) and three teachers (rows). Our original data matrix,

$$F = \begin{bmatrix} 1 & 3 & 6 \\ 3 & 5 & 2 \\ 6 & 3 & 0 \end{bmatrix},$$

can now be represented in the 29×6 matrix F^*,

$$
\begin{array}{c}
\qquad\qquad \text{Set 1 (categories)} \qquad\qquad \text{Set 2 (teachers)} \\
\qquad\qquad \text{good} \quad \text{average} \quad \text{poor} \qquad\qquad 1 \quad 2 \quad 3
\end{array}
$$

$$
F^*_{29\times 6} = \begin{bmatrix}
1 & 0 & 0 & 1 & 0 & 0 \\
0 & 1 & 0 & 1 & 0 & 0 \\
0 & 1 & 0 & 1 & 0 & 0 \\
0 & 1 & 0 & 1 & 0 & 0 \\
0 & 0 & 1 & 1 & 0 & 0 \\
\vdots & \vdots & \vdots & \vdots & \vdots & \vdots \\
0 & 1 & 0 & 0 & 0 & 1
\end{bmatrix}
$$

$$
= \Big[\underset{29\times 3}{F_1^*} , \underset{29\times 3}{F_2^*} \Big], \text{ say.} \tag{3.56}
$$

In F^*, we see one row of pattern (100 100), indicating the combination 'good, Teacher 1,' three of pattern (010 100) of combination 'average, Teacher 1,' six of pattern (001 100) of combination 'poor, Teacher 1,' and so on. The last pattern is (010 001), of combination 'average, Teacher 3'; there are three of this pattern. Consider linear combinations of variables for the two sets, namely, $F_1^*\mathbf{x}$ and $F_2^*\mathbf{y}$. If we indicate vector $[\mathbf{x}',\mathbf{y}']$ by $\mathbf{z}'$, the cross-product of the weighted responses, $\mathbf{z}'F^{*\prime}F^*\mathbf{z}$, can be expressed as

$$
\mathbf{z}'F^{*\prime}F^*\mathbf{z} = [\mathbf{x}',\mathbf{y}'] \begin{bmatrix} F_1^{*\prime} \\ F_2^{*\prime} \end{bmatrix} [F_1^*, F_2^*] \begin{bmatrix} \mathbf{x} \\ \mathbf{y} \end{bmatrix}
$$

$$
= [\mathbf{x}',\mathbf{y}'] \begin{bmatrix} D & F' \\ F & D_n \end{bmatrix} \begin{bmatrix} \mathbf{x} \\ \mathbf{y} \end{bmatrix}, \tag{3.57}
$$

where F, D, and D_n are as defined before. When the original data in F^* are adjusted to the origin zero, D corresponds to the variance-covariance matrix of the first set of variables, Σ_{11} say, and D_n to that of the second set of variables, Σ_{22}. F' corresponds to the covariance matrix of the two sets of variables, Σ_{12}. Thus, in terms of the ordinary notation for variables adjusted to the zero origin, (3.57) presents the familiar quadratic form with partitioned variance-covariance matrix, that is,

$$
[\mathbf{x}',\mathbf{y}'] \begin{bmatrix} \Sigma_{11} & \Sigma_{12} \\ \Sigma_{21} & \Sigma_{22} \end{bmatrix} \begin{bmatrix} \mathbf{x} \\ \mathbf{y} \end{bmatrix}. \tag{3.58}
$$

Our objective is to determine $\mathbf{x}$ and $\mathbf{y}$ so as to maximize the canonical correlation. This is nothing but the problem of maximizing $\mathbf{x}'\Sigma_{12}\mathbf{y}$ (or $\mathbf{y}'\Sigma_{21}\mathbf{x}$) under some constraints on $\mathbf{x}$ and $\mathbf{y}$. If we use the conditions $\mathbf{f}'\mathbf{x}=\mathbf{g}'\mathbf{y}=0$ and $\mathbf{x}'D\mathbf{x}=\mathbf{y}'D_n\mathbf{y}=f_t$, and note that $\mathbf{y}'\Sigma_{21}\mathbf{x}=\mathbf{y}'F\mathbf{x}$, then it is obvious that the problem is exactly the same as in the bivariate correlation approach. The two

Lagrange multipliers in the canonical correlation approach turn out to be canonical correlations, and they are identical. As is obvious from the previous section, the squared canonical correlation, γ^2 say, turns out to be equal to η^2. Therefore, in dual scaling, the relation

$$\eta^2 = \rho^2 = \gamma^2 \tag{3.59}$$

holds, and the three criteria, η^2, ρ, and γ, lead to identical results.

3.2.4 *Simultaneous linear regression approach*

Hirschfeld (1935) posed the question: 'Is it always possible to introduce new variates for the rows and the columns of a contingency table such that *both* regressions are linear?' In reply, he derived the formulas of dual scaling discussed earlier. This approach was later employed by Lingoes (1963, 1968), who was obviously unaware of the Hirschfeld study, but noted that the approach would use the basic theory and equations worked out by Guttman (1941). Since the formulas of dual scaling have been thoroughly discussed, we shall present here merely an example to illustrate the procedure. Suppose that from a survey on the use of sleeping pills 28 subjects were randomly sampled from each of the following five groups: those strongly against the use, those moderately against, those neutral, those moderately for, those strongly for. Suppose that the subjects representing these five groups were asked the question: 'Do you have nightmares in sleep?' Suppose the responses were as given in Table 3.1. Let us now assign the following scores to the categories: x_1 (never) $= -2$, x_2 (rarely) $= -1$, x_3 (sometimes) $= 0$, x_4 (often) $= 1$, x_5 (always) $= 2$; y_1 (strongly against) $= -2$, y_2 (against) $= -1$, y_3 (neutral) $= 0$, y_4 (for) $= 1$, y_5 (strongly for) $= 2$. Given these values for the columns and the rows, we can calculate the means of the rows and of the columns. For example, the mean

TABLE 3.1

	Nightmares						
Sleeping pills	Never	Rarely	Sometimes	Often	Always	Total	Score
Strongly against	15	8	3	2	0	28	−2
Against	5	17	4	0	2	28	−1
Neutral	6	13	4	3	2	28	0
For	0	7	7	5	9	28	1
Strongly for	1	2	6	3	16	28	2
Total	27	47	24	13	29	140	
Score	−2	−1	0	1	2		

value of 'never' is calculated from the five frequencies (15, 5, 6, 0, 1) in column 'never' and the corresponding weights as

$$[15\times(-2)+5\times(-1)+6\times0+0\times1+1\times2]/27 = -1.2 = x_1^*, \text{ say.}$$

The mean value of 'strongly against' can be calculated from the response frequencies (15, 8, 3, 2, 0) and the corresponding weights as

$$[15\times(-2)+8\times(-1)+3\times0+2\times1+0\times2]/28 = -1.3 = y_1^*, \text{ say.}$$

In this way we obtain the two sets of mean scores

$$\mathbf{x}^* = D^{-1}F'\mathbf{y} = \begin{bmatrix} -1.2 \\ -0.5 \\ 0.4 \\ 0.5 \\ 1.3 \end{bmatrix}, \qquad \mathbf{y}^* = D_n^{-1}F\mathbf{x} = \begin{bmatrix} -1.3 \\ -0.8 \\ -0.6 \\ 0.6 \\ 1.1 \end{bmatrix}.$$

Figure 3.1 shows the relations between **x** and $D^{-1}F'\mathbf{y}$ and between **y** and $D_n^{-1}F\mathbf{x}$. Let us call these two relations regression of **x** on **y** and regression of **y** on **x**, respectively. Both relations are non-linear. Under the scoring scheme (−2, −1, 0, 1, 2), the product-moment correlation between the two variables (nightmares and sleeping pills) is 0.6277. The correlation increases as the plot

Figure 3.1

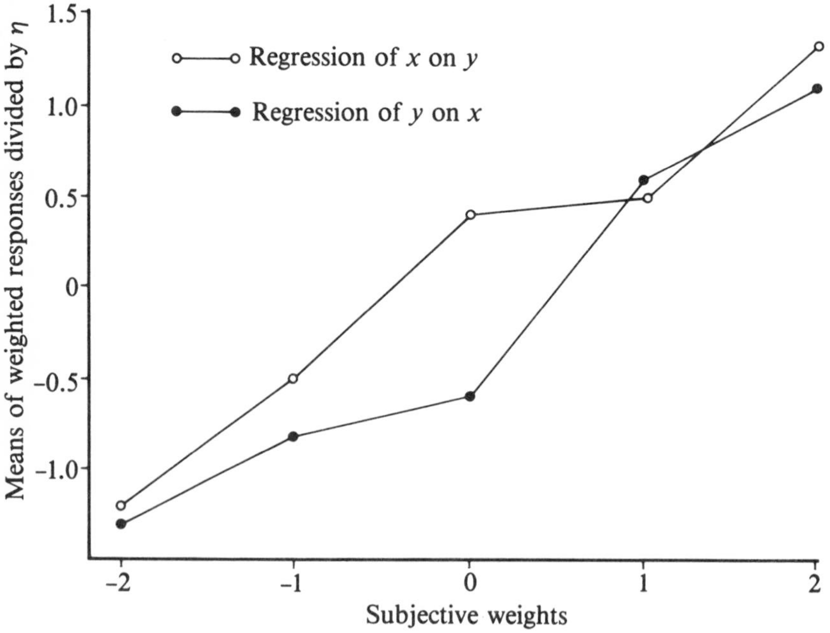

approaches a straight line and attains its maximum when all the means fall on a straight line. The simultaneous linear regression approach then considers stretching and compressing the category intervals of both variables so that the plots of means against the category scores may be both linear. Once this process is translated into mathematical operations, we obtain the same formulas as those derived for dual scaling. To illustrate the validity of this approach, we simply present the dual scaling results for our example in Table 3.2 and Figure 3.2. There are a few interesting points to note. First, the product-moment correlation is now increased to 0.6472 from 0.6277. Second, the two lines in Figure 3.1 are now merged into a single straight line. In other words, all the mean values now fall on a common straight line. This is an interesting effect of our choice of the constant of proportionality, $1/\eta$ (see (2.75) and (3.8)). If $1/\eta$ is not used, there will be two straight lines, one for **x** and the other for **y**. Another interesting effect of constant $1/\eta$ is that the slope of the plot in Figure 3.2 corresponds to the product-moment correlation between the two variables, that is, 0.6472.

As illustrated by our example, dual scaling indeed attains simultaneous linear regressions, and the formulas of dual scaling in turn can be derived from the simultaneous linear regression point of view, as demonstrated by

Figure 3.2

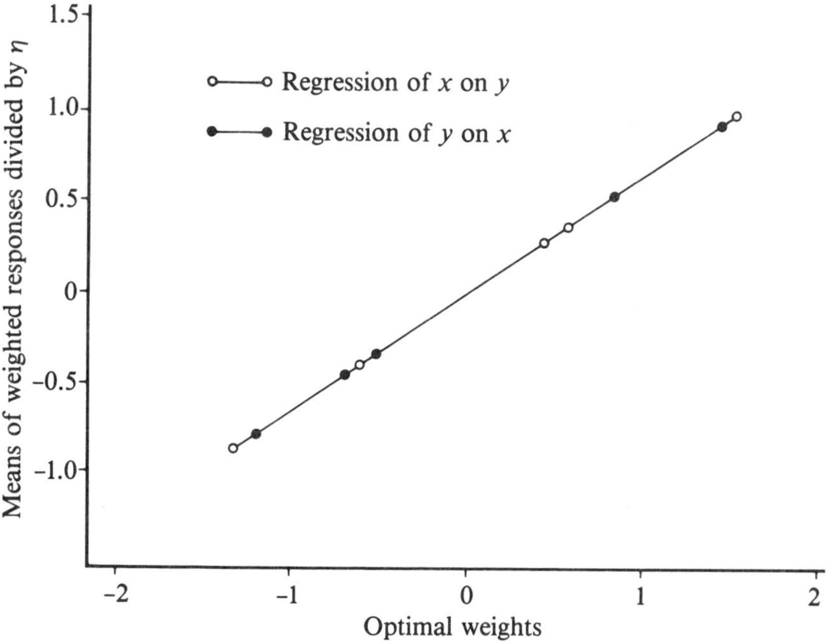

TABLE 3.2

	Optimal weight x	Mean x*		Optimal weight y	Mean y*
Never	−1.30	−0.84	Strongly against	−1.20	−0.78
Rarely	−0.59	−0.38	Against	−0.64	−0.42
Sometimes	0.43	0.28	Neutral	−0.49	−0.32
Often	0.58	0.38	For	0.87	0.56
Always	1.55	1.00	Strongly for	1.47	0.95

$\eta_1^2=0.4189$, correlation $=\sqrt{\eta_1^2}=0.6472$.

Hirschfeld (1935). The simultaneous linear regressions approach suggests the possibility of developing a simple graphical method for obtaining a coarse approximation to the optimal solution. Note also that such a plot as Figure 3.1 may offer some ideas about the relative goodness of any empirical scoring scheme.

3.3 BASIC CHARACTERISTICS AND MATHEMATICAL STRUCTURE

As described in this chapter, dual scaling quantifies data in a two-way classification table so as to maximize *simultaneously* the between-row and the between-column sums of squares, relative to the total sum of squares. Its procedure can be formulated in several ways, which seems to indicate the versatility of the technique.

In addition to its interesting and important property of duality, it has other noteworthy characteristics:

1. As was seen above, it does not employ any assumptions about the distribution of responses.
2. It explains the joint responses in the data matrix only in terms of weights (scores) assigned to the marginals, and hence offers a data-reduction technique for categorical data.
3. It is a method for constructing an 'optimal' composite of several categorical variables.
4. It can determine the weights for categorical variables so as to discriminate maximally among several groups of subjects, and as such it provides discriminant function analysis of categorical variables.

Furthermore, as will be seen in Chapter 5 and other chapters,

5. It maximizes the generalized Kuder-Richardson reliability of multiple-choice data.
6. It provides a simple, but effective, means of item selection in test construction.
7. It makes effective use of item-difficulty levels in arriving at test scores for individual subjects.
8. It can capture partial knowledge subjects might have of the topics in multiple-choice testing, through differential weighting of response options.
9. In applications to other statistical models, it can quantify categorical data to suit such analyses as factor analysis and the analysis of variance.
10. It provides multidimensional analysis as warranted by the data, thus without jeopardizing the validity of quantification. Many scaling techniques, in contrast, are valid only under the assumption of a unidimensional distribution of the 'attribute' under investigation.
11. As applied to general scaling problems, it takes into consideration individual differences in judgment, and thus is capable of isolating homogeneous subgroups of subjects in terms of their distinct sets of response patterns. This approach of individual differences scaling also suggests that dual scaling determines scale values of objects so as to maximize the reproducibility of responses of individual subjects in terms of the scale values.

In regard to its mathematical structure, many interesting relations have been noted, in particular those formulas used in the theorem on duality, that is, (3.28) to (3.31). For illustrative purposes, it is often helpful to express data matrix F in terms of its derived components. To conclude our discussion, let us derive such an expression. Consider formula (3.41), that is,

$$F = \sum \frac{F_j \mathbf{x}_j \mathbf{y}_j' F_j}{\mathbf{y}_j' F_j \mathbf{x}_j}. \tag{3.41}$$

In the bivariate correlation approach, we obtained the relations

$$F_j \mathbf{x}_j = \rho_j D_n \mathbf{y}_j, \tag{3.60}$$

$$F_j' \mathbf{y}_j = \rho_j D \mathbf{x}_j. \tag{3.61}$$

Noting that $\mathbf{y}_j' D_n \mathbf{y}_j = f_t$, we obtain from (3.60)

$$\mathbf{y}_j' F_j \mathbf{x}_j = \rho_j \mathbf{y}_j' D_n \mathbf{y}_j = f_t \rho_j. \tag{3.62}$$

Substituting (3.60), (3.61), and (3.62) for the corresponding parts of (3.41), we obtain

$$F = D_n\left(\sum \rho_j \mathbf{y}_j \mathbf{x}_j'\right) D / f_t. \tag{3.63}$$

F is now expressed in terms of $\mathbf{x}$, $\mathbf{y}$, and the correlation. Suppose that $r(F) = t+1$, and indicate by X and Y matrices of the entire sets of vectors $\mathbf{x}_j$

and $\mathbf{y}_j$, arranged as follows:

$$X = \begin{bmatrix} \mathbf{x}_0' \\ \mathbf{x}_1' \\ \vdots \\ \mathbf{x}_t' \end{bmatrix}, \qquad Y = \begin{bmatrix} \mathbf{y}_0' \\ \mathbf{y}_1' \\ \vdots \\ \mathbf{y}_t' \end{bmatrix}. \tag{3.64}$$

Let us also indicate the diagonal matrix of the $t+1$ values of ρ_j by Λ,

$$\Lambda = \begin{bmatrix} 1 & 0 & 0 & \cdots & 0 \\ 0 & \rho_1 & 0 & \cdots & 0 \\ 0 & 0 & \rho_2 & \cdots & 0 \\ \vdots & \vdots & \vdots & \ddots & \vdots \\ 0 & 0 & 0 & \cdots & \rho_t \end{bmatrix}. \tag{3.65}$$

This corresponds to the Λ defined for (3.31). Note that $\rho_0 = 1$, that is, the value corresponding to the trivial solution. Using (3.64) and (3.65), we can express (3.63) as

$$F = D_n Y'\Lambda X D / f_t. \tag{3.66}$$

As we discussed earlier, the matrices of orthogonal vectors which correspond to X and Y are given respectively by

$$W = XD^{1/2}, \tag{3.67}$$

$$U = YD_n^{1/2}. \tag{3.68}$$

Thus data matrix F can be expressed in terms of W and U:

$$F = D_n^{1/2} U'\Lambda W D^{1/2} / f_t. \tag{3.69}$$

Therefore the typical element of F,f_{ik}, can be expressed as

$$f_{ik} = \frac{f_{i.} f_{.k}}{f_t}\left(1 + \sum_{j=1}^{t} \rho_j y_{ji} x_{jk}\right) \tag{3.70}$$

$$= \frac{\sqrt{f_{i.} f_{.k}}}{f_t}\left(1 + \sum_{j=1}^{t} \rho_j u_{ji} w_{jk}\right), \tag{3.71}$$

where $f_{i.}$ and $f_{.k}$ are the ith and the kth diagonal elements of D_n and D, respectively, and similarly subscripts i and k for y_{ji}, u_{ji}, x_{jk}, and w_{jk} are the ith and kth elements of the corresponding vectors $\mathbf{y}_j$, $\mathbf{u}_j$, $\mathbf{x}_j$, and $\mathbf{w}_j$. The standardized data matrix, B, can be expressed as

$$B = D_n^{-1/2} F D^{-1/2} = D_n^{1/2} Y'\Lambda X D^{1/2} / f_t \tag{3.72}$$

$$= U'\Lambda W / f_t. \tag{3.73}$$

The typical element of B, therefore, can be expressed as

$$b_{ik} = \frac{f_{ik}}{\sqrt{f_{i.}f_{.k}}} = \frac{\sqrt{f_{i.}f_{.k}}}{f_t}\left(1+\sum_{j=1}^{t}\rho_j y_{ji} x_{jk}\right) \tag{3.74}$$

$$= \frac{1}{f_t}\left(1+\sum_{j=1}^{t}\rho_j u_{ji} w_{jk}\right). \tag{3.75}$$

These formulas for F and B hold for all the approaches discussed in this chapter. Under appropriate constraints on the component matrices, the formulas can be used to solve the dual scaling problem. In this sense, these equations of F and B, expressed in terms of Λ, X, Y, W, and U, can be regarded as the major part of the structural model of dual scaling. From a practical point of view, it would be of some interest to use the relation (3.63) and to define the following matrices: *the first-order approximation* of data matrix F, which is given by

$$F_{(1)} = D_n(\mathbf{1}\mathbf{1}'+\rho_1\mathbf{y}_1\mathbf{x}_1')D/f_t, \tag{3.76}$$

and *the residual matrix*

$$F_{\mathrm{res}(1)} = F - F_{(1)}; \tag{3.77}$$

similarly, the *kth order approximation*

$$F_{(k)} = \frac{1}{f_t}D_n\left(\mathbf{1}\mathbf{1}'+\sum_{j=1}^{k}\rho_j\mathbf{y}_j\mathbf{x}_j'\right)D, \tag{3.78}$$

and the corresponding residual matrix

$$F_{\mathrm{res}(k)} = F - F_{(k)}. \tag{3.79}$$

These matrices can be used, together with statistic δ, defined by (2.90) and χ^2 tests of η_j^2, to illustrate how much information contained in the data matrix can be explained by scaling. The percentage of the total variance accounted for by the first k non-trivial solutions can be expressed as

$$\delta_k = \sum_{j=1}^{k}\eta_j^2\times 100\% \Big/ \operatorname{tr}(C_1). \tag{3.80}$$

One of the perennial problems in scaling involves the 'quality' of the scaled quantity. This is a difficult, but very important, problem which has a direct bearing on the interpretability of the derived measurement. One way to handle it is to consider metric properties of the scaled data. Given a finite set of numbers, G, a distance (metric) is defined as a function, d, on the Cartesian product, $G\times G$, satisfying the following axioms:

1. The distance between any point and itself is zero,
2. The distance between any two points is non-negative,

3. The distance from point i to point j is the same as the distance from point j to point i,
4. The sum of the length of any two sides of a triangle created by joining three points cannot be shorter than the length of the remaining side.

The Euclidean distance is an example of a metric that satisfies these so-called metric axioms. Greenacre (1978a, b) discusses several variants of dual scaling to see whether or not the square of the biplot distance in each case can generate the squared distance calculated from the coordinates of the kth-order approximation. His discussion can be summarized as follows.

Consider the Eckart-Young decomposition of data matrix F, that is,

$$F_{p\times q} = P\Lambda Q', \tag{3.81}$$

where P is the $p\times r$ matrix of the r normalized eigenvectors of FF' in the rows, Q is the $q\times r$ matrix of the r normalized eigenvectors of $F'F$ in the rows, and Λ is the diagonal matrix of positive square roots of the eigenvalues of FF', which is the same as for $F'F$. In other words,

$$FF' = P\Lambda^2P', \quad F'F = Q\Lambda^2Q', \quad P'P = Q'Q = I_r. \tag{3.82}$$

One can rewrite (3.81) as

$$F = P\Lambda Q' = KL', \tag{3.83}$$

where K and L are matrices of weights (scores) for the rows and the columns, respectively, of F in r dimensions. K and L may be defined as

$$K = P\Lambda^{1/2}, \qquad L = Q\Lambda^{1/2}, \tag{3.84}$$

$$K = P, \qquad L = Q\Lambda, \tag{3.85}$$

or

$$K = P\Lambda, \qquad L = Q. \tag{3.86}$$

Suppose that one is interested in the variable related to the columns of F. Then the squared distances among columns in the kth-order approximation can be expressed as

$$F_k'F_k = L_kK_k'K_kL_k'. \tag{3.87}$$

Thus, if we define L by (3.85), then $K_k'K_k = I_k$, so that

$$F_k'F_k = L_kL_k'. \tag{3.88}$$

In other words, $F_k'F_k$ can be calculated only from the weights (scores) for the columns, and hence in this instance (3.85) seems to be the best choice of the three. At the same time, this choice of K and L makes the meaning of the squared distance between rows somewhat ambiguous, for then one cannot obtain such a simple relation between F_k and K_k as the relation between F_k and L_k indicated by (3.88).

In dual scaling as formulated here, the weights for the rows and those for the columns are scaled symmetrically, and the problem of distance is somehow more complicated than discussed above. The presence of diagonal matrices D_n and D, as seen in (3.66), makes it rather difficult to derive such a simple expression as (3.88). More work is needed on metric properties of measurement derived by dual scaling. Greenacre's work, for example, would serve as a good introduction to this problem and further research.

4

Contingency (two-item) and response-frequency tables

4.1
SPECIAL PROBLEMS

In Chapter 1, the response-frequency table was defined in such a way as to distinguish it from the contingency table. Except for some points, however, these two kinds of tables can be analysed using the same formulation, so they are both considered in this chapter.

4.1.1 *Contingency (two-item) table*

Contingency tables are often used in the behavioural sciences as a means of summarizing joint responses to a set of variables. Consider a 2×3 contingency table, Table 4.1. The data are collected by asking two questions: (1) Are you a member of the union? – (yes, no). (2) Would you accept the salary proposal? – (accept, reject, abstain). If the elements of this table were not frequencies but sets of normal variates, the table would ordinarily be subjected to the analysis of variance, using a two-way crossed design. With frequencies in the cells, however, this type of table is often handled by the analysis of variance, using a log linear model (e.g., Dyke and Patterson 1952;

TABLE 4.1

Contingency table

	Salary proposal			
Membership	Accept	Reject	Abstain	Total
Union	f_{11}	f_{12}	f_{13}	$f_{1.}$
Non-union	f_{21}	f_{22}	f_{23}	$f_{2.}$
Total	$f_{.1}$	$f_{.2}$	$f_{.3}$	f_t

Bock 1965, 1968; Nishisato 1966, 1970; Bishop 1969; Grizzle, Starmer, and Koch 1969; Fienberg 1970; Goodman 1971; Haberman 1974; Bishop, Fienberg, and Holland 1975). Logarithmic transforms of frequencies are thus subjected to the analysis of variance, guarded with all the assumptions normally employed in the analysis. The results are then summarized in a table similar to that used in the usual analysis of variance. This approach will be considered again in Chapter 8.

One may wonder whether the contingency table can be directly subjected to dual scaling, without requiring any assumptions or transformations such as logarithmic. The answer is affirmative, and it is indeed a straightforward application. There are, however, several interesting problems in the application, which we shall explore below.

The 2×3 data matrix in Table 4.1 is given as

$$F = \begin{bmatrix} f_{11} & f_{12} & f_{13} \\ f_{21} & f_{22} & f_{23} \end{bmatrix}. \tag{4.1}$$

The problem is to determine weights $x_{1j}(j=1,2)$ for question 1 and $x_{2k}(k=1,2,3)$ for question 2. As mentioned in Chapter 1, the contingency table can also be represented in the format of a response-pattern table. The 2×3 matrix F can therefore be represented, without loss of information, in the response-pattern format as follows:

$$\begin{array}{cc} & \begin{array}{ccccc} \multicolumn{2}{c}{\text{Q.1}} & \multicolumn{3}{c}{\text{Q.2}} \\ 1 & 2 & 1 & 2 & 3 \end{array} \\ \underset{f_t \times 5}{F_a} = & \left[\begin{array}{ccccc} 1 & 0 & 1 & 0 & 0 \\ \vdots & \vdots & \vdots & \vdots & \vdots \\ 1 & 0 & 1 & 0 & 0 \\ 1 & 0 & 0 & 1 & 0 \\ \vdots & \vdots & \vdots & \vdots & \vdots \\ 1 & 0 & 0 & 1 & 0 \\ \vdots & \vdots & \vdots & \vdots & \vdots \\ 0 & 1 & 0 & 0 & 1 \\ \vdots & \vdots & \vdots & \vdots & \vdots \\ 0 & 1 & 0 & 0 & 1 \end{array}\right] \begin{array}{l} \left.\begin{array}{l} \\ \\ \\ \end{array}\right\} f_{11} \\ \left.\begin{array}{l} \\ \\ \\ \end{array}\right\} f_{12} \\ \quad\vdots \\ \left.\begin{array}{l} \\ \\ \\ \end{array}\right\} f_{23} \end{array} \end{array} \tag{4.2}$$

The first f_{11} rows consist of pattern (10100), the next f_{12} rows consist of pattern (100010), and so on. There is thus redundancy in F_a, and F_a can be reduced, without loss of information, to the matrix

$$\underset{6\times 5}{F_b} = \begin{array}{c} \begin{array}{ccccc} \multicolumn{2}{c}{\text{Q.1}} & \multicolumn{3}{c}{\text{Q.2}} \\ 1 & 2 & 1 & 2 & 3 \end{array} \\ \begin{bmatrix} f_{11} & 0 & f_{11} & 0 & 0 \\ f_{12} & 0 & 0 & f_{12} & 0 \\ f_{13} & 0 & 0 & 0 & f_{13} \\ 0 & f_{21} & f_{21} & 0 & 0 \\ 0 & f_{22} & 0 & f_{22} & 0 \\ 0 & f_{23} & 0 & 0 & f_{23} \end{bmatrix} \end{array}. \tag{4.3}$$

It is known (Benzécri 1969; Nishisato 1973a) that if two columns (rows) are proportional, then the optimal weights of the two columns (rows) are identical. Thus the optimal weight vector of F_a,$\mathbf{x}_a$, is identical with the optimal weight vector of F_b, $\mathbf{x}_b$, and the optimal scores for the corresponding patterns in F_a and F_b (e.g., [1,0,1,0,0] and $[f_{11},0,f_{11},0,0]$) are identical. Noting this type of equivalence in structure between F_a and F_b, we now turn our attention to comparing F and F_a (or F_b). As we recall, F_a was constructed from F without loss of information. It is not difficult to see that F can be constructed from F_a, too. However, the two matrices are not equivalent in terms of the scaling results. There are at least two points of difference between F and F_a. The first is that the f_t rows of F_a are imbedded in F. This circumstance suggests that an extra step is necessary to calculate f_t optimal scores from F, which correspond to the scores for the rows of F_a. The second point is that the ranks of the two matrices are not the same. If F is $n\times m$ and F_a is therefore $f_t\times(n+m)$, and if f_t is much larger than $n+m$, then

$$r(F) \leq \min(n,m), \tag{4.4}$$

$$r(F_a) \leq n+m-1, \tag{4.5}$$

where $\min(n,m)$ indicates 'the smaller value of n and m.' Thus the number of solutions, excluding trivial ones, is at most $\min(n,m)-1$ for F and $n+m-2$ for F_a. We note that the latter becomes larger than the former when n and m are greater than one. Then one may ask how the extra solutions from F_a are generated.

To clarify the relation between these two representations of the same data, we first introduce the following notation:

n = the number of rows of F,
m = the number of columns of F,
$\mathbf{y}$ = the $n\times 1$ vector of weights for the rows of F (e.g., question 1),
$\mathbf{x}$ = the $m\times 1$ vector of weights for the columns of F (e.g., question 2),
$\mathbf{z} = \begin{bmatrix} \mathbf{z}_1 \\ \mathbf{z}_2 \end{bmatrix}$ = the $(n+m)\times 1$ vector of weights for the columns of F_a (e.g., question 1 + question 2),

D_n = the $n \times n$ diagonal matrix of the row totals of F,
D = the $m \times m$ diagonal matrix of the column totals of F,
D_N = the $f_t \times f_t$ diagonal matrix of the row totals of F_a,
D_a = the $(n+m) \times (n+m)$ diagonal matrix of the column totals of F_a,
η^2 = the squared correlation ratio for F,
η_a^2 = the squared correlation ratio for F_a,
f_t = the sum of the elements of F,
f_t^* = the sum of the elements of F_a.

Note the relations

$$f_t^* = 2f_t, \tag{4.6}$$

$$D_N = 2I. \tag{4.7}$$

Let us consider dual scaling of F_a. η_a^2 can be expressed in terms of the weight vector $\mathbf{z}$ as

$$\eta_a^2 = \frac{\mathbf{z}'F_a'D_N^{-1}F_a\mathbf{z}}{\mathbf{z}'D_a\mathbf{z}} = \frac{\frac{1}{2}\mathbf{z}'F_a'F_a\mathbf{z}}{\mathbf{z}'D_a\mathbf{z}}. \tag{4.8}$$

That is,

$$\frac{\mathbf{z}'F_a'F_a\mathbf{z}}{\mathbf{z}'D_a\mathbf{z}} = 2\eta_a^2. \tag{4.9}$$

Note that

$$F_a'F_a = \begin{bmatrix} \underset{n\times n}{D_n} & \underset{n\times m}{F} \\ \underset{m\times n}{F'} & \underset{m\times m}{D} \end{bmatrix}, \tag{4.10}$$

$$D_a = \begin{bmatrix} D_n & O \\ O' & D \end{bmatrix}, \tag{4.11}$$

where O is the $n \times m$ null matrix. We maximize $\mathbf{z}'F_a'F_a\mathbf{z}$ with respect to $\mathbf{z}$, subject to the condition that $\mathbf{z}'D_a\mathbf{z} = f_t^*$. Define

$$Q = \mathbf{z}'F_a'F_a\mathbf{z} - \lambda(\mathbf{z}'D_a\mathbf{z} - f_t^*), \tag{4.12}$$

where λ is a Lagrange multiplier, and solve

$$\partial Q/\partial \mathbf{z} = \mathbf{0} \quad \text{and} \quad \partial Q/\partial \lambda = 0. \tag{4.13}$$

After a few algebraic manipulations, we find that $\lambda = 2\eta_a^2$ and obtain the equations

$$\begin{bmatrix} D_n & F \\ F' & D \end{bmatrix}\begin{bmatrix} \mathbf{z}_1 \\ \mathbf{z}_2 \end{bmatrix} = 2\eta_a^2\begin{bmatrix} D_n & O \\ O' & D \end{bmatrix}\begin{bmatrix} \mathbf{z}_1 \\ \mathbf{z}_2 \end{bmatrix}, \tag{4.14}$$

$$[\mathbf{z}_1'\mathbf{z}_2']\begin{bmatrix} D_n & O \\ O' & D \end{bmatrix}\begin{bmatrix} \mathbf{z}_1 \\ \mathbf{z}_2 \end{bmatrix} = 2f_t. \tag{4.15}$$

These two equations can be rewritten as

$$F\mathbf{z}_2 = (2\eta_a^2 - 1)D_n\mathbf{z}_1, \qquad F'\mathbf{z}_1 = (2\eta_a^2 - 1)D\mathbf{z}_2, \tag{4.16}$$

$$\mathbf{z}_1'D_n\mathbf{z}_1 + \mathbf{z}_2'D\mathbf{z}_2 = 2f_t, \tag{4.17}$$

respectively. Compare the two equations of (4.16) with (3.45) and (3.44) derived for the bivariate correlation approach, that is,

$$F\mathbf{x} - \rho D_n\mathbf{y} = \mathbf{0}, \qquad F'\mathbf{y} - \rho D\mathbf{x} = \mathbf{0}, \tag{4.18}$$

where ρ is substituted for λ and μ in (3.44) and (3.45). As is clear from our discussion in Chapter 3, (4.18) also defines the stationary equations for optimal vectors of the contingency table. Thus (4.16) and (4.18) are the equations for F_a and F, respectively. Note that if we set $\mathbf{x}=\mathbf{z}_2$, $\mathbf{y}=\mathbf{z}_1$, and

$$\rho = 2\eta_a^2 - 1, \tag{4.19}$$

(4.16) and (4.18) become identical. Since we can consider only the case in which ρ is non-negative, (4.19) indicates that: (1) $\eta_a^2 \geq \rho$, (2) $\eta_a^2 = \rho$ only when $\rho = 1$, and (3) $1 \geq \eta_a^2 \geq 0.5$ when $1 \geq \rho \geq 0$. Note, however, that these relations between ρ and η_a^2 hold only for the $\min(n,m)-1$ corresponding solutions. For the remaining extra solutions associated with F_a, the eigenvalues can be smaller than 0.5.

In regard to the unit of weights, simple algebraic manipulations of (4.16) lead to the following interesting relation:

$$\mathbf{z}_1'D_n\mathbf{z}_1 = \mathbf{z}_2'D\mathbf{z}_2 = f_t. \tag{4.20}$$

Since we impose the same conditions on $\mathbf{x}$ and $\mathbf{y}$ for F (i.e., $\mathbf{y}'D_n\mathbf{y} = \mathbf{x}'D\mathbf{x} = f_t$), we see that the two analyses use the same scale unit.

When F is analysed, one may wish to maximize η^2 instead of ρ. Then, noting that $\eta^2 = \rho^2$, we can obtain from (4.19) the relation between the two correlation ratios, one from F and the other from F_a, namely,

$$\eta^2 = (2\eta_a^2 - 1)^2, \tag{4.21}$$

which is the same as the result reported by de Leeuw (1973). This relation holds for pairs of eigenvalues, (η_j^2, η_{ja}^2), say, $j = 0, 1, \ldots, \min(n,m)-1$. As for (4.19), (4.21) indicates that $1 \geq \eta_a^2 \geq 0.5$ when $1 \geq \eta^2 \geq 0$ and $\eta_a^2 \geq \eta^2$ with equality holding when $\eta^2 = 1$. For the extra solutions associated with F_a, the eigenvalues can be smaller than 0.5.

In summary, we conclude that vectors $\mathbf{x}$ and $\mathbf{y}$ obtained from F are identical respectively with $\mathbf{z}_2$ and $\mathbf{z}_1$ obtained from F_a, that is,

$$\mathbf{z} = \begin{bmatrix} \mathbf{z}_1 \\ \mathbf{z}_2 \end{bmatrix} = \begin{bmatrix} \mathbf{y} \\ \mathbf{x} \end{bmatrix}, \tag{4.22}$$

and that the two corresponding correlation ratios are related by (4.21). It may be of interest to know that duality of the contingency table analysis, as inferred from (4.16), leads to the relations

$$\mathbf{z} = \begin{bmatrix} \mathbf{z}_1 \\ \mathbf{z}_2 \end{bmatrix} = \frac{1}{2\eta_a^2 - 1} \begin{bmatrix} D_n^{-1} & O_1 \\ O_1' & D^{-1} \end{bmatrix} \begin{bmatrix} F & O_2 \\ O_3 & F' \end{bmatrix} \begin{bmatrix} \mathbf{z}_2 \\ \mathbf{z}_1 \end{bmatrix}, \tag{4.23}$$

where the orders of the null matrices O_1, O_2, and O_3 are $n \times m$, $n \times n$, and $m \times m$, respectively.

Let us now consider the scores for the rows of F_a, which can be calculated in the same way as discussed in Chapter 2, that is,

$$D_N^{-1} F_a \mathbf{z}/\eta_a = F_a \mathbf{z}/2\eta_a = \mathbf{v}_a, \text{ say.} \tag{4.24}$$

The corresponding score of f_{ij} subjects falling in the ith row and the jth column of F can be defined (Nishisato 1976) as

$$v_{ij} = (y_i + x_j)/2\eta_a. \tag{4.25}$$

Noting, however, that η_a^2 can be expressed as

$$\eta_a^2 = (\eta + 1)/2, \tag{4.26}$$

we obtain

$$v_{ij} = (y_i + x_j)/\sqrt{2(\eta+1)}\ . \tag{4.27}$$

v_{ij} is now expressed in terms of quantities obtained from F. Vector $\mathbf{v}$, with elements v_{ij} arranged in the $f_t \times 1$ form, is then identical with $\mathbf{v}_a$ obtained by (4.24). Score vectors $\mathbf{v}$ and $\mathbf{v}_a$ satisfy the conditions

$$\sum \sum f_{ij} v_{ij}^2 = f_t, \tag{4.28}$$

$$\mathbf{v}_a' D_N \mathbf{v}_a = 2 f_t, \tag{4.29}$$

where f_t is the total number of observations in F and f_{ij} is the element in the ith row and jth column of F. It is interesting to note that v_{ij} in (4.27) consists of only additive components, y_i and x_j, that is to say, no multiplicative component such as $y_i x_j$ is involved. It is in this context that the additivity aspect of dual scaling was discussed in Chapter 3.

A perfect relationship also exists for orthogonal vectors calculated from $\mathbf{x}$, $\mathbf{y}$, $\mathbf{z}$, $\mathbf{v}$, and $\mathbf{v}_a$. The orthogonal vectors, which correspond to $\mathbf{x}$ and $\mathbf{y}$, are indicated by $\mathbf{w}$ and $\mathbf{u}$, respectively, and are defined by

$$\mathbf{w} = D^{1/2}\mathbf{x}, \qquad \mathbf{u} = D_n^{1/2}\mathbf{y}. \tag{4.30}$$

The orthogonal vector corresponding to $\mathbf{z}$ is indicated by $\mathbf{t}$ and is given by

$$\mathbf{t} = \begin{bmatrix} D_n^{1/2} & O \\ O & D^{1/2} \end{bmatrix} \mathbf{z} = \begin{bmatrix} D_n^{1/2}\mathbf{y} \\ D^{1/2}\mathbf{x} \end{bmatrix} = \begin{bmatrix} \mathbf{u} \\ \mathbf{w} \end{bmatrix}. \tag{4.31}$$

The orthogonal vector corresponding to $\mathbf{v}_a$ is indicated by $\mathbf{s}_a$ and is defined by

$$\mathbf{s}_a = D_N^{1/2}\mathbf{v}_a = \sqrt{2}\,\mathbf{v}_a \tag{4.32}$$

$$= \sqrt{2}\,\mathbf{v} = \mathbf{s}, \text{ say,} \tag{4.33}$$

where $\mathbf{s}$ is the orthogonal vector corresponding to $\mathbf{v}$.

The last problem relates to the discrepancy arising from analyses of F and F_a. As mentioned earlier, the rank of F_a is in general greater than that of F and the former cannot be smaller than the latter. All the foregoing discussion on the equality of results from the two analyses then implies that the entire results obtained from dual scaling of contingency table F constitute a subset of the results obtained from F_a. What are the remaining solutions obtained from F_a? Where do they come from? One interpretation may be given in the context of the analysis of variance. F_a can be regarded as a three-way classification table, where the variables are subjects, items, and item options. Scaling results then may reflect variations due to the three variables. But F may also be regarded as a two-way classification table, constructed from F_a by ignoring one variable, subjects. Dual scaling of F then would not reflect any variations due to the subjects. Therefore the extra solutions from F_a must arise from such variations. This interpretation seems to be reasonable and is perhaps correct. It is interesting to know, however, that unlike in the usual analysis of variance, the two-way table F contains complete information on F_a. In other words, it is true not only that F can be constructed from F_a, but also that F_a can be constructed from F.

The χ^2 test of η^2, obtained from contingency table F, may be regarded as a test for between-row and between-column discrimination, but more appropriately it is a test for association between the two scaled variates. As we have seen in Chapter 3, dual scaling maximizes the canonical correlation between the two sets of weighted categories. (2.87) can be used as a test statistic.

We have introduced many symbols and relations in this section. Table 4.2 summarizes some of them. (a) and (c) are for contingency table F, and (b) and (d) are for response-pattern representation F_a of F. Note the relations

$$\mathbf{y}'F\mathbf{x} = \mathbf{u}'D_n^{-1/2}FD^{-1/2}\mathbf{w}, \tag{4.34}$$

$$\mathbf{v}_a'F_a\mathbf{z} = \mathbf{s}_a'D_N^{-1/2}F_a\begin{bmatrix} D_n^{-1/2} & O \\ O & D^{-1/2} \end{bmatrix}\mathbf{t}. \tag{4.35}$$

From these, one can find the relations between the two vectors in each of pairs $(\mathbf{x},\mathbf{w})$, $(\mathbf{y},\mathbf{u})$, $(\mathbf{z},\mathbf{t})$, and $(\mathbf{v}_a,\mathbf{s}_a)$. The relation $\mathbf{z}'=(\mathbf{y}',\mathbf{x}')$ leads to the relation $\mathbf{t}'=(\mathbf{y}'D_n^{1/2},\mathbf{x}'D^{1/2})=(\mathbf{u}',\mathbf{w}')$. We have already demonstrated the equality relations, $\mathbf{v}_a=\mathbf{v}$ and $\mathbf{s}_a=\mathbf{s}$.

In multidimensional analysis, it is convenient to scale all the vectors by

TABLE 4.2

Symbols and their relations

Contingency table F	Response-pattern table F_a
(a) Columns: $\mathbf{x}'$; rows: $\mathbf{y}$; $\underset{n\times m}{F}$, $\mathbf{v}=(v_{ij})$ by (4.25)	(b) Columns: $\mathbf{z}'=(\mathbf{y}',\mathbf{x}')$; rows: $\mathbf{v}_a$; $\underset{N\times(n+m)}{F_a}$
(c) Columns: $\mathbf{w}'$; rows: $\mathbf{u}$; $\underset{n\times m}{D_n^{-1/2}FD^{-1/2}}$, $\mathbf{s}=(s_{ij})$ $=\mathbf{s}_a$	(d) Columns: $\mathbf{t}'=(\mathbf{u}',\mathbf{w}')$; rows: $\mathbf{s}_a$; $\underset{N\times(n+m)}{D_N^{-1/2}F_a\begin{bmatrix} D_n^{-1/2} & 0 \\ 0 & D^{-1/2}\end{bmatrix}}$
(e) $\eta_j^2=(2\eta_{aj}^2-1)^2$	(f) $\eta_{aj}^2=\frac{1}{2}(\eta_j+1)$

respective eigenvalues, η_j^2 and η_{aj}^2. The scaled vectors are indicated by asterisks, and given by

$$\left.\begin{array}{l} \mathbf{x}_j^*=\eta_j\mathbf{x}_j, \qquad \mathbf{w}_j^*=\eta_j\mathbf{w}_j, \\ \mathbf{y}_j^*=\eta_j\mathbf{y}_j, \quad \mathbf{u}_j^*=\eta_j\mathbf{u}_j, \quad \mathbf{v}_j^*=\eta_j\mathbf{v}_j; \end{array}\right\} \tag{4.36}$$

$$\left.\begin{array}{ll} \mathbf{z}_j^*=\eta_{aj}\mathbf{z}_j, & \mathbf{t}_j^*=\eta_{aj}\mathbf{t}_j, \\ \mathbf{v}_{aj}^*=\eta_{aj}\mathbf{v}_{aj}, & \mathbf{s}_{aj}^*=\eta_{aj}\mathbf{s}_{aj}. \end{array}\right\} \tag{4.37}$$

Those in (4.36) are for F and those in (4.37) are for F_a. Since the corresponding eigenvalues are in general different, some of the identity relations between vectors are now changed to collinear relations.

4.1.2 *Response-frequency table*

In Chapter 1 the response-frequency table was defined in distinction from the contingency table. The difference between the two may not appear obvious, and will therefore be illustrated with small examples. Suppose that six subjects were asked two questions: (1) In what country were you born? (2)

Are you a Christian? Suppose that the following data were obtained:

$$F_c = \begin{array}{c} \text{Q.2} \\ \begin{array}{cc} \text{yes} & \text{no} \end{array} \\ \left[\begin{array}{cc} 2 & 1 \\ 1 & 2 \end{array}\right] \end{array} \begin{array}{l} \\ \\ \left.\begin{array}{l} \text{country A} \\ \text{country B} \end{array}\right\} \text{Q.1} \end{array} \tag{4.38}$$

Suppose now that three of the subjects were asked the question 'Do you like apples?' and that the other three were asked 'Do you like pears?' Suppose that the following data were obtained:

$$F_r = \begin{array}{c} \begin{array}{cc} \text{yes} & \text{no} \end{array} \\ \left[\begin{array}{cc} 2 & 1 \\ 1 & 2 \end{array}\right] \end{array} \begin{array}{l} \\ \text{Q.1} \\ \text{Q.2} \end{array} \tag{4.39}$$

(4.38) is an example of a contingency table, and (4.39) is an example of a response-frequency table. Their respective representations in the response-pattern format are:

$$F_c = \left[\begin{array}{cc|cc} \multicolumn{2}{c|}{\text{Country}} & \multicolumn{2}{c}{\text{Christian}} \\ \text{A} & \text{B} & \text{yes} & \text{no} \\ \hline 1 & 0 & 1 & 0 \\ 1 & 0 & 1 & 0 \\ 1 & 0 & 0 & 1 \\ 0 & 1 & 1 & 0 \\ 0 & 1 & 0 & 1 \\ 0 & 1 & 0 & 1 \end{array}\right], \tag{4.40}$$

$$F_r = \left[\begin{array}{cc|cc|cc} \multicolumn{2}{c|}{\text{Group}} & \multicolumn{2}{c|}{\text{Fruit}} & \multicolumn{2}{c}{\text{Category}} \\ 1 & 2 & \text{apple} & \text{pear} & \text{yes} & \text{no} \\ \hline 1 & 0 & 1 & 0 & 1 & 0 \\ 1 & 0 & 1 & 0 & 1 & 0 \\ 1 & 0 & 1 & 0 & 0 & 1 \\ 0 & 1 & 0 & 1 & 1 & 0 \\ 0 & 1 & 0 & 1 & 0 & 1 \\ 0 & 1 & 0 & 1 & 0 & 1 \end{array}\right]. \tag{4.41}$$

F_r is now different from F_c. Note that the patterns for 'group' in F_r coincide with those for 'fruit.' Therefore the canonical correlation between 'group' and 'category' is identical with the canonical correlation between 'fruit' and 'category.' Our interest, however, lies only in the latter correlation. In other words, we would like to attribute the maximized canonical correlation only to the relation between 'fruit' and 'category,' and would like to determine optimal weights for the categories and optimal scores for the fruit. For this purpose, subjects in the response-frequency table are randomly assigned to groups to minimize the difference due to the grouping. In dual scaling, the dimension 'group' is consequently suppressed. However, groups in the contingency table ('country' in (4.40)) represent one of the variables, of which scaled variates should have the maximum dispersion.

Another way to distinguish between F_c and F_r is in terms of the number of judgments that subjects are asked to make. The contingency table can always be interpreted as arising from responses to two questions. Of course, there are occasions in which the investigator classifies the subjects on an a priori basis, such as by nationality. Then the subjects are asked only one question. Even then, however, the same data can be obtained by asking the subjects two questions, one the same question as above and the other about their group membership such as by nationality. In contrast, the repsonse-frequency table is obtained by asking subjects only one question, and hence no joint responses are generated. This restriction in turn allows the investigator to introduce more than one stimulus (objects, 'fruits' in (4.41)) in data collection.

In spite of these differences between F_c and F_r, we can analyse (4.38) and (4.39) by using the same procedure of dual scaling to determine weight vectors and score vectors for the columns and rows. This procedural match is obvious, considering that if we delete 'group' from (4.41) the rest is identical with (4.40), and that these two representations are the forms to which the canonical approach is applied. In other words, dual scaling results are the same for both F_c and F_r if F_c of form (4.38) is equal to F_r of form (4.39).

Because of the nature of the response-frequency table, however, we are not interested in the scores for subjects, which if wanted are the same as those for the rows of F_r. The subjects in the contingency table are imbedded in the table, however, and calculation of their scores requires an additional step, which was discussed earlier in this chapter. Thus, as far as the computational procedure is concerned, there is nothing new to discuss here. One has simply to follow the procedure presented in Chapter 2.

4.2 APPLICATIONS

We have discussed some interesting aspects of the contingency table. To verify numerically some of the relations, the first two examples below are presented. Then three more examples of the contingency table and one example of the response-frequency table are provided to facilitate discussion of other aspects of application.

Example 4.1 This example will show that the analysis of contingency table F provides the same optimum results as the analysis of the same data represented in the two formats of response-pattern tables F_a and F_b.

Suppose that 13 people were asked two questions: 'Do you smoke?' (yes, no) and, 'Do you prefer coffee to tea?' (yes, not always, no), and that the following contingency table was obtained:

$$F = \begin{array}{c} \text{Q.2} \\ \begin{array}{ccc} \text{yes} & \text{not always} & \text{no} \end{array} \\ \left[\begin{array}{ccc} 3 & 2 & 1 \\ 1 & 2 & 4 \end{array}\right] \end{array} \begin{array}{l} \\ \\ \left.\begin{array}{l} \text{yes} \\ \text{no} \end{array}\right\} \text{Q.1} \end{array}$$

Then

$$D = \begin{bmatrix} 4 & 0 & 0 \\ 0 & 4 & 0 \\ 0 & 0 & 5 \end{bmatrix}, \qquad D_n = \begin{bmatrix} 6 & 0 \\ 0 & 7 \end{bmatrix}, \qquad f_t = 13.$$

Let us first obtain the weight vector for the options of question 2, that is **x**. The matrix subjected to the analysis of eigenvalues and eigenvectors, C_c say, is

$$C_c = D^{-1/2}F'D_n^{-1}FD^{-1/2} - \frac{1}{f_t}D^{1/2}\mathbf{1}\mathbf{1}'D^{1/2}$$

$$= \begin{bmatrix} 0.1030 & 0.0137 & -0.1044 \\ 0.0137 & 0.0018 & -0.0139 \\ -0.1044 & -0.0139 & 0.1059 \end{bmatrix}, \tag{4.42}$$

and

$$\operatorname{tr}(C_c) = 0.1030 + 0.0018 + 0.1059 = 0.2107.$$

C_c is now subjected to the iterative method, and we obtain

$$\eta^2 = 0.2107, \qquad \delta = \frac{100\eta^2}{\operatorname{tr}(C_c)} = 100, \qquad \mathbf{x} = \begin{bmatrix} 1.2605 \\ 0.1681 \\ -1.1429 \end{bmatrix}.$$

The corresponding vector for the options of question 1, **y**, is

$$\mathbf{y} = \frac{1}{\eta}D^{-1}F\mathbf{x} = \begin{bmatrix} 1.0801 \\ -0.9258 \end{bmatrix}.$$

Therefore the optimal weight vector for the five options is

$$\begin{bmatrix} \mathbf{y} \\ \mathbf{x} \end{bmatrix} = \begin{bmatrix} 1.0801 \\ -0.9258 \\ 1.2605 \\ 0.1681 \\ -1.1429 \end{bmatrix}. \tag{4.43}$$

Note that $\eta_1{}^2 = \text{tr}(C_c) = 0.2107$, which is reflected in the value of δ, 100. In other words, the optimum solution accounts for the entire variation in the data. The exhaustive analysis can also be inferred from the fact that $r(C_c) = \min(2,3) - 1 = 1$. In consequence, it would be meaningless to calculate the orthogonal vectors corresponding to **y** and **x**, that is, **u** and **w**, respectively.

The significance of discriminability of this solution, or the significance of association between the two scaled variables, can be tested by our χ^2 approximation (2.87),

$$\chi^2 = -\left[13-1-(2+3-1)/2\right]\log_e(1-0.2107) = 2.3661,$$

$$\text{df} = 2+3-1-2 = 2.$$

This value is not significant at the 0.05 level. The failure to secure a significant value may have been caused by the small value of f_t. Anyway, this is a good example to show that δ and χ^2 provide different pieces of information. It is not right to conjecture that if the total variance is completely accounted for by one solution ($\delta_1 = 100$), then the solution must provide a significantly discriminative weight vector. δ is independent of η^2.

The optimal scores for cell (i, j) can be calculated by (4.25). In our case, this multiplier turns out to be 0.5854. If we calculate only distinct scores, that is, scores for the six cells of F, we obtain

$$\mathbf{v} = \begin{bmatrix} v_{11} \\ v_{12} \\ v_{13} \\ v_{21} \\ v_{22} \\ v_{23} \end{bmatrix} = 0.5854 \begin{bmatrix} y_1+x_1 \\ y_1+x_2 \\ y_1+x_3 \\ y_2+x_1 \\ y_2+x_2 \\ y_2+x_3 \end{bmatrix} = \begin{bmatrix} 1.3702 \\ 0.7307 \\ -0.0368 \\ 0.1959 \\ -0.4436 \\ -1.2110 \end{bmatrix}. \qquad (4.44)$$

If one wishes to start with finding the optimal weights for the rows, then

$$C_c = D_n{}^{-1/2}FD^{-1}F'D_n{}^{-1/2} - \frac{1}{f_t}D_n{}^{1/2}\mathbf{11}'D_n{}^{1/2}$$

$$= \begin{bmatrix} 0.1135 & -0.1050 \\ -0.1050 & 0.0973 \end{bmatrix}. \qquad (4.45)$$

Analysis of this 2×2 matrix leads by virtue of duality to the same results as obtained from the 3×3 matrix of (4.42). Computationally, (4.45) is simpler to use than (4.42), and is preferred for eigenvalue decomposition.

Let us now analyse F_a and F_b, where

$$\underset{13\times 5}{F_a} = \begin{array}{c} \begin{array}{cc} \underline{\text{Q.1}} & \underline{\text{Q.2}} \end{array} \\ \left[\begin{array}{ccccc} 1 & 0 & 1 & 0 & 0 \\ 1 & 0 & 1 & 0 & 0 \\ 1 & 0 & 1 & 0 & 0 \\ 1 & 0 & 0 & 1 & 0 \\ 1 & 0 & 0 & 1 & 0 \\ 1 & 0 & 0 & 0 & 1 \\ 0 & 1 & 1 & 0 & 0 \\ 0 & 1 & 0 & 1 & 0 \\ 0 & 1 & 0 & 1 & 0 \\ 0 & 1 & 0 & 0 & 1 \\ 0 & 1 & 0 & 0 & 1 \\ 0 & 1 & 0 & 0 & 1 \\ 0 & 1 & 0 & 0 & 1 \end{array}\right] \end{array} \begin{array}{l} \text{Frequency} \\ \left.\begin{array}{c} \\ \\ \\ \end{array}\right\} 3 \\ \left.\begin{array}{c} \\ \\ \end{array}\right\} 2 \\ \quad 1 \\ \quad 1 \\ \left.\begin{array}{c} \\ \\ \end{array}\right\} 2 \\ \left.\begin{array}{c} \\ \\ \\ \\ \end{array}\right\} 4 \end{array}$$

$$\underset{6\times 5}{F_b} = \begin{bmatrix} 3 & 0 & 3 & 0 & 0 \\ 2 & 0 & 0 & 2 & 0 \\ 1 & 0 & 0 & 0 & 1 \\ 0 & 1 & 1 & 0 & 0 \\ 0 & 2 & 0 & 2 & 0 \\ 0 & 4 & 0 & 0 & 4 \end{bmatrix}.$$

Note that f_t for each of F_a and F_b is now 26, twice f_t for F. The diagonal matrix of column totals, D, for F_a is the same as for F_b. D_n for F_a is $2I$, where I is the 13×13 identity matrix, whereas D_n for F_b is 6×6 with diagonal elements (3,2,1,2,4). Both F_a and F_b, however, yield the same matrix C,

$$C_a = C_b = D^{-1/2}F_k'D_n^{-1}F_kD^{-1/2} - D^{1/2}\mathbf{11}'D^{1/2}/f_t, \qquad k = a,b,$$

$$= \begin{bmatrix} 0.2692 & & & & (\text{symmetric}) \\ -0.2493 & 0.2308 & & & \\ 0.1178 & -0.1090 & 0.3462 & & \\ 0.0157 & -0.0145 & -0.1539 & 0.3462 & \\ -0.1194 & 0.1105 & -0.1720 & 0.1720 & 0.3077 \end{bmatrix}. \tag{4.46}$$

This equality of C_a and C_b illustrates why F_a and F_b provide identical correlation ratios and weight vectors for the item options. By subjecting C_a (C_b) to the iterative method, we obtain

$$\eta_a^2 = 0.7295, \qquad \delta_a = 48.6, \qquad \mathbf{z} = \begin{bmatrix} 1.0801 \\ -0.9258 \\ 1.2605 \\ 0.1681 \\ -1.1429 \end{bmatrix}. \tag{4.47}$$

Note that $(2\eta_a^2 - 1)^2 = (2\times 0.7295 - 1)^2 = 0.2107$, which is equal to η^2 obtained from F. This is the relation indicated by (4.21). We also note that $\mathbf{z}'$ of (4.47) is equal to (y′,x′) of (4.43), as was expected from (4.22).

The optimal score vectors for F_a and F_b,

$$\mathbf{y}_a = \begin{bmatrix} 1.3702 \\ 1.3702 \\ 1.3702 \\ 0.7307 \\ 0.7307 \\ -0.0368 \\ 0.1959 \\ -0.4436 \\ -0.4436 \\ -1.2110 \\ -1.2110 \\ -1.2110 \\ -1.2110 \end{bmatrix} \begin{matrix} \left.\begin{matrix} \\ \\ \\ \end{matrix}\right\} 3 \\ \left.\begin{matrix} \\ \\ \end{matrix}\right\} 2 \\ 1 \\ 1 \\ \left.\begin{matrix} \\ \\ \end{matrix}\right\} 2 \\ \left.\begin{matrix} \\ \\ \\ \\ \end{matrix}\right\} 4 \end{matrix}, \qquad \mathbf{y}_b = \begin{bmatrix} 1.3702 \\ 0.7307 \\ -0.0368 \\ 0.1959 \\ -0.4436 \\ -1.2110 \end{bmatrix},$$

are different in their dimensions, but are equivalent in terms of their information content. $\mathbf{y}_b$ is identical with $\mathbf{v}$ of (4.44), obtained from F.

The rank of $C_a(C_b)$ is $2+3-2=3$, and the trace is $0.2692+0.2308+0.3462+0.3462+0.3077=1.5001$. Since $\delta_a=48.6$, approximately 51% of the total variance is distributed over the remaining two dimensions. Recall that the contingency table was exhaustively accounted for by the optimal scale alone. Following the procedure discussed in Chapter 2, we obtain the remaining two eigenvalues of $F_a(F_b)$,

$$\eta_{a2}^2 = 0.5000\ (\delta_{a2}=82.0), \qquad \eta_{a3}^{\ 2} = 0.2705\ (\delta_{a3}=100.0).$$

Note that $\eta_a^{\ 2}+\eta_{a2}^{\ 2}+\eta_{a3}^{\ 3}=\mathrm{tr}(C_a)$, indicating exhaustive analysis of data, that is, $\delta_{a3}=100$.

Example 4.2 This example is a case where $r(C_c)$ is greater than one, so that the multidimensional decomposition of the contingency table may be compared with the corresponding analysis of the same data represented in the same format as F_a or F_b. Suppose that 50 subjects were administered an aptitude test and a vocabulary test, and that scores of each test were then categorized into four groups. The bivariate distribution of the two sets of scores was therefore summarized and presented as a 4×4 contingency table:

$$\begin{array}{c} \text{Aptitude test} \\ \text{level} \\ \hline \end{array}$$

$$F_{4\times4} = \begin{array}{c} \begin{matrix} \text{A} & \text{B} & \text{C} & \text{D} \end{matrix} \\ \begin{bmatrix} 6 & 4 & 3 & 1 \\ 3 & 5 & 5 & 2 \\ 0 & 1 & 6 & 3 \\ 1 & 2 & 4 & 4 \end{bmatrix} \end{array} \left.\begin{matrix} a \\ b \\ c \\ d \end{matrix}\right\} \begin{array}{c} \text{vocabulary test} \\ \text{level} \end{array}$$

This contingency table can also be represented in the formats of F_a and F_b. Since F_a and F_b provide the same results, let us consider the smaller matrix, F_b:

$$\underset{15\times 8}{F_b} = \begin{array}{c} \overbrace{\begin{array}{cccc} a & b & c & d \end{array}}^{\text{Vocabulary}} \; \overbrace{\begin{array}{cccc} \text{A} & \text{B} & \text{C} & \text{D} \end{array}}^{\text{Aptitude}} \\ \left[\begin{array}{cccccccc}
6 & 0 & 0 & 0 & 6 & 0 & 0 & 0 \\
4 & 0 & 0 & 0 & 0 & 4 & 0 & 0 \\
3 & 0 & 0 & 0 & 0 & 0 & 3 & 0 \\
1 & 0 & 0 & 0 & 0 & 0 & 0 & 1 \\
0 & 3 & 0 & 0 & 3 & 0 & 0 & 0 \\
0 & 5 & 0 & 0 & 0 & 5 & 0 & 0 \\
0 & 5 & 0 & 0 & 0 & 0 & 5 & 0 \\
0 & 2 & 0 & 0 & 0 & 0 & 0 & 2 \\
0 & 0 & 1 & 0 & 0 & 1 & 0 & 0 \\
0 & 0 & 6 & 0 & 0 & 0 & 6 & 0 \\
0 & 0 & 3 & 0 & 0 & 0 & 0 & 3 \\
0 & 0 & 0 & 1 & 1 & 0 & 0 & 0 \\
0 & 0 & 0 & 2 & 0 & 2 & 0 & 0 \\
0 & 0 & 0 & 4 & 0 & 0 & 4 & 0 \\
0 & 0 & 0 & 4 & 0 & 0 & 0 & 4
\end{array}\right] \end{array}$$

Note that the row of F_b which corresponds to the third row and the first column of F is missing because the frequency is zero.

Using the same formulas (4.45) and (4.46), we obtain

$$C_c = \begin{bmatrix} 0.1262 & & & \text{(symmetric)} \\ 0.0453 & 0.0328 & & \\ -0.0709 & -0.0319 & 0.0491 & \\ -0.0808 & -0.0384 & 0.0693 & 0.0765 \end{bmatrix},$$

$$C_b = \begin{bmatrix} 0.36 & & & & & & & \text{(symmetric)} \\ -0.14 & 0.35 & & & & & & \\ -0.12 & -0.12 & 0.40 & & & & & \\ -0.12 & -0.13 & -0.10 & 0.39 & & & & \\ 0.14 & 0.00 & -0.10 & -0.06 & 0.40 & & & \\ 0.02 & 0.05 & -0.06 & -0.03 & -0.11 & 0.38 & & \\ -0.06 & -0.01 & 0.09 & 0.00 & -0.13 & -0.15 & 0.32 & \\ -0.08 & -0.04 & 0.05 & 0.09 & -0.10 & -0.11 & -0.13 & 0.40 \end{bmatrix}.$$

Note that $\mathrm{tr}(C_c)=0.2774$, $\mathrm{tr}(C_b)=3.0000$, $r(C_c)=3$, and $r(C_b)=6$. C_c and C_b are now subjected to the iterative method. Some of the results from F and F_b are summarized in Tables 4.3 and 4.4, respectively. It can be seen from the tables that the three solutions from F correspond to the first three solutions

TABLE 4.3

Analysis of contingency table F

$\eta_1^2=0.2428$ ($\delta_1=87.5\%$)	$\eta_2^2=0.0198$ ($\delta_2=94.6\%$)	$\eta_3^2=0.0147$ ($\delta_3=100\%$)

$$\mathbf{y}_1=\begin{bmatrix}1.28\\0.30\\-1.37\\-0.79\end{bmatrix}\quad \mathbf{y}_2=\begin{bmatrix}-0.42\\1.01\\0.77\\-1.54\end{bmatrix}\quad \mathbf{y}_3=\begin{bmatrix}0.87\\-1.10\\1.24\\-0.73\end{bmatrix}$$

$$\mathbf{x}_1=\begin{bmatrix}1.58\\0.62\\-0.68\\-1.10\end{bmatrix}\quad \mathbf{x}_2=\begin{bmatrix}-0.73\\0.63\\0.88\\-1.61\end{bmatrix}\quad \mathbf{x}_3=\begin{bmatrix}0.99\\-1.54\\0.73\\-0.45\end{bmatrix}$$

$$\mathbf{y}_1^*=\begin{bmatrix}0.63\\0.15\\-0.68\\-0.39\end{bmatrix}\quad \mathbf{y}_2^*=\begin{bmatrix}-0.06\\0.14\\0.11\\-0.22\end{bmatrix}\quad \mathbf{y}_3^*=\begin{bmatrix}0.11\\-0.13\\0.15\\-0.09\end{bmatrix}$$

$$\mathbf{x}_1^*=\begin{bmatrix}0.78\\0.31\\-0.34\\-0.54\end{bmatrix}\quad \mathbf{x}_2^*=\begin{bmatrix}-0.10\\0.09\\0.12\\-0.23\end{bmatrix}\quad \mathbf{x}_3^*=\begin{bmatrix}0.12\\-0.19\\0.09\\-0.05\end{bmatrix}$$

TABLE 4.4

Analysis of response-pattern table F_b

$\eta_{b1}^2=0.7464$ ($\delta_1=24.9\%$)	$\eta_{b2}^2=0.5704$ ($\delta_2=43.9\%$)	$\eta_{b3}^2=0.5606$ ($\delta_3=62.6\%$)

$$\mathbf{z}_1=\begin{bmatrix}1.28\\0.30\\-1.37\\-0.79\\1.58\\0.62\\-0.68\\-1.10\end{bmatrix}\quad \mathbf{z}_2=\begin{bmatrix}-0.42\\1.01\\0.77\\-1.54\\-0.73\\0.63\\0.88\\-1.61\end{bmatrix}\quad \mathbf{z}_3=\begin{bmatrix}0.87\\-1.10\\1.24\\-0.73\\0.99\\-1.54\\0.73\\-0.45\end{bmatrix}$$

$$\mathbf{z}_1^*=\begin{bmatrix}1.11\\0.26\\-1.18\\-0.68\\1.37\\0.54\\-0.58\\-0.95\end{bmatrix}\quad \mathbf{z}_2^*=\begin{bmatrix}-0.32\\0.76\\0.58\\-1.16\\-0.55\\0.48\\0.66\\-1.22\end{bmatrix}\quad \mathbf{z}_3^*=\begin{bmatrix}0.65\\-0.82\\0.93\\-0.55\\0.74\\-1.15\\0.55\\-0.34\end{bmatrix}$$

from F_b, namely,

$$\begin{bmatrix} \mathbf{y}_j \\ \mathbf{x}_j \end{bmatrix} = \mathbf{z}_j, \qquad j = 1,2,3.$$

The corresponding eigenvalues from F_b, η_{bj}^2, are all greater than 0.5 as expected. The relation between $\mathbf{t}'$ and $(\mathbf{u}', \mathbf{w}')$ can also be verified to be (4.31). Similarly, the relation between $\mathbf{v}$ and $\mathbf{v}_b$ and the relation between $\mathbf{s}$ and $\mathbf{s}_b$ can be numerically verified.

It seems important to note that the distributions of η_j^2 and η_{bj}^2, $j=1,2,3$, are quite different in spite of the perfect correspondence between the two sets of vectors associated with them. The first solution of F accounts for 87.5% of the total variance and is highly dominant. The remaining two solutions account for more or less negligible portions of the total variance. In contrast, the first three solutions of F_b account for 24.9%, 19.0%, and 18.7%, respectively, of the total variance and leave 37.4% unaccounted for. These comparisons suggest that in data analysis one can obtain a much more discriminative result by ignoring some of the variables (e.g., 'subjects' in the above example of F). The interesting point here is the fact that F and F_b provided identical sets of vectors for the first three solutions but that the accountabilities of the two sets of solutions differ. In other words, discarding 'subjects' from the analysis did not alter the weight (score) vectors, but increased the dominance of the optimum solution, η_1^2, over the remaining solutions. The χ^2 statistic for the significance of η_1^2 is

$$\chi^2 = -\left[50-1-\tfrac{1}{2}(4+4-1)\right]\log_e(1-0.2428) = 12.65,$$

$$\text{df} = 4+4-1-2 = 5.$$

This is significant at the 0.05 level ($\chi^2_{0.05,5}=11.07$). The same statistic for η_2^2 is less than one with three degrees of freedom, which is not significant.

The eigenvalues and δ_j of the remaining three solutions from F_b are

$$\eta_{b4}^2=0.4393, \qquad \eta_{b5}^2=0.4296, \qquad \eta_{b6}^2=0.2563$$
$$(77.2\%) \qquad (91.5\%) \qquad (100\%).$$

It is interesting to note that the eigenvalues obtained from analysis of F_b are rather evenly distributed. The way the additional dimension of 'subjects' affects the outcome of dual scaling, as we have seen in the present example, may be a worthwhile topic for future investigation.

Example 4.3 (*Hollingshead* 1949) Hollingshead (1949) found that the members of a small Middle Western community in the United States divided themselves into five social classes. He investigated his prediction that adolescents in the different social classes would enrol in different curricula (college preparatory, general, commercial) at a high school. Table 4.5 is adopted from Hollingshead's study. Social classes I and II were merged because the frequencies were small. (As a general strategy, it is advisable to consider

TABLE 4.5

Different curricula and social classes (Hollingshead 1949)

Curricula	Social classes I&II	III	IV	V	Total
College preparatory	23	40	16	2	81
General	11	75	107	14	207
Commercial	1	31	60	10	102
Total	35	146	183	26	390

merging categories of very small frequencies into their adjacent categories. Otherwise, because of their 'uniqueness' [i.e., very small frequencies] in the data matrix, such 'unpopular' categories may receive exceedingly large or small weights.) 390 students were classified in terms of their social classes and actual curriculum enrolments. Since the data table is 3×4, we should first obtain three scores for the curricula. If we specify data matrix F as

$$F = \begin{bmatrix} 23 & 11 & 1 \\ 40 & 75 & 31 \\ 16 & 107 & 60 \\ 2 & 14 & 10 \end{bmatrix},$$

then

$$C_c = \begin{bmatrix} 0.1334 & & (\text{symmetric}) \\ -0.0369 & 0.0107 & \\ -0.0662 & 0.0177 & 0.0338 \end{bmatrix},$$

$\mathrm{tr}(C_c) = 0.1779$, and $r(C_c) = 2$. The results for the optimum solution are summarized in Table 4.6. The optimum solution accounts for more than 99%

TABLE 4.6

Optimum solution
$\eta_1{}^2 = 0.1765$ ($\delta_1 = 99.2\%$), $\chi^2 = 75.61$, df $= 4$ ($p < 0.001$)

Curricula	x	Social classes	y
College preparatory	1.9068	I&II	2.6785
General	−0.3290	III	0.4133
Commercial	−0.8465	IV	−0.7216
		V	−0.8476

of the variance. The chi-square statistic is highly significant, indicating clear-cut differences in the pattern of curriculum selection by the four groups of social classes. This result confirms Hollingshead's prediction. There are distinct tendencies for the college preparatory curriculum and the commercial curriculum to be chosen by students from social classes I&II and V, respectively. These can be inferred from the comparison of elements of **x** and **y**. Since most of the information is accounted for by the optimum solution, the analysis ends at this point.

Example 4.4 200 children of five ethnic groups were asked to indicate their preferred subject out of mathematics, history, and music (Table 4.7). The table is 3×5. Again we start with determining the smaller number of weights, that is, three, by specifying the data matrix to be 5×3 instead of 3×5. The corresponding matrix C_c is given by

$$C_c = \begin{bmatrix} 0.1760 & & \text{(symmetric)} \\ -0.0760 & 0.0336 & \\ -0.1185 & 0.0505 & 0.0804 \end{bmatrix}.$$

The results are summarized in Table 4.8. Since $\delta_1 = 99.5\%$, the only optimum solution was extracted. The significant χ^2 indicates that there is a clear-cut pattern of preference for the three subjects by the five groups. Note that groups C and E received the same score, reflecting the characteristic of dual scaling that columns (rows) which are proportional are evaluated as equivalent in terms of their scale values. Comparisons of optimal weights and scores suggest that group B is the most oriented towards mathematics.

Example 4.5 28 students from each of five countries were asked to indicate their preferred dish out of five (Table 4.9). The results of dual scaling are

TABLE 4.7

Preferred subject chosen by children of five ethnic groups

Subjects	Ethnic groups A	B	C	D	E	Total
Mathematics	30	30	5	2	10	77
History	10	5	10	8	20	53
Music	10	5	15	10	30	70
Total	50	40	30	20	60	200

TABLE 4.8

Optimum solution
$\eta_1^2 = 0.2886$ ($\delta_1 = 99.5\%$), $\chi^2 = 66.57$, df = 5 ($p < 0.001$)

Subjects	x	Groups	y
Mathematics	1.2586	A	0.8311
History	−0.6543	B	1.3980
Music	−0.8891	C	−0.8429
		D	−1.0803
		E	−0.8429

given in Table 4.10. The first two solutions were significant, but the third was not. Thus the third and fourth solutions are not listed in Table 4.10. The first solution accounts for 75% of the variance and the second for 19%. Inspection of Table 4.9 and 4.10 reveals that the optimal scale reflects the patterns of dishes 1 and 4 and countries A and E. The corresponding frequency patterns are

$$\begin{cases} \text{dish 1: } 16, 2, 9, 2, 0 & (x_1 = 1.55), \\ \text{dish 4: } 1, 6, 0, 5, 15 & (x_4 = -1.30); \end{cases}$$

$$\begin{cases} \text{country A: } 16, 6, 2, 1, 3 & (y_A = 1.47), \\ \text{country E: } 0, 3, 8, 15, 2 & (y_E = -1.20). \end{cases}$$

These two dishes and two countries are the main determiners of the first solution, which can clearly be seen if we construct the 2×2 contingency table

TABLE 4.9

Preferred dishes chosen by people from five countries

	Dishes					
Countries	1	2	3	4	5	Total
A	16	6	2	1	3	28
B	2	4	13	6	3	28
C	9	7	7	0	5	28
D	2	4	17	5	0	28
E	0	3	8	15	2	28
Total	29	24	47	27	13	140

TABLE 4.10

Two solutions

$\eta_1^2=0.4189$ ($\delta_1=75\%$)		$\eta_2^2=0.1082$ ($\delta_2=94\%$)	
$\chi^2=73.01$		$\chi^2=15.40$	
df = 7		df = 5	
($p<0.001$)		($p<0.01$)	
$\mathbf{x}_1$	$\mathbf{w}_1$	$\mathbf{x}_2$	$\mathbf{w}_2$
1.55	8.32	−0.44	−2.38
0.43	2.32	0.16	0.77
−0.59	−4.04	1.19	8.18
−1.30	−6.75	−1.54	−8.04
0.58	2.86	−0.41	−1.46
$\mathbf{y}_1$	$\mathbf{u}_1$	$\mathbf{y}_2$	$\mathbf{u}_2$
1.47	7.77	−0.71	−3.73
−0.49	−2.60	0.52	2.73
0.87	4.59	0.38	1.98
−0.64	−3.41	1.33	7.06
−1.20	−6.35	−1.52	−8.05

consisting of them:

	Dish		
Country	1	4	
A	16	1	17
E	0	15	15
	16	16	

The first solution, however, fails to capture subdominant patterns, which are reflected in the second solution. From Table 4.10, we find that the second dimension is mainly contributed by dishes 3 and 4 and countries D and E. The 2×2 contingency table which consists of them is

	Dish		
Country	3	4	
D	17	5	22
E	8	15	23
	25	20	

Again we see strong association among these variables.

The patterns in the two dimensions can also be observed for orthogonal vectors $\mathbf{w}_j$ and $\mathbf{u}_j$. In the present example, $\mathbf{y}_j = \mathbf{u}_j/\sqrt{28}$, and hence $\mathbf{y}_1$ and $\mathbf{y}_2$ are orthogonal to one another. Thus there is nonorthogonality only with respect to $\mathbf{x}_1$ and $\mathbf{x}_2$, created by inequalities among the column totals of F. It is therefore expected that $\mathbf{w}_j$ and $\mathbf{u}_j$ would reveal patterns similar to those for $\mathbf{x}_j$ and $\mathbf{y}_j$.

Example 4.6 Let us now look at an example of a response-frequency table. Suppose that 1,000 high school students were randomly assigned to five groups of equal size. Five well-trained teachers were then assigned to the five groups and taught their respective groups using five distinct teaching methods. After two months, the students were asked to evaluate the different teaching methods in terms of categories excellent, good, mediocre, and boring. The data are listed in Table 4.11. Data matrix F_r is 5×4, with elements arranged as in Table 4.11. The other quantities we need are $f_t = 1{,}000$, $D_n = 200I$, $\mathbf{f}' = (168, 317, 280, 235)$, and $D = \text{diag}[168, 317, 280, 235]$. Using the standard formula, we obtain matrix C_1 as

$$C_1 = \begin{bmatrix} 0.1561 & & & \text{(symmetric)} \\ 0.1204 & 0.1570 & & \\ -0.0925 & -0.0772 & 0.1028 & \\ -0.1709 & -0.1999 & 0.0556 & 0.3160 \end{bmatrix}.$$

The maximum non-trivial eigenvalue is 0.5906. Since $\text{tr}(C_1) = 0.7319$,

$$\delta_1 = 0.5906 \times 100/0.7319 = 80.7(\%).$$

The discriminability of the weighted responses can be tested by (2.87). In our example,

$$\chi^2 = -\left[1000 - 1 - \tfrac{1}{2}(5+4-1)\right]\log_e(1-0.5906) = 888.60,$$

$$\text{df} = 5 + 4 - 1 - 2 = 6.$$

TABLE 4.11

Evaluation of teaching methods

Methods	Categories Excellent	Good	Mediocre	Boring	Total
1	45	126	24	5	200
2	87	93	19	1	200
3	0	0	52	148	200
4	36	68	74	22	200
5	0	30	111	59	200
Total	168	317	280	235	1000

This is highly significant. In other words, the students perceive that the five teaching methods are significantly different. The optimal weights for the four categories and the optimal scores for the teaching methods are

Excellent	1.1277	Method 1	0.9283
Good	0.8748	Method 2	1.1011
Mediocre	−0.4627	Method 3	−1.5382
Boring	−1.4349	Method 4	0.2230
		Method 5	−0.7142

In terms of these scores, method 2 is the best and method 3 is the most boring.

This example presents a few problems which require discussion. The first is that the response categories are ordered in that excellent > good > mediocre > boring. The analysis has revealed that the order of the optimal weights conforms to this a priori order. (The order of category weights is judged only by $\mathbf{x}_1$, not by $\mathbf{w}_1$.) This finding is, however, a matter of coincidence, for dual scaling does not have any built-in procedure to guarantee that the optimal weights of ordered categories are ordered in the same way. The signs of weights may be completely reversed, or the weight for 'good' may turn out to be larger than the weight for 'excellent.' This last case may be difficult to understand, considering that all the subjects are mature enough to know the order relations among the response categories. However, it is not the subjects' understanding of the response categories that is responsible for disordered weights. The real cause is multidimensionality of the attribute of stimulus objects (e.g., goodness of individual teaching methods). The multidimensionality may appear in the form of several dominant criteria for good teaching methods such as the motivational aspect versus the actual achievement, the traditional classroom teaching versus individualistic teaching, or the laissez-faire versus authoritarian teaching. The multidimensionality may also appear in the form of several patterns of judgment. Under these circumstances, the principle of internal consistency no longer generates results which have face validity. To obtain interpretable results, therefore, one must consider constrained optimization. More specifically, one should determine weight vector $\mathbf{x}$ under the condition that the elements of $\mathbf{x}$ are ordered in a specified manner. This problem of constrained optimization will be discussed in Chapter 8.

The second problem is related to the first. Because of the way dual scaling of F_r is formulated, the imposition of the constraint of complete order on the categories has the effect of restricting the number of solutions to one, no matter whether the attribute of the objects is multidimensional or not. In our example, the optimum solution, which also satisfies the order relations among the categories, has left approximately 19% of the variance unaccounted for. Since $r(C_1)=3$, the two additional solutions will account for this portion of the variance. Using the procedure discussed in Chapter 2, we obtain $\eta_2{}^2=0.1102$ ($\delta_2=95.8\%$) and $\eta_3{}^2=0.0311$ ($\delta=100\%$). The chi-square statistics for $\eta_2{}^2$ and $\eta_3{}^2$ are respectively 116.17 (df=4) and 31.44 (df=2), which are both significant. As stated above, however, the weights of the response categories from these two solutions are not correctly ordered. Thus these solutions may be of little practical importance. The significance of the chi squares in the current example is likely to be due to the large value of f_t rather than the values of η^2.

As a general problem, one should bear in mind the interpretability of the results. There are certain objective criteria which serve as an indicator of when to stop extracting further solutions, such as the chi-square statistic, the cumulative percentage of the total variance accounted for, or the average eigenvalue (i.e., $\text{tr}(C_1)/r(C_1)$). In practice, however, the investigator has to decide how many solutions to report. He may be interested only in the optimal vectors, $\mathbf{x}_1$ and $\mathbf{y}_1$, irrespective of what the significance levels of the remaining solutions are. He may wish to use only the optimum solution, but may report the values of the remaining eigenvalues and their chi-square values as a supplementry piece of information about the data structure. Or he may wish to use all the significant solutions and may strive to find clues to interpret them. Anyway, what seems to be important is for the investigator to be able to interpret, clarify, and explain all the results to be reported. The effective use of analysis depends heavily on the investigator's judgment. In this regard, the value of face validity cannot be unduly discounted.

5

Response-pattern table: multiple-choice (n-item) data

5.1

SPECIAL PROBLEMS

Response-pattern tables are often obtained from sociological surveys and psychological tests. One of the most popular procedures adopted in such surveys and tests is the so-called multiple-choice method. The investigator presents a set of response options (alternatives) for each question and asks subjects to choose only one option per question. In consequence, if the questionnaire consists of n multiple-choice items, each subject will provide n responses (choices). This is probably the most popular type of multiple-choice data, and will therefore be discussed first.

Table 5.1 is a small example of multiple-choice data, where eight subjects answered three multiple-choice questions. The first two items have two options each, and the third has three options. Since each subject was instructed to choose only one option per item, the sum of coded responses within each item is constant across subjects and items. Note that if we take any two items, we can construct a contingency table. In this regard, the response-pattern table is a compact form of the multidimensional (n-dimensional) contingency table, and can be called 'n-item data,' as opposed to 'two-item data,' a name given to the contingency table in Chapter 1. Let us define the notation:

n = the number of multiple-choice items;
m_k = the number of options for item k, $k=1,2,\ldots,n$;
$m=\Sigma m_k$ = the total number of options of n items;
N = the number of subjects;
F = the $N\times m$ data matrix of 1's and 0's;
$\mathbf{f}$ = the $m\times 1$ vector of column totals of F;
$\mathbf{g}$ = the $N\times 1$ vector of row totals of $F=n\mathbf{1}$;
D = the $m\times m$ diagonal matrix of column totals of F;
D_N = the $N\times N$ diagonal matrix of row totals of $F=nI$;
f_t = the sum of elements of $F=nN$;
$\mathbf{x}$ = the $m\times 1$ vector of weights;
$\mathbf{y}$ = the $N\times 1$ vector of weights (scores).

TABLE 5.1

	Item 1		Item 2		Item 3			
Subjects	1*	2	1	2	1	2	3	Total
1	0	1	0	1	0	1	0	3
2	0	1	1	0	0	0	1	3
3	1	0	0	1	1	0	0	3
4	0	1	0	1	0	0	1	3
5	1	0	1	0	1	0	0	3
6	0	1	1	0	0	1	0	3
7	0	1	0	1	0	0	1	3
8	1	0	1	0	1	0	0	3
Total	3	5	4	4	3	2	3	24
Total	8		8		8			

*Options.

With these definitions and the substitution of D_N for D_n, we can carry out dual scaling in almost the same way as described in Chapter 2. There are only a few points one should note:

1. The condition that

$$\mathbf{f}'\mathbf{x} = 0 \tag{5.1}$$

now implies that the sum of weighted responses within *each item* is equal to zero (see interesting discussions on this point in Guttman [1941] and McDonald, Torii, and Nishisato [1979]). This implication is due to the restriction that one is allowed to choose only one option per item.

2. Assuming that N is much larger than m,

$$r(F) = m - n + 1. \tag{5.2}$$

3. Because of the restriction imposed on the data structure, our χ^2 statistic is modified as

$$\chi^2 = -\left[f_t - 1 - \tfrac{1}{2}(N + m - n - 1)\right]\log_e\left(1 - \eta_j^2\right), \tag{5.3}$$

$$\text{df} = N + m - n - 1 - 2j, \tag{5.4}$$

$$\eta_1^2 > \eta_2^2 > \eta_3^2 > \cdots$$

4. Noting that $D_N = nI$, one can simplify some of the formulas. For instance, under the condition that $\mathbf{f}'\mathbf{x} = \mathbf{g}'\mathbf{y} = 0$,

$$\eta^2 = \frac{\mathbf{x}'F'D_N^{-1}F\mathbf{x}}{\mathbf{x}'D\mathbf{x}} = \frac{\mathbf{x}'F'F\mathbf{x}/n}{\mathbf{x}'D\mathbf{x}}, \tag{5.5}$$

$$\mathbf{y} = \frac{1}{\eta} D_N{}^{-1}F\mathbf{x} = \frac{1}{n\eta} F\mathbf{x}, \tag{5.6}$$

$$\eta^2 = \frac{\mathbf{y}'FD^{-1}F'\mathbf{y}}{\mathbf{y}'D_N\mathbf{y}} = \frac{\mathbf{y}'FD^{-1}F'\mathbf{y}}{n\mathbf{y}'\mathbf{y}}, \tag{5.7}$$

$$C_1 = D_N{}^{-1/2}FD^{-1}F'D_N{}^{-1/2} - \frac{1}{f_t} D_N{}^{1/2}\mathbf{1}\mathbf{1}'D_N{}^{1/2}$$

$$= \frac{1}{n} FD^{-1}F' - \frac{n}{f_t}\mathbf{1}\mathbf{1}'. \tag{5.8}$$

In spite of these differences, duality of analysis is preserved, and one can start scaling with either rows or columns.

Considering that most data that fit into this restricted format are obtained from multiple-choice tests, the concept of reliability often becomes a central concern for the investigators. In 1958 Lord demonstrated that maximization of η^2 leads to maximization of the generalized Kuder-Richardson reliability, which is also referred to as Cronbach's alpha, indicated by α (Cronbach 1951). Given the $N \times n$ (subjects-by-items) data matrix, Lord considered the partitioning of the total sum of squares, ss_t, into the sum of squares between items, ss_n, the sum of squares between subjects, ss_b, and the residual sum of squares, ss_e, that is,

$$ss_t = ss_n + ss_b + ss_e. \tag{5.9}$$

In terms of these quantities, α and η^2 can be expressed as

$$\alpha = \frac{n}{n-1}\left(1 - \frac{ss_t - ss_n}{n ss_b}\right), \tag{5.10}$$

$$\eta^2 = ss_b/ss_t. \tag{5.11}$$

As stated in Chapter 2, dual scaling typically employs the condition that $\mathbf{f}'\mathbf{x} = 0$. As stated in note 1 above, this condition implies that the sum of weighted responses of every item is zero. It then follows that

$$ss_n = 0. \tag{5.12}$$

Under this condition, we obtain

$$\eta^2 = \frac{1}{1 + (n-1)(1-\alpha)}, \tag{5.13}$$

or

$$\alpha = 1 - \frac{1-\eta^2}{(n-1)\eta^2}. \tag{5.14}$$

This formula indicates that α *is maximized by maximizing* η^2, as Lord (1958) concluded.

As inferred from (5.14), α can be negative. In practice, however, it is difficult to interpret negative reliability. Therefore, if we restrict the range of α to

$$1 \geq \alpha \geq 0, \tag{5.15}$$

we find that η^2 must be bounded:

$$1 \geq \eta^2 \geq 1/n. \tag{5.16}$$

When $\eta^2 = 1/n$, $\alpha = 0$. Thus, for most practical purposes, the value $\eta^2 = 1/n$ may be used as one of the critical values for terminating further extraction of solutions.

Equation (5.12) indicates that the mean contributions of items to the total score are all equal. This condition serves to make effective use of item difficulty in the calculation of the total score. For example, consider that 100 subjects answered two dichotomous questions in a mathematics test and that the following results were obtained: 90 subjects passed the first item, but only 10 subjects passed the second item. Thus item 1 was very easy and item 2 was very difficult for this group of subjects. Assuming that the two items were equally discriminative, we can consider typical weights that the four options could have got, for example,

Item 1		*Item 2*	
Correct	0.2	Correct	1.8
Incorrect	−1.8	Incorrect	−0.2

These weights satisfy (5.12). An immediate implication of (5.12) is now obvious: if a student fails an easy item, the penalty is great (e.g., −1.8); if he passes a difficult item, the reward is great (e.g., 1.8). Herein lies an advantage of dual scaling over the traditional (1 for correct and 0 for incorrect) scoring method which ignores the effect of item difficulty on discrimination among subjects.

Let us indicate by x_{jk} the weight for option k of item j, and by s_j the standard deviation of option weights of item j, that is,

$$s_j = \sqrt{\frac{\sum_{k=1}^{m_j} (x_{jk} - \bar{x}_j)^2}{m_j - 1}},$$

where m_j is the total number of options of item j and $\bar{x}_j$ is the mean of the m_j option weights. It is known (Mosier 1946) that s_j is proportional to the correlation between that item and the total scores, r_{jt} say. Thus the relative contribution of item j to the generalized Kuder-Richardson reliability of a test can be inferred from the magnitude of s_j. Since the computation of s_j is simpler than that of r_{jt}, s_j may be preferred to r_{jt} for item selection in test construction. The recommended procedure then is to retain those items with relatively large values of s_j and discard the others. This would guarantee selecting a set of most internally consistent items from the given pool of items.

This procedure of item selection may appear paradoxical to the known fact that the optimal level of item difficulty of a dichotomous item for the Kuder-Richardson reliability is 0.5. In other words, collecting items with large values of s_j may appear to be in conflict with collecting items which have item difficulty levels near 0.5. This concern does not seem to be a problem, however, for s_j is a function not only of item difficulty but, more importantly, of the internal consistency of that item. For instance, if an item is not at all consistent internally, all the options of the item will have almost identical weights, indicating that it does not matter which option one may choose. Then s_j is nearly zero, whatever the item difficulty level may be. This is an example in which the effect of internal consistency overrides that of item difficulty. The interactive effect of s_j and item difficulty on reliability, however, seems to require further investigation.

Let us now consider a more general case where one does not impose the restriction that only one option per item should be chosen. Subjects are then allowed to omit some items or may choose more than one option per item. One may have a definite idea about association between specific attitudes of subjects and their tendencies to omit responses or tendencies to provide multiple responses. If one is interested in that kind of association, the data matrix should be reorganized. For example, one may create an additional option, called omission, for each item, or one may divide the subjects into two groups, one for those without multiple responses and the other for those with multiple responses. In practice, however, omissions or multiple responses are quite commonly observed whether one is interested in them or not. Without knowing what is responsible for omissions or multiple responses, the best way to deal with them seems to be to subject the given response-pattern table to analysis, using the general formulation. In other words, one should use D_N^{-1} for $1/n$, where the diagonal elements of D_N are the numbers of actual (observed) responses. This procedure is at least consistent with the idea of maximizing η^2.

Another problem which one often encounters in practice is the analysis of multiple-choice data in which some of the options have weights assigned already on an a priori basis. Since this problem requires new procedural considerations and presents a different set of formulas, we shall discuss it in a separate section. The approach is referred to as partially optimal scaling (Nishisato and Inukai 1972).

5.2 PARTIALLY OPTIMAL SCALING

Nishisato and Torii (1971) described decompositions of weights into several components. Let x_{jk} stand for the weight for option k of item j. x_{jk} can be decomposed as follows:

(a) *Crossed case* The crossed case is characterized by the fact that all the multiple-choice items share a *common set of options*. In other words, subjects are asked to answer all the items in terms of the same set of response categories, for example, 'never, sometimes, often, always.' Then x_{jk} can be expressed as

$$x_{jk} = \mu + \alpha_j + \beta_k + \gamma_{jk}, \tag{5.17}$$

where μ is the grand mean, α_j the contribution of item j, β_k the contribution of option k, and γ_{jk} the contribution of the 'item j'×'option k' interaction.

(b) *Nested case* The nested case is one where each item has a different set of options. In this case x_{jk} can be expressed as

$$x_{jk} = \mu + \alpha_j + \beta_{k(j)}, \tag{5.18}$$

where $\beta_{k(j)}$ is the contribution of option k within item j.

So far we have considered a procedure which determines x_{jk} without singling out any particular component, μ, α_j, β_k, γ_{jk}, or $\beta_{k(j)}$. Suppose now that the investigator employs some a priori option weights, that is, β_k in (5.17) and $\beta_{k(j)}$ in (5.18) are now fixed. We can no longer control γ_{jk} in (5.17) in any direct manner. However, γ_{jk} is a component which contributes to the unreliability of the scoring scheme (Nishisato and Torii 1971) and hence can be ignored in our consideration of optimization. μ is not of any particular interest to our scaling problem since it applies to all the items and options equally. What is left for the optimization problem is then the variation in α_j. Nishisato and Inukai (1972) considered weighting subjects' responses in terms of only α_j, given option weights, and called this procedure *partially optimal scaling* for obvious reasons. They also considered the case of mixed data, that is, some items with and others without preassigned option weights. To describe the procedure, let us define the following notation:

$A^* = (a^*_{ij})$ = the $N \times n$ (subject-by-item) data matrix, expressed in terms of scores, preassigned to the options;

$a_{ij} = a^*_{ij} - a^*_{.j}/N$ = the deviation score from the column mean of subject i and item j;

$A=(a_{ij})=$ the $N\times n$ matrix of deviation scores;

$D_a=\text{diag}\left(\sum_{i=1}^{N} a_{ij}^2\right)=$ the $n\times n$ diagonal matrix of the column sums of squares of deviation scores;

$D_N=$ the $N\times N$ diagonal matrix of the *numbers* of responses from N subjects $=nI$, if there is no omitted response or multiple response;

$\mathbf{x}=$ the $n\times 1$ vector of weights for n items;

$\mathbf{a}=$ the $n\times 1$ vector of the column totals of $A=\mathbf{0}$;

$f_t=$ the total number of responses in the data matrix $=nN$, if there is no omitted response or multiple response.

By the definition of **a**, the origin of the derived measurement is zero, i.e.,

$$\mathbf{a}'\mathbf{x} = 0. \tag{5.19}$$

Thus the correlation ratio of weighted responses is

$$\eta^2 = \frac{\mathbf{x}'A'D_N^{-1}A\mathbf{x}}{\mathbf{x}'D_a\mathbf{x}} \tag{5.20}$$

$$= \frac{\mathbf{w}'D_a^{-1/2}A'D_N^{-1}AD_a^{-1/2}\mathbf{w}}{\mathbf{w}'\mathbf{w}}, \tag{5.21}$$

where

$$\mathbf{w} = D_a^{1/2}\mathbf{x}, \quad \text{or} \quad \mathbf{x} = D_a^{-1/2}\mathbf{w}. \tag{5.22}$$

Maximization of (5.21), subject to the constraint that $\mathbf{w}'\mathbf{w}=\mathbf{x}'D_a\mathbf{x}=f_t$, results in the equations

$$\left(D_a^{-1/2}A'D_N^{-1}AD_a^{-1/2}-\eta^2 I\right)\mathbf{w} = \mathbf{0}, \tag{5.23}$$

$$\mathbf{w}'\mathbf{w} = f_t. \tag{5.24}$$

Equation (5.23) has only one trivial solution, $\mathbf{w}=\mathbf{0}$. The second trivial solution does not exist since A is a matrix of deviation scores. Using the iterative method, we can successively extract the solutions, $(\eta_1^2,\mathbf{w}_1)$, $(\eta_2^2,\mathbf{w}_2)$, and so on. **w** can be transformed to **x** by (5.22). The score vectors which correspond to optimal and conditionally optimal weight vectors $\mathbf{x}_j$ and those which correspond to orthogonal weight vectors $\mathbf{w}_j$ can be calculated, respectively, by

$$\mathbf{y}_j = D_N^{-1}A\mathbf{x}_j/\eta_j, \tag{5.25}$$

$$\mathbf{u}_j = D_N^{-1/2}A\mathbf{x}_j/\eta_j = D_N^{-1/2}AD_a^{-1/2}\mathbf{w}_j/\eta_j. \tag{5.26}$$

The discriminability of the jth solution can be tested by

$$\chi^2 = -\left[f_t-1-\tfrac{1}{2}(N+n)\right]\log_e\left(1-\eta_j^2\right), \tag{5.27}$$

$$\text{df} = N+n-2j, \tag{5.28}$$

$$\eta_1^2 > \eta_2^2 > \cdots > \eta_j^2.$$

Let us now consider the case of mixed data. Mixed data refer to a collection containing a mixture of qualitative and quantitative data, or to the use of a priori weights for only some of the multiple-choice items. A question which has then to be answered is how to combine dual scaling and partially optimal scaling. Nishisato and Inukai (1972; see also Inukai 1972) have presented a method to deal with this problem. They considered the case in which all the items are of multiple-choice type.

Let h be the number of items, of which options have preassigned values (i.e., the format of partially optimal scaling), t be the total number of options of these h items, and m be the total number of options of n items. Then the $m-t$ options of the remaining $n-h$ items have unknown weights (i.e., the format of dual scaling). The data matrix then is a mixture of A^* and F. Define the notation:

$G^* = (g_{ij}^*) =$ the $N \times (h+m-t)$ response matrix in mixed format, where the first h columns represent the h items with preassigned option weights as elements and the remaining $m-t$ columns represent the responses to the $m-t$ options of the $n-h$ items, coded as 1's and 0's;

$$g_{ij} = \begin{cases} g_{ij}^* - \sum_{i=1}^{N} g_{ij}^*/N, & \text{if } g_{ij}^* \text{ are preassigned values,} \\ g_{ij}^*, & \text{otherwise, thus leaving } g_{ij} \text{ to be binary numbers;} \end{cases}$$

$G = (g_{ij}) =$ the $N \times (h+m-t)$ matrix of deviation scores and binary numbers;

$\mathbf{g} =$ the $(h+m-t) \times 1$ vector of the column totals of G;

$D_g = \operatorname{diag}\left(\sum_{i=1}^{N} g_{ij}^2\right) =$ the diagonal matrix of order $h+m-t$;

$D_N =$ the $N \times N$ diagonal matrix of the *numbers* of responses $= nI$, if there is no omitted response or multiple response;

$\mathbf{x} =$ the vector of weights for the h items and the $m-t$ options;

$$c_k = \begin{cases} 0, & \text{if the } k\text{th column of } G^* \text{ has a preassigned value,} \\ 1, & \text{otherwise;} \end{cases}$$

$\mathbf{c} =$ the $(h+m-t) \times 1$ vector of c_k;

$f_t =$ the total number of scores in $G^* = nN$, if there is no omitted response or multiple response.

For the mixed response matrix, G, we want to determine $\mathbf{x}$ so as to maximize the ratio of the between-subject sum of squares to the total sum of squares. Using the condition

$$\mathbf{g}'\mathbf{x} = 0, \tag{5.29}$$

we can express the correlation ratio as

$$\eta^2 = \frac{\mathbf{x}'G'D_N^{-1}G\mathbf{x}}{\mathbf{x}'D_g\mathbf{x}}. \tag{5.30}$$

Following the same procedure as before, we obtain

$$\left(D_g^{-1/2}G'D_N^{-1}GD_g^{-1/2}-\eta^2 I\right)\mathbf{w}=\mathbf{0}, \tag{5.31}$$

$$\mathbf{w}'\mathbf{w}=f_t, \tag{5.32}$$

where

$$\mathbf{w}=D_g^{1/2}\mathbf{x}, \quad \text{or} \quad \mathbf{x}=D_g^{-1/2}\mathbf{w}. \tag{5.33}$$

Equation (5.31) has two trivial solutions, namely, (i) $\mathbf{w}=\mathbf{0}$ and (ii) $\mathbf{w}=D_g^{1/2}\mathbf{c}$, that is, $\mathbf{x}=\mathbf{c}$. Note that the second trivial solution does not satisfy (5.29), for

$$\mathbf{g}'\mathbf{x}=\mathbf{g}'\mathbf{c}=N(n-h)>0 \quad \text{since } n>h. \tag{5.34}$$

Furthermore, the second trivial solution satisfies (5.31), whatever matrix G happens to be. The eigenvalue which corresponds to this solution is given by

$$\eta^2=\frac{\mathbf{c}'G'D_N^{-1}G\mathbf{c}}{\mathbf{c}'D_g\mathbf{c}}=\frac{\mathbf{c}'G'G\mathbf{c}/n}{N(n-h)}$$

$$=\frac{N(n-h)^2/n}{N(n-h)}=1-\frac{h}{n}. \tag{5.35}$$

It is interesting to note that $\eta^2=1$ if $h=0$ (i.e., all the responses are in the response-pattern format) and that $\eta^2=0$ if $h=n$ (i.e., all the responses have preassigned scores as in partially optimal scaling). Let us now indicate by C_1 the matrix obtained after eliminating the second trivial solution. This matrix is given by

$$C_1=D_g^{-1/2}G'D_N^{-1}GD_g^{-1/2}-\left(1-\frac{h}{n}\right)\frac{D_g^{1/2}\mathbf{c}\mathbf{c}'D_g^{1/2}}{\mathbf{c}'D_g\mathbf{c}}$$

$$=D_g^{-1/2}G'D_N^{-1}GD_g^{-1/2}-\frac{1}{f_t}D_g^{-1/2}\mathbf{c}\mathbf{c}'D_g^{1/2} \tag{5.36}$$

since

$$\mathbf{c}'D_g\mathbf{c}=N(n-h) \quad \text{and} \quad f_t=nN.$$

Using the iterative method, we solve the equation

$$\left(C_1-\eta^2 I\right)\mathbf{w}=\mathbf{0}, \tag{5.37}$$

and rescale $\mathbf{w}$ so that (5.32) is satisfied.

The vectors which correspond to $\mathbf{x}_j$ and $\mathbf{w}_j$ (the jth solution) are given respectively by

$$\mathbf{y}_j=D_N^{-1}G\mathbf{x}_j/\eta_j, \tag{5.38}$$

$$\mathbf{u}_j=D_N^{-1/2}GD_g^{-1/2}\mathbf{w}_j/\eta_j=D_N^{-1/2}G\mathbf{x}_j/\eta_j. \tag{5.39}$$

The discriminability of solution η_j^2 can be tested by a χ^2 approximation,

$$\chi^2 = -\left[f_t - 1 - \tfrac{1}{2}(N+m+h-t-1)\right]\log_e\left(1-\eta_j^2\right), \tag{5.40}$$

$$\mathrm{df} = N+m+h-t-1-2j, \tag{5.41}$$

$$\eta_1^2 > \eta_2^2 > \cdots > \eta_j^2.$$

5.3 COMPARISONS OF OPTION WEIGHTING AND ITEM WEIGHTING

5.3.1 *General case*

There are certain similarities and dissimilarities between dual scaling of multiple-choice options and partially optimal scaling. The former may be referred to as *option weighting* and the latter as *item weighting*. A comparison of the two will provide some useful guidelines for their application.

When one considers item weighting, given a priori option weights, the reduction in computation time is enormous in comparison with option weighting. The minimum difference in computation time between the two is obtained when all the items are dichotomous. Option weighting then deals with the eigenequation of a $2n \times 2n$ matrix, derived from $N \times 2n$ data matrix F, while item weighting, given $2n$ a priori option weights, handles the eigenequation of an $n \times n$ matrix, derived from $N \times n$ data matrix A^*. Assuming that the computation time is proportional to the number of elements in the matrix, the ratio of the time needed in item weighting to that needed in option weighting is roughly $1/k^2$, where k is the average number of options per multiple-choice item.

Another problem related to computation is the size problem. Suppose that the questionnaire consists of 50 items with seven response options per item. In item weighting, the investigator uses a priori weights for the seven options of each item, and his task is to determine optimal weights for the 50 items. In option weighting, however, the objective is to determine optimal weights for the 350 (50×7) options. Most available computers can handle an eigenequation of a 50×50 matrix, but not necessarily an eigenequation of a 350×350 matrix.

It is important to note, however, that item weighting is based on a priori or preassigned weights for the options. Thus the validity of the scaling outcome depends heavily on the preassigned weights. In regard to the measure of the internal consistency, option weighting provides in general a better result than item weighting. Suppose that multiple-choice data are first subjected to option weighting, and that the same data are then subjected to item weighting by assigning a priori weights to the options. If we indicate by η_o^2 and η_p^2 the correlation ratios obtained from option weighting and item weighting, respectively, then

$$\eta_o^2 \geq \eta_p^2. \tag{5.42}$$

The inequality relation is understandable since more degrees of freedom are in general available for option weighting than for item weighting. Equality can be obtained under two conditions. The first is when the preassigned option weights are proportional to the optimal option weights, which is, of course, rather unlikely to happen in practice. The second is when all the items are *dichotomous*, in which case, the equality of (5.42) holds whatever a priori option weights are assigned. Recall that item weighting is computationally much faster than option weighting.

When the options are ordered (e.g., never, sometimes, often, always), item weighting often provides a more practical method than option weighting. In the former, one can always assign a set of ordered values to the options, prior to scaling. For example, one might assign 0, 1, 2, and 3 to options never, sometimes, often, and always, respectively, which guarantees that subjects' scores are at least related to the a priori order of the options. Option weighting, in contrast, does not have any built-in procedure to guarantee that the optimal weights satisfy the a priori order of the options. It could happen that the optimal weight for 'always' may lie between those for 'sometimes' and 'often.' To guarantee the a priori order of options, option weighting requires an additional step to impose order constraints on options, and such a procedure makes dual scaling somewhat complex (see Bradley, Katti, and Coons 1962; Kruskal 1965; Nishisato and Arri 1975). Thus, admitting that the procedure of partially optimal scaling may not be satisfactory from a theoretical point of view, it is nevertheless often preferred to that of option weighting with order constraints, a procedure that will be discussed in Chapter 8.

Another problem related to ordered options is the observation that when the complete order constraint is imposed on the options certain techniques (e.g., Nishisato 1973b; Nishisato and Arri 1975) provide only one solution. Then the search for a multidimensional structure in the data has to be abandoned. Nevertheless partially optimal scaling with a set of completely ordered a priori weights is capable of providing a multidimensional account of data since the pattern of *item weights* is still free to vary. This is another aspect of partially optimal scaling which may appeal to some investigators.

Nishisato and Inukai (1972) conducted a Monte Carlo study on the use of the two weighting schemes, and provided the following tentative suggestions:

1. Item weighting should be used whenever the items are dichotomous.
2. When the number of options for each item is greater than two, option weighting is preferred to item weighting. However, if the total number of

options is too large to handle, those items of relatively few options should be modified into the format of partially optimal scaling (item weighting) and the combined scaling method based on data matrix G^* should be used.
3. When the number of subjects is small, say $N<50$, option weighting should be used. When N is large, item weighting may provide a result very close to that obtained by option weighting.
4. If a reduction in computer time is crucial, item weighting should be used.
5. If one wishes to preserve the order of options of an item, such an item should be modified into the format of partially optimal scaling and the combined method of scaling should be used.

These are tentative guidelines; further empirical studies are required before more explicit recommendations can be made.

As we have seen, option weighting operates under the condition that the item means of all the weighted responses are equal, and item weighting works under the condition that the option weights are fixed. It seems possible to develop a method which starts with item weighting with given option weights, but which modifies the preassigned option weights successively in an iterative manner, keeping the dimension of matrix C $n\times n$. This idea suggests a possibility of developing a method which may possess the merits of both option weighting and item weighting. The details remain to be worked out.

There is one aspect of partially optimal scaling which tends to be overlooked. It is the fact that the method is formulated for continuous data rather than categorical. Partially optimal scaling is therefore nothing but the familiar component analysis formulated within the context of dual scaling. Its application is not restricted to multiple-choice data. It is more suited to general types of questionnaire data with continuous scores.

A comparison between option weighting and item weighting presents interesting relations when all the items are dichotomous. The relations are important both theoretically and practically, and will therefore be discussed in detail in the next section.

5.3.2 *Dichotomous case*

In the previous section, it was noted that the equality in (5.42) would hold under two conditions. One was when items were dichotomous. Our question now is whether or not one can obtain the same information from the two analyses, option weighting and item weighting. To simplify the comparison, let us assign 1 and 0 as the a priori weights to the two options of each item. Then the data matrix with a priori weights A^*, is the same as the response-pattern matrix, F, with one of the two columns of each item being struck out. For each item, option weighting based on F provides two weights, x_1 and x_2, and item weighting based on A^* determines just one weight, x^*. The sign of

x^* is altered depending on the choice of the column to be deleted, that is, on whether it is one associated with x_1 or x_2. However, the absolute value of x^* is not affected by the choice of the column to be deleted. As we know, the two columns for each item in F are completely redundant. This redundancy suggests some clear-cut relations between the pair x_1 and x_2, and x^*. Some of the interesting comparisons (Nishisato 1976) are:

1. $\eta_o^2 = \eta_p^2$ with respect to all correlation ratios.
2. $x_1 x_2 \leq 0$, but x^* can be greater than, less than, or equal to zero.
3. The absolute value of x^* is equal to the sum of the absolute values of x_1 and x_2,

$$|x^*| = |x_1| + |x_2|. \tag{5.43}$$

This relation holds when we use the familiar (1,0) option weights as the a priori weights for partially optimal scaling, that is, item weighting. The above relation, however, does not hold for orthogonal weights, $\mathbf{w}_1$, $\mathbf{w}_2$, and $\mathbf{w}^*$.

4. Let p_i be the proportion of 1's for option i, $i=1,2$. Then $p_1 x_1 + p_2 x_2 = 0$, where $p_1 + p_2 = 1$. Therefore, using (5.43), one can recover x_1 and x_2 from x^* and the item difficulty, p, by the relations

$$x_1 = \pm p_2 x^*, \qquad x_2 = \mp p_1 x^*. \tag{5.44}$$

5. Since all the n items are dichotomous, the total number of options, m, is $2n$. Thus, assuming that the number of subjects is substantially larger than the number of items, we obtain

$$r(A^*) = r(A) = n, \tag{5.45}$$

$$r(F) = m - n + 1 = n + 1. \tag{5.46}$$

However, the ranks of the corresponding matrices for eigenequations, C_p and C_1 say, are the same, that is,

$$r(C_p) = r(C_1) = n. \tag{5.47}$$

Note that the rank of F is reduced by one when the contribution of the second trivial solution is eliminated.

6. It is also known that

$$\mathrm{tr}(C_p) = \mathrm{tr}(C_1) = 1. \tag{5.48}$$

Thus one can see from (5.47) and (5.48) that both analyses must provide multidimensional accounts of data which are equivalent in terms of their information content.

The choice of a priori weights for dichotomous options does not change the value of η_p^2, but affects the value of x^*. This change in x^*, however, is quite systematic, and it is immaterial what a priori option weights one may adopt.

More specifically:
(a) Weights $(1,0),(2,1),(3,2),\ldots,(k+1,k)$ all yield identical values of x^*, say x_1^*.
(b) Weights $(2,0),(3,1),(4,2),\ldots,(k+2,k)$ all yield identical values of x^*, say x_2^*, and $x_2^* = x_1^*/2$.
(c) Similarly, $x_s^* = x_1^*/s$, where x_s^* is the partially optimal weight associated with a priori weights $(k+s,k)$.

Therefore, optimal weights x_1 and x_2 can be recovered from any choice of a priori weights.

The relation between the item weights and a set of option weights of the general polychotomous item has not been fully investigated. However, considering Mosier's remark (1946) about the relation between the item-total correlation, r_{it}, and the standard deviation of the options of item i, s_i, it is likely that a relation similar to (5.43) must be found between item weight x^* and a set of option weights $(x_1, x_2, \ldots)$.

5.4
APPLICATIONS

We shall now discuss numerical examples of comparisons between option weighting and item weighting, scaling of dichotomous items, and effects of different sets of a priori weights on the internal consistency.

Example 5.1: *Dichotomous items* In presenting this example, several points discussed in the previous sections will be verified numerically. Let us consider a small example of 10 subjects answering three dichotomous questions. Suppose that the following response-pattern table was obtained:

$$
\begin{array}{cccccc}
\multicolumn{2}{c}{\underline{\text{Q.1}}} & \multicolumn{2}{c}{\underline{\text{Q.2}}} & \multicolumn{2}{c}{\underline{\text{Q.3}}}
\end{array}
$$

$$F = \begin{bmatrix} 1 & 0 & 1 & 0 & 0 & 1 \\ 0 & 1 & 0 & 1 & 0 & 1 \\ 1 & 0 & 1 & 0 & 1 & 0 \\ 0 & 1 & 0 & 1 & 1 & 0 \\ 0 & 1 & 1 & 0 & 0 & 1 \\ 0 & 1 & 1 & 0 & 1 & 0 \\ 1 & 0 & 1 & 0 & 0 & 1 \\ 1 & 0 & 0 & 1 & 1 & 0 \\ 1 & 0 & 1 & 0 & 1 & 0 \\ 0 & 1 & 0 & 1 & 0 & 1 \end{bmatrix}, \tag{5.49}$$

Then the matrix subjected to the iterative method is

$$C_1 = \begin{bmatrix} 0.167 & & & & & \text{(symmetric)} \\ -0.167 & 0.167 & & & & \\ 0.061 & -0.061 & 0.133 & & & \\ -0.075 & 0.075 & -0.163 & 0.200 & & \\ 0.033 & -0.033 & 0.000 & 0.000 & 0.167 & \\ -0.033 & 0.033 & 0.000 & 0.000 & -0.167 & 0.167 \end{bmatrix}.$$

Dual scaling of C_1 results in three solutions as listed in Table 5.2. Note that $r(C_1)=3$ and

$$\sum_{s=1}^{3} \eta_s^2 = \text{tr}(C_1) = 1.0000.$$

Since the reliability coefficients of the second and third solutions are respectively zero and negative, only those vectors associated with the first solution are listed.

TABLE 5.2

Analysis of C_1: option weighting

	Solution 1	Solution 2	Solution 3
η^2	0.4849	0.3333	0.1818
Reliability (α)	0.4688	0.0000	(−1.2503)
χ^2	15.26	9.32	4.61
df	10	8	6
p	>0.10		

	$\mathbf{x}_1$	$\mathbf{w}_1$	$\mathbf{y}_1$	$\mathbf{u}_1$
Item 1	1.225	2.739	0.758	1.313
	−1.225	−2.739	−1.489	−2.579
			1.274	2.207
Item 2	0.898	2.200	−0.973	−1.685
	−1.347	−2.694	−0.414	−0.718
			0.102	0.176
Item 3	0.539	1.205	0.758	1.313
	−0.539	−1.205	0.199	0.346
			1.274	2.207
			−1.489	−2.579

The same data are now subjected to partially optimal scaling under the following conditions:

1. A priori weights (1,0),(1,0),(1,0), that is, the response to the first option of each item is given as 1, and the response to the second option as 0. Thus A_1^* consists of columns 1, 3, and 5 of F in (5.49),

$$A_1^* = \begin{bmatrix} 1 & 1 & 0 \\ 0 & 0 & 0 \\ 1 & 1 & 1 \\ 0 & 0 & 1 \\ 0 & 1 & 0 \\ 0 & 1 & 1 \\ 1 & 1 & 0 \\ 1 & 0 & 1 \\ 1 & 1 & 1 \\ 0 & 0 & 0 \end{bmatrix}.$$

2. A priori weights (1,0),(1,0),(0,1), that is, A_2^* consists of columns 1, 3, and 6 of F.

3. A priori weights (0,1),(0,1),(0,1), that is A_3^* consists of columns 2, 4, and 6 of F.

4. A priori weights for A_4^* (2,1),(2,1),(2,1), that is,

$$A_4^* = \begin{bmatrix} 2 & 2 & 1 \\ 1 & 1 & 1 \\ 2 & 2 & 2 \\ 1 & 1 & 2 \\ 1 & 2 & 1 \\ 1 & 2 & 2 \\ 2 & 2 & 1 \\ 2 & 1 & 2 \\ 2 & 2 & 2 \\ 1 & 1 & 1 \end{bmatrix}.$$

5. A priori weights for A_5^* (3,1),(3,1),(3,1).

6. A priori weights for A_6^* (1,0),(2,0),(3,0).

The six matrices, A_k^*, $k=1,2,\ldots,6$, are now subjected to partially optimal scaling. Matrices C_p for cases 1, 3, 4, 5, and 6 are identical,

$$C_p = \begin{bmatrix} 0.333 & \text{(symmetric)} & \\ 0.136 & 0.333 & \\ 0.067 & 0.000 & 0.333 \end{bmatrix} \tag{5.50}$$

and C_p for case 2 is the same as (5.50) except that c_{31} is -0.067 instead of 0.067.

TABLE 5.3

Analysis of C_p for six cases: item weighting

		1	2	3	4	5	6
A priori weights		(1,0)	(1,0)	(0,1)	(2,1)	(3,1)	(1,0)
		(1,0)	(1,0)	(0,1)	(2,1)	(3,1)	(2,0)
		(1,0)	(0,1)	(0,1)	(2,1)	(3,1)	(3,0)
$\mathbf{x}_1^* =$	x_1^*	2.449	2.449	2.449	2.449	1.225	2.449
	x_2^*	2.245	2.245	2.245	2.245	1.123	1.123
	x_3^*	1.078	−1.078	1.078	1.078	0.539	0.359
$\mathbf{w}_1 =$	w_1	3.873	3.873	3.873	3.873	3.873	3.873
	w_2	3.478	3.478	3.478	3.478	3.478	3.478
	w_3	1.704	−1.704	1.704	1.704	1.704	1.704
η^2		The six cases yield identical values: $\eta_1^2=0.4849$, $\eta_2^2=0.3333$, $\eta_3^2=0.1818$					
$\mathbf{y}_1$		The six cases yield identical values, except for the reversed sign for 3, and they are equal to $\mathbf{y}_1$ listed in Table 5.2					
$\mathbf{u}_1$		The six cases yield identical values, except for the reversed sign for 3, and they are equal to $\mathbf{u}_1$ listed in Table 5.2					

As expected, partially optimal scaling of the six cases results in correlation ratios identical with those obtained from option weighting, that is, $\eta_1^2=0.4849$, $\eta_2^2=0.3333$, and $\eta_3^2=0.1818$. As in Table 5.2, we list in Table 5.3 only those vectors associated with the first solutions of option weighting and six cases of item weighting.

Partially optimal weights present several interesting points. First, the effect of choosing a priori weights (0, 1) instead of (1,0) is reflected in a reversal in the sign of the item weight, as illustrated in x_3^* of cases 1 and 2. Second, however, the direction of scaling is arbitrary and hence the sign of the entire vector may be reversed, as illustrated by the relation between 1 and 3. Third, a priori weights $(k+1,k)$ yield identical optimal weights to those associated with (1,0), as illustrated in 4. Fourth, a priori weights $(k+s,k)$ yield weights which are one sth of those associated with (1,0), as illustrated in 5 and 6. Fifth, one can choose a priori weights for each item, independently of the others, without changing the outcome of scaling, as illustrated in 6. Sixth, although x_i reflect the choice of a priori weights as indicated by 5 and 6, the corresponding orthogonal weights, w_i, are invariant under the choice of a priori weights. Finally, as stated earlier, the partially optimal item weight x^* and the optimal option weights x_1 and x_2 of each item are related by (5.43).

For instance, the first two items of our example provide the following values:

x^*	x_1	x_2	$\lvert x_1\rvert+\lvert x_2\rvert$
2.4495	1.2247	−1.2247	2.4494
2.2450	0.8980	−1.3470	2.2450

When partially optimal weights in cases 1 and 2 are compared, it may appear as though the two weighting schemes could provide partially optimal scores of different origins. It should be noted, however, that the column totals of A are all zero, and that in consequence the sum of partially optimal scores is always zero, thus leading to the same origin, irrespective of what a priori weights are chosen.

In summary, partially optimal scaling of dichotomous items is not affected by the choice of a priori weights. Furthermore, option weighting and item weighting provide the same scores. This statement does not, of course, hold in general when items are not dichotomous.

Example 5.2: *Polychotomous items* Once the number of options for each item in a multiple-choice questionnaire becomes greater than two, a discrepancy between option weighting and item weighting arises. In the present example the magnitude of the discrepancy changes depending on the choice of a priori weights. The data are given in Table 5.4. In addition to option weighting,

TABLE 5.4

	Q.1			Q.2			Q.3		
Subjects	1	2	3	1	2	3	1	2	3
1	0	0	1	0	1	0	0	0	1
2	1	0	0	1	0	0	1	0	0
3	0	1	0	0	1	0	1	0	0
4	1	0	0	0	1	0	1	0	0
5	0	1	0	0	1	0	0	1	0
6	0	0	1	0	0	1	0	0	1
7	1	0	0	1	0	0	1	0	0
8	1	0	0	0	1	0	0	1	0
9	0	0	1	0	0	1	0	0	1
10	0	1	0	0	0	1	0	1	0
11	1	0	0	0	1	0	0	0	1
12	0	0	1	1	0	0	0	1	0
13	0	0	1	1	0	0	0	1	0
14	0	1	0	1	0	0	1	0	0
15	1	0	0	0	0	1	1	0	0

three kinds of a priori weights were used for item weighting. It was assumed that the options of each item were ordered.

(a) *Traditional multiple-choice scores* We assign 1 to the correct (best) option and 0 to the remaining options. In our example, the weights for the nine options are (1,0,0),(1,0,0),(1,0,0).

(b) *Integer weights* We assign the score of 3 to the best option, 2 to the second option, and 1 to the worst option. In our example, the weights are (3,2,1),(3,2,1),(3,2,1).

(c) *Likert weights* When options of each item are ordered, so-called Likert weights (Likert 1932) are sometimes used as option weights. In calculating the weights, one assumes that the attribute the item is supposed to measure is normally distributed, and that the relative frequencies of the ordered options can therefore be interpreted as partitioned areas under the normal distribution. The weight of each option is then defined as a representative value of the corresponding area which may be the mean or the median of the area. Let us indicate by f_i the total number of responses for the ith option, where all the options are now arranged according to the order. Let p_k be the cumulative proportion of responses from the lowest (worst) option up to the kth option, that is,

$$p_k = \sum_{i=1}^{k} f_i \Big/ N, \tag{5.51}$$

where N is the number of subjects.

(i) *Mean approach* Let z_k be the normal deviate corresponding to p_k and $f(z_k)$ be the ordinate of the unit normal distribution at z_k. Then the Likert weight for option k, x_k, is given by

$$\begin{aligned} x_k &= \frac{\int_{z_{k-1}}^{z_k} zf(z)\,dz}{\int_{z_{k-1}}^{z_k} f(z)\,dz} = \frac{f(z_{k-1})-f(z_k)}{p_k-p_{k-1}} \\ &= \frac{N}{f_k}\left[f(z_{k-1})-f(z_k)\right], \qquad f_k \neq 0, \end{aligned} \tag{5.52}$$

where $f(z_0)=0$, $p_0=0$, and $x_k=0$ if $f_k=0$.

(ii) *Median approach* If we define $p_0=0$, then

$$x_k = \text{the normal deviate corresponding to } \tfrac{1}{2}(p_{k-1}+p_k). \tag{5.53}$$

The mean approach provides a slightly larger value of the variance of the option weights than the median approach, but the latter is computationally simpler than the former. Since the difference between the two is likely to be negligible, we shall use the simpler, median approach. In our example, the marginal frequencies of the three options of each item are:

Q.1 (6,4,5); Q.2 (5,6,4); Q.3 (6,5,4).

It is convenient to prepare a computational table for each item as follows. For item 1,

Option	f_k	p_k	$\frac{1}{2}(p_{k-1}+p_k)$
1	6	$(5+4+6)/15$	$(5+4+5+4+6)/30 = 0.800$
2	4	$(5+4)/15$	$(5+5+4)/30 = 0.4667$
3	5	$5/15$	$(0+5)/30 = 0.1667$

The normal deviates corresponding to the values in the last column can be obtained from a table of the unit normal distribution, and they are the Likert weights. In our example,

$$x_1 = 0.84, \qquad x_2 = -0.08, \qquad x_3 = -0.97.$$

Similarly, the Likert weights for items 2 and 3 can be calculated. The weights thus obtained are

Q.1 (0.84, −0.08, −0.97),

Q.2 (0.97, −0.08, −1.11),

Q.3 (0.84, −0.17, −1.11).

The results of option weighting and item weighting with three sets of a priori option weights are listed in Table 5.5. Let us first compare the option weights and the item weights. The three sets of item weights under (a), (b), and (c) indicate the consistent pattern that item 3 and item 2 have respectively the largest and the smallest weights. The magnitudes of item weights indicate the relative importance of the items in generating subjects' scores. A similar indicator for option weighting is the one mentioned by Mosier (1946), that is, the standard deviation of *option weights* of each item. In our example, the standard deviations of the three items are

$$s_1 = 2.22, \qquad s_2 = 1.67, \qquad s_3 = 2.53.$$

We note that s_j indicate the same pattern as the item weights.

In terms of the reliability coefficient, α, the values obtained from (b) and (c) are close to the optimum value, namely, the value obtained from option weighting. For evaluating differences in the scaling results, α may not always be an appropriate statistic, since it is relatively insensitive to the effects of changes in weight x_j. χ^2 associated with η^2 may not be appropriate either, especially considering that degrees of freedom for option weighting and item weighting are different. One statistic which is relatively sensitive to an increase in $\alpha(\eta^2)$ is an index of the *effectiveness in test length due to scaling*, which can be calculated by the general Spearman-Brown formula

TABLE 5.5

Analysis of polychotomous data

	Option weighting			Item weighting with a priori weights		
Item	1*	2	3	(a) (1,0)	(b) Integer	(c) Likert
x′ 1	0.78	0.68	−1.49	2.10	1.17	1.29
2	0.68	0.22	−1.17	1.42	1.01	0.98
3	1.19	−0.05	−1.72	2.50	1.47	1.51
η_1^2		0.6040		0.5100	0.5836	0.5842
α		0.6722		0.5196	0.6432	0.6441
χ^2		31.50		24.97	30.66	30.71
df		18		16	16	16
p		<0.05		>0.05	<0.05	<0.05
η_2^2		0.4344		(0.3333)	(0.3034)	(0.3030)
η_3^2		0.3892		(0.1567)	(0.1130)	(0.1128)
η_4^2		(0.3094)†				
η_5^2		(0.2042)				
η_6^2		(0.0588)				
Trace		2.0000		1.0000	1.0000	1.0000

*Options.
†Values in parentheses provide negative values of α.

$$\alpha_s = \frac{h\alpha}{1+(h-1)\alpha}, \tag{5.54}$$

where α is the reliability coefficient of the original test, h is the number of parallel tests, and α_s is the reliability of the lengthened test (h times the original test). The statistic of our interest is h, that is,

$$h = \frac{\alpha_s(1-\alpha)}{\alpha(1-\alpha_s)}. \tag{5.55}$$

If we define α to be the reliability of the non-scaled (1,0) scoring scheme and α_s to be that of scaled scores, we can calculate by (5.55) the effect of option weighting and item weighting on the equivalent test length. This statistic, h, can be interpreted in such a way that a particular weighting scheme has the effect of providing h times the amount of information transmitted by the (1,0) scoring scheme. For example, suppose the number of items is 100 and $h=2$. Then one may say that this particular weighting scheme provides the amount of information that 200 statistically equivalent items would transmit under the (1,0) scoring scheme. Or one may say that this weighting scheme would require only 50 (n/h) items to transmit the amount of information conveyed by 100 equivalent items under the (1,0) scoring scheme. In our example, we can calculate α for the traditional scoring method of (a) *without* item weighting, using formula (5.10). From Table 5.4, we obtain $ss_n = 0.0444$, $ss_b = 5.2444$, $ss_t = 10.5778$, and $n = 3$, so that $\alpha = 0.4958$. The values of α_s are given in Table 5.5. Thus the effects of option weighting and item weighting are

Option weighting		$h = 2.09$
Item weighting	(a)	$h = 1.06$
	(b)	$h = 1.83$
	(c)	$h = 1.84$

Since the same data are used, the differences in h are due to the scoring methods. Note that this example shows the advantage of option weighting and item weighting (b) and (c) over the traditional (1,0) non-scaled scoring scheme. For example, the traditional (1,0) scoring scheme would require roughly double the number of items in order to achieve the same degree of internal consistency reliability as could be achieved by option weighting $(h = 2.09)$.

The index h was proposed to rectify the insensitivity of α to the effects of changes in weight x_j. Another problem with the use of α for comparison stems from the fact that α has the ceiling unity, and hence has a skewed sampling distribution (i.e., the scale of α is not composed of equal intervals). In other words, it is more difficult to raise α, for example, from 0.90 to 0.95 than from 0.10 to 0.15 through weighting. Since a negative value of α is meaningless as a measure of reliability, and an α of unity is unattainable in practice, one can consider only $0 < \alpha < 1$. For this restricted range of α, Nishisato (1979c) proposed another index for the effectiveness of scaling, T_α, which approximates an equal interval scale of α,

$$T_\alpha = -\tfrac{1}{2}\log_e(1-\alpha). \qquad (5.56)$$

This is a variance-homogenizing transformation, and should have the effect of separating large values of α more widely than α does. If we take the (1,0) non-scaled case as the base, and divide α and T_α by the base values of α and T_α, respectively, we obtain

		$\alpha/0.4958$	$T_\alpha/0.34$
Option weighting		1.3	1.7
Item weighting	(a)	1.0	1.1
	(b)	1.3	1.4
	(c)	1.3	1.5

We can see the effects of transformation T_α very clearly.

6
Rank order and paired comparison tables

6.1
SPECIAL PROBLEMS

Rank order data are one of the most popular types we see in our daily life. For instance, we rank TV programs, restaurants, beauty contestants, or election candidates according to the order of our preference. When there are n objects to rank, one often assigns score n to the first choice, $n-1$ to the second choice, and in general $(n+1)-k$ to the kth choice. The total or average rank score is then calculated for each object and is used as the score of the object. This scoring scheme appears reasonable and simple enough to use. However, it is not always appropriate, and can often be quite misleading. It may provide a set of meaningless scores when the attribute of the objects is multidimensional, or when subjects are heterogeneous with respect to their rank order judgments. To illustrate, let us consider choosing a group representative by ranking 10 candidates, A, B, C, D, E, F, G, H, I, and J. Suppose that the group consists of two distinct factions, and that their rankings of the candidates are as shown in Table 6.1. According to the above scoring scheme of average ranks, the group representative for this example is the candidate no one ranked at higher than the fifth choice. Furthermore, the last choice is the candidate no one ranked at lower than the sixth. In terms of the original rankings, these two candidates appear to be the closest pair, but the scoring scheme puts one at the top and the other at the bottom of the 10 candidates. This is, of course, an extreme case, but can nevertheless happen.

In practice, one often collects incomplete ranking data by asking judges to rank, for example, as many candidates as they like. If a judge chooses only one candidate, this candidate receives score n and the others 0. If a judge chooses two candidates, the first and the second candidates receive scores n and $n-1$, respectively, and the other candidates receive 0. These are typical arrangements for incomplete ranking when the above scoring scheme is used. It is not too difficult to see that these arrangements would by no means mitigate the problem. Rather they would militate against the scoring scheme.

In 1931 Thurstone presented a method to transform rank order proportions to paired-comparison proportions and to analyse the latter, using the method

TABLE 6.1

Ranking of 10 candidates by a group consisting of two factions

Candidate	Faction I	Faction II	Average rank
A	1	10	5.5
B	2	9	5.5
C	3	8	5.5
D	4	7	5.5
E	5	5	5
F	6	6	6
G	7	4	5.5
H	8	3	5.5
I	9	2	5.5
J	10	1	5.5

NOTE: Group representative is F, with score 6; last candidate is E, with score 5.

of paired comparisons, which was introduced by Fechner in 1860 and developed by Thurstone in 1927. His method deals with a set of complete rank orders and provides only a unidimensional solution. Guilford (1954) reviews the rank-score method mentioned above, Thurstone's method, and other methods such as the normalized-rank method, the method of the first k ranks, the method of first choices, and the method of triad. The paired-comparison approach to ranking data has been further investigated by Bradley and Terry (1952), using the proportions of first choice, and by Luce (1959), who generalized the Bradley-Terry model. Block and Marschak (1960) examined thoroughly the logical interrelations of various paired and multiple comparison models.

Regarding the traditional method of paired comparisons, one can find excellent treatments by Torgerson (1958), David (1963), and Bock and Jones (1968). Extensive review of recent studies on ranking and paired comparison are given by Bradley (1976) and Nishisato (1977).

These approaches are more or less categorized as distinct from the dual scaling approach, where the direct quantification of response categories is involved. As will be seen shortly, the dual scaling approach quantifies differentially the responses from different subjects, while those mentioned above regard each individual as a random sample from a single population and construct a scale of stimuli based on the *average* judgments. The dual scaling approach to ranking and paired comparison data was pioneered by Guttman (1946). His principle is that of internal consistency. Guttman states that his method 'seems to differ from previous approaches in at least two important respects: (a) it is based on but one simple principle, namely that the quantification shall be the one best able to *reproduce the judgment of each person in the population on each comparison*; and, as a consequence, (b) the approach yields solutions not only to the traditional case of ordinary comparisons, but also to more complex cases that do not seem to have been discussed previously' (Guttman 1946, p. 145). Approaches similar to Guttman's have been investigated by several people, among others, Slater (1960a), Tucker (1960), Hayashi (1964, 1967), Carroll and Chang (1964, 1968), de Leeuw (1968, 1973), Bechtel, Tucker, and Chang (1971), Carroll (1972), and Nishisato (1976, 1978a, c).

In this chapter, we discuss a few approaches within the framework of dual scaling. As stated in Chapter 1, two types of rank order data are considered, paired comparison data and general ranking data. Basic differences between these two types of data are: (i) paired comparison of n stimuli requires $n(n-1)$, or $n(n-1)/2$, judgments, while ranking of the same stimuli requires fewer judgments; (ii) it is often more difficult to judge (rank) n stimuli simultaneously than to compare two stimuli at a time, and (iii) comparing two stimuli at a time allows inconsistencies to appear within judgments of an individual (e.g., A is preferred to B; B is preferred to C; C is preferred to A), while ranking n stimuli provides a single *transitive* rank order.

6.2 PAIRED COMPARISON DATA AND INDIVIDUAL DIFFERENCES SCALING

The data are collected in the following way. First, the investigator prepares $n(n-1)/2$ possible pairs of n stimuli, (X_j, X_k), where $j<k$, $j=1,2,\ldots,n-1$, and $k=2,3,\ldots,n$. Each pair is then presented to N subjects, and the subjects are asked 'Which stimulus do you think is greater (higher, more attractive, preferred, better, heavier, etc.), X_j or X_k?' Since each subject answers the question for $n(n-1)/2$ pairs, the investigator collects $Nn(n-1)/2$ responses from N subjects. The main objective of such a paired comparison experiment is to determine the so-called scale values (preference values) of the n stimuli. There may be occasions in which pairs (X_j, X_k) and (X_k, X_j) must be treated as different. For example, in a taste experiment the so-called *order effect* ('Cake A is tasted first and then cake B' versus 'Cake B is tasted first and then cake A') may need to be considered, or in other experiments the so-called *position effect* (e.g., when a subject may tend to prefer the stimulus presented on the right-hand side of the screen to the one on the left-hand side) may arise. In such cases, one should employ $n(n-1)$ possible pairs of different stimuli, (X_j, X_k) and (X_k, X_j), rather than $n(n-1)/2$ pairs (X_j, X_k), $j=1,2,\ldots,n-1$, $k=2,3,\ldots,n$, $j<k$.

In this section, we shall discuss four distinct formulations of the dual

TABLE 6.2

Sample data of paired comparisons

Subjects	(X_1,X_2)	(X_1,X_3)	(X_2,X_3)
1	$X_1>X_2$	$X_1>X_3$	$X_2<X_3$
2	$X_1>X_2$	$X_1<X_3$	$X_2<X_3$
3	$X_1<X_2$	$X_1>X_3$	$X_2<X_3$
4	$X_1>X_2$	$X_1>X_3$	$X_2>X_3$
5	$X_1>X_2$	$X_1<X_3$	$X_2<X_3$

scaling approach, those by Guttman (1946), Nishisato (1978a, c), Slater (1960a), and Tucker and Carroll (Tucker 1960; Carroll 1972). To illustrate differences in these procedures, we shall use a small example with three stimuli and five subjects (i.e., $n=2, N=5$), as given in Table 6.2.

6.2.1 *Guttman's method*

The basic principle of this approach requires that the scale values of stimuli judged "greater" than others by a subject should be as different as possible from the scale values of stimuli judged "smaller" than others by the same subject. For pair (X_j, X_k), define

$$ {}_ie_{jk} = \begin{cases} 1 & \text{if subject } i \text{ judges } X_j > X_k, \\ 0 & \text{if subject } i \text{ judges } X_j < X_k, \end{cases} \tag{6.1} $$

$$ i = 1,2,\ldots,N;\ j = 1,2,\ldots,n; $$
$$ k = 1,2,\ldots,n;\ j \neq k. $$

The responses in Table 6.2 can be represented in terms of ${}_ie_{jk}$ as follows:

$$ ({}_ie_{jk}) = \begin{bmatrix} 1 & 0 & 1 & 0 & 0 & 1 \\ 1 & 0 & 0 & 1 & 0 & 1 \\ 0 & 1 & 1 & 0 & 0 & 1 \\ 1 & 0 & 1 & 0 & 1 & 0 \\ 1 & 0 & 0 & 1 & 0 & 1 \end{bmatrix}, \tag{6.2} $$

where the rows indicate the subjects and the columns pairs corresponding to (X_1,X_2), (X_2,X_1), (X_1,X_3), (X_3,X_1), (X_2,X_3), and (X_3,X_2). Define

$$ f_{ij} = \sum_{k=1}^{n} {}_ie_{jk}, \tag{6.3} $$

$$ g_{ij} = \sum_{k=1}^{n} {}_ie_{kj}, \tag{6.4} $$

where n is the number of stimuli. f_{ij} is the frequency of X_j judged to be greater than other stimuli by subject i and g_{ij} is the frequency of X_j judged smaller than other stimuli by subject i. Denote by F^* and G^* the $N \times n$ matrices with elements f_{ij} and g_{ij}, respectively. In our example, the elements of F^* and G^* can be obtained from (6.2), or more easily from Table 6.2:

$$F^* = (f_{ij}) = \begin{bmatrix} 2 & 0 & 1 \\ 1 & 0 & 2 \\ 1 & 1 & 1 \\ 2 & 1 & 0 \\ 1 & 0 & 2 \end{bmatrix}, \qquad G^* = (g_{ij}) = \begin{bmatrix} 0 & 2 & 1 \\ 1 & 2 & 0 \\ 1 & 1 & 1 \\ 0 & 1 & 2 \\ 1 & 2 & 0 \end{bmatrix}. \tag{6.5}$$

Note that

$$f_{ij} + g_{ij} = n - 1 \quad \text{for every } i \text{ and } j, \tag{6.6}$$

that is,

$$F^* + G^* = (n-1)\mathbf{1}_N\mathbf{1}_n' \tag{6.7}$$

and

$$\sum_{j=1}^{n} f_{ij} = \sum_{j=1}^{n} g_{ij} = n(n-1)/2. \tag{6.8}$$

Hence,

$$F^*\mathbf{1}_n = G^*\mathbf{1}_n = [n(n-1)/2]\mathbf{1}_N. \tag{6.9}$$

In our example, (6.5) gives us the relation $f_{ij} + g_{ij} = 2$ for all i and j, which verifies (6.6). The row sums of F^* and G^* of (6.5) are all equal to 3, which verifies (6.8) and (6.9). Let x_j be the scale value of X_j, t_i and s_i be the means of the x-values of objects subject i judged to be, respectively, higher and lower than other objects, and p_i and q_i be the sums of squares of the x-values of objects subject i judged to be, respectively, higher and lower than other objects. That is,

$$t_i = \frac{2}{n(n-1)} \sum_k x_k f_{ik}, \tag{6.10}$$

$$s_i = \frac{2}{n(n-1)} \sum_k x_k g_{ik}, \tag{6.11}$$

$$p_i = \sum_k (x_k - t_i)^2 f_{ik} = \sum_k x_k^2 f_{ik} - \frac{n(n-1)}{2} t_i^2, \tag{6.12}$$

$$q_i = \sum_k (x_k - s_i)^2 g_{ik} = \sum_k x_k^2 g_{ik} - \frac{n(n-1)}{2} s_i^2, \tag{6.13}$$

Let r be the mean of all the x-values, i.e.

$$r = \frac{1}{Nn(n-1)} \sum_k x_k N(n-1) = \frac{1}{n} \sum x_k. \tag{6.14}$$

Considering that each of N subjects provides $2n$ statistics (f_{ij}, g_{ij}), Guttman defines the between-subject sum of squares, ss_b, the total sum of squares, ss_t, and the within-subject sum of squares, ss_w, as follows:

$$ss_b = \sum_{i=1}^{N} \left[(t_i - r)^2 + (s_i - r)^2 \right] \frac{n(n-1)}{2} \tag{6.15}$$

$$= \frac{n(n-1)}{2} \sum_i (t_i^2 + s_i^2) - r^2 Nn(n-1), \tag{6.16}$$

$$ss_t = N \sum_k (x_k - r)^2 (n-1) \tag{6.17}$$

$$= N(n-1) \sum_k x_k^2 - r^2 Nn(n-1), \tag{6.18}$$

$$ss_w = ss_t - ss_b. \tag{6.19}$$

Our objective is to determine the values of x_j so as to maximize the squared correlation ratio,

$$\eta^2 = ss_b / ss_t. \tag{6.20}$$

Since the correlation ratio is invariant under a shift in the origin of measurement, we set

$$r = 0, \quad \text{or equivalently} \quad \mathbf{1}'\mathbf{x} = 0, \tag{6.21}$$

where $\mathbf{x}$ is the $n \times 1$ vector of scale values. Under condition (6.21), we obtain from (6.16) and (6.18)

$$\eta^2 = \frac{2}{Nn(n-1)^2} \frac{\mathbf{x}'(F^{*\prime}F^* + G^{*\prime}G^*)\mathbf{x}}{\mathbf{x}'\mathbf{x}} \tag{6.22}$$

$$= \frac{\mathbf{x}'H_g\mathbf{x}}{\mathbf{x}'\mathbf{x}}, \text{ say,} \tag{6.23}$$

where

$$H_g = \frac{2}{Nn(n-1)^2}(F^{*\prime}F^* + G^{*\prime}G^*). \tag{6.24}$$

In our example,

$$H_g = \frac{2}{5 \times 3 \times (3-1)^2}(F^{*\prime}F^* + G^{*\prime}G^*)$$

$$= \frac{1}{30}\begin{bmatrix} 14 & 8 & 8 \\ 8 & 16 & 6 \\ 8 & 6 & 16 \end{bmatrix} = \begin{bmatrix} 0.4667 & 0.2667 & 0.2677 \\ 0.2667 & 0.5333 & 0.2000 \\ 0.2667 & 0.2000 & 0.4667 \end{bmatrix}.$$

We now want to determine $\mathbf{x}$ so as to maximize η^2. As discussed in Chapter 2, this amounts to solving the eigenequation

$$(H_g - \eta^2 I)\mathbf{x} = \mathbf{0}, \tag{6.25}$$

under the condition that $\mathbf{x}'\mathbf{x}$ is fixed, say,

$$\mathbf{x}'\mathbf{x} = n. \tag{6.26}$$

As before, the eigenequation has a trivial solution arising from the condition indicated by (6.21). This trivial solution is

$$\eta^2 = 1, \qquad \mathbf{x} = \mathbf{1}, \tag{6.27}$$

which satisfies (6.25) irrespective of what F^* and G^* are. To illustrate the triviality, suppose that $\mathbf{x}=\mathbf{1}$. $\mathbf{x}$ is $n\times 1$, and hence $\mathbf{x}'\mathbf{x}=n$, satisfying (6.26). We also obtain $\Sigma x_k f_{ik} = \Sigma x_k g_{ik} = \Sigma f_{ik} = \Sigma g_{ik} = n(n-1)/2$ for every i. Then (6.10) and (6.11) imply that $t_i = s_i = 1$ for every i, which means that

$$\mathbf{1}'F^{*\prime}F^*\mathbf{1} = \mathbf{1}'G^{*\prime}G^*\mathbf{1} = \left[\tfrac{1}{2}n(n-1)\right]^2 \times N.$$

Putting the results together, we obtain

$$\frac{2}{Nn(n-1)^2}\,\frac{\mathbf{1}'(F^{*\prime}F^* + G^{*\prime}G^*)\mathbf{1}}{\mathbf{1}'\mathbf{1}} = \frac{2}{Nn(n-1)^2}\,\frac{\left[\tfrac{1}{2}n(n-1)\right]^2 \times 2N}{n} = 1,$$

which is η^2 corresponding to $\mathbf{x}=\mathbf{1}$.

To eliminate this trivial solution, we define the residual matrix in the same way as discussed in previous chapters. That is, residual matrix C is given by

$$C = H_g - \eta^2\frac{\mathbf{x}\mathbf{x}'}{\mathbf{x}'\mathbf{x}} = H_g - \frac{1}{n}\mathbf{1}\mathbf{1}'. \tag{6.28}$$

In our example,

$$C = H_g - \tfrac{1}{3}\mathbf{1}\mathbf{1}' = \begin{bmatrix} 0.1333 & -0.0667 & -0.0667 \\ -0.0667 & 0.2000 & -0.1333 \\ -0.0667 & -0.1333 & 0.2000 \end{bmatrix}.$$

Our problem is now to solve

$$(C - \eta^2 I)I\mathbf{x} = \mathbf{0}, \qquad \mathbf{x}'\mathbf{x} = n. \tag{6.29}$$

The iterative method discussed in Chapter 2 can be used to find the largest correlation ratio and the corresponding eigenvector satisfying (6.29). The elements of the eigenvector are scale values of the n objects. If one wants to carry out multidimensional analysis, one may follow the same procedure as discussed in Chapter 2. Since the column marginals of $F^* + G^*$ are constant, optimal and conditionally optimal solution vectors are mutually orthogonal.

6.2.2 *Nishisato's method*

Let us define response variable ${}_if_{jk}$ by

$$
{}_if_{jk} = \begin{cases} 1 & \text{if subject } i \text{ judges } X_j > X_k, \\ -1 & \text{if subject } i \text{ judges } X_j < X_k, \end{cases} \tag{6.30}
$$

$i = 1,2,\ldots,N; j = 1,2,\ldots,n-1;\ k = 2,3,\ldots,n; j < k.$

Let us indicate by F the $N \times n(n-1)/2$ data matrix consisting of ${}_if_{jk}$. In our example,

$$
F = \begin{bmatrix} 1 & 1 & -1 \\ 1 & -1 & -1 \\ -1 & 1 & -1 \\ 1 & 1 & 1 \\ 1 & -1 & -1 \end{bmatrix}.
$$

The rows of F represent subjects and the columns indicate paired stimuli. '1' in column (j,k) means that the contrast of the perceived scale values of stimuli X_j and X_k, that is, $x_j - x_k$, is greater than zero, and '-1' indicates that this contrast is negative. If F is subjected to dual scaling, we shall obtain a set of most discriminative weights for the subjects and a set of most discriminative weights for the pairs of stimuli. Our problem, however, is to find scale values, x_j, for the n stimuli rather than for the $n(n-1)/2$ contrasts. Thus we would like to transform F to the subject-by-stimulus ($N \times n$) matrix so that dual scaling of the transformed matrix will provide two sets of most discriminative weights for subjects and stimuli.

Let us indicate by $\mathbf{z}$ the $n(n-1)/2 \times 1$ vector of weights for the columns of F, and by $\mathbf{x}$ the $n \times 1$ vector of stimulus scale values. Since the (j,k) element of $\mathbf{z}$ has the structure $x_j - x_k$, we can relate $\mathbf{z}$ to $\mathbf{x}$ by the familiar design matrix used in the traditional method of paired comparison. In our example,

$$
\mathbf{z} = \begin{bmatrix} x_1 - x_2 \\ x_1 - x_3 \\ x_2 - x_3 \end{bmatrix} = \begin{bmatrix} 1 & -1 & 0 \\ 1 & 0 & -1 \\ 0 & 1 & -1 \end{bmatrix} \begin{bmatrix} x_1 \\ x_2 \\ x_3 \end{bmatrix} = A\mathbf{x}, \text{ say.} \tag{6.31}
$$

The design matrix for n stimuli can be expressed as

$$
\underset{\frac{1}{2}n(n-1)\times n}{A} = \begin{bmatrix} 1 & -1 & 0 & 0 & \cdots & 0 & 0 \\ 1 & 0 & -1 & 0 & \cdots & 0 & 0 \\ 1 & 0 & 0 & -1 & \cdots & 0 & 0 \\ \vdots & \vdots & \vdots & \vdots & \ddots & \vdots & \vdots \\ 1 & 0 & 0 & 0 & \cdots & 0 & -1 \\ 0 & 1 & -1 & 0 & \cdots & 0 & 0 \\ 0 & 1 & 0 & -1 & \cdots & 0 & 0 \\ \vdots & \vdots & \vdots & \vdots & \ddots & \vdots & \vdots \\ 0 & 0 & 0 & 0 & \cdots & 1 & -1 \end{bmatrix}. \tag{6.32}
$$

For example, if $n=5$, then

$$\underset{10\times 5}{A} = \begin{bmatrix} 1 & -1 & 0 & 0 & 0 \\ 1 & 0 & -1 & 0 & 0 \\ 1 & 0 & 0 & -1 & 0 \\ 1 & 0 & 0 & 0 & -1 \\ 0 & 1 & -1 & 0 & 0 \\ 0 & 1 & 0 & -1 & 0 \\ 0 & 1 & 0 & 0 & -1 \\ 0 & 0 & 1 & -1 & 0 \\ 0 & 0 & 1 & 0 & -1 \\ 0 & 0 & 0 & 1 & -1 \end{bmatrix}, \quad \text{so that} \quad A\mathbf{x} = \begin{bmatrix} x_1 - x_2 \\ x_1 - x_3 \\ x_1 - x_4 \\ x_1 - x_5 \\ x_2 - x_3 \\ x_2 - x_4 \\ x_2 - x_5 \\ x_3 - x_4 \\ x_3 - x_5 \\ x_4 - x_5 \end{bmatrix}.$$

As was seen in Chapter 2, dual scaling of F determines vectors $\mathbf{y}$ and $\mathbf{z}$ so as to maximize $\mathbf{y}'F\mathbf{z}$, given constraints on $\mathbf{y}$ and $\mathbf{z}$. Using (6.31), dual scaling of paired comparison data can be expressed as

$$\max_{\mathbf{y},\mathbf{z}} (\mathbf{y}'F\mathbf{z}) = \max_{\mathbf{y},\mathbf{x}} (\mathbf{y}'FA\mathbf{x}) = \max_{\mathbf{y},\mathbf{x}} (\mathbf{y}'E\mathbf{x}), \text{ say,} \tag{6.33}$$

subject to the constraints on $\mathbf{y}$, $\mathbf{z}$ and $\mathbf{x}$, where

$$\underset{N\times n}{E} = (e_{ij}) = FA. \tag{6.34}$$

Matrix E is the required subject-by-stimulus matrix, and is obtained by post-multiplying the data matrix by the design matrix. In our example,

$$E = FA = \begin{bmatrix} 1 & 1 & -1 \\ 1 & -1 & -1 \\ -1 & 1 & -1 \\ 1 & 1 & 1 \\ 1 & -1 & -1 \end{bmatrix} \begin{bmatrix} 1 & -1 & 0 \\ 1 & 0 & -1 \\ 0 & 1 & -1 \end{bmatrix} = \begin{bmatrix} 2 & -2 & 0 \\ 0 & -2 & 2 \\ 0 & 0 & 0 \\ 2 & 0 & -2 \\ 0 & -2 & 2 \end{bmatrix}.$$

Matrix E has two interesting properties:

$$\sum_j e_{ij} = 0 \quad \text{for every } i; \tag{6.35}$$

$$E = F^* - G^*, \tag{6.36}$$

where F^* and G^* are Guttman's matrices (f_{ij}) and (g_{ij}). Property (6.35) leads to the structure that all the marginals of $E'E$ are zero; in other words, $E'E$ is double centred. Property (6.36) serves to establish that the scale values obtained by Nishisato's method are identical with those obtained by Guttman's method; this comparison will be discussed shortly.

Dual scaling of E requires the determination of the number of responses involved in its rows and columns, for *unlike the data matrices discussed so far matrix E has negative elements*. In this case, neither the marginals of E nor the sums of the absolute values of e_{ij} indicate the actual number of responses involved in the rows and the columns of E. Note that up to this point D and

D_N (or D_n) were diagonal matrices of column totals and row totals of the data matrix, respectively. This simple procedure does not apply to dual scaling of E. To specify the elements of D and D_N, let us look at our example. If we consider responses weighted by $\mathbf{z}$ and $\mathbf{x}$, the data corresponding to F and E can be expressed, respectively, as

$$\begin{bmatrix} x_1-x_2 & x_1-x_3 & -(x_2-x_3) \\ x_1-x_2 & -(x_1-x_3) & -(x_2-x_3) \\ -(x_1-x_2) & x_1-x_3 & -(x_2-x_3) \\ x_1-x_2 & x_1-x_3 & x_2-x_3 \\ x_1-x_2 & -(x_1-x_3) & -(x_2-x_3) \end{bmatrix} = F(z), \text{ say,}$$

$$\begin{bmatrix} x_1+x_1 & -x_2-x_2 & -x_3+x_3 \\ x_1-x_1 & -x_2-x_2 & x_3+x_3 \\ -x_1+x_1 & x_2-x_2 & -x_3+x_3 \\ x_1+x_1 & -x_2+x_2 & -x_3-x_3 \\ x_1-x_1 & -x_2-x_2 & x_3+x_3 \end{bmatrix} = E(x), \text{ say.}$$

Note that matrix E consists of the coefficients for x_1, x_2, and x_3 of $E(x)$. From $E(x)$, we can generalize that the number of responses involved in each row of $N\times n$ matrix E is $2\times[n(n-1)/2]=n(n-1)$, and that the number of responses for each column is $N(n-1)$. More precisely, $n(n-1)$ can be arrived at by considering that each of $n(n-1)/2$ columns has two scale values in the form of a contrast, and $N(n-1)$ indicates that N subjects compare each stimulus with $n-1$ remaining stimuli. As $F(z)$ and $E(x)$ show, the total number of scale values involved in the data matrix is $Nn(n-1)$. Thus, for the paired comparison case, we have

$$D = \text{diag}\left[N(n-1)\right] = N(n-1)I, \tag{6.37}$$

$$D_N = \text{diag}\left[n(n-1)\right] = n(n-1)I. \tag{6.38}$$

Because $E'E$ is double centred, the so-called correction terms of ss_b and ss_t vanish, resulting in the expressions

$$ss_b = \mathbf{x}'E'D_N^{-1}E\mathbf{x} = \frac{1}{n(n-1)}\mathbf{x}'E'E\mathbf{x}, \tag{6.39}$$

$$ss_t = \mathbf{x}'D\mathbf{x} = N(n-1)\mathbf{x}'\mathbf{x}. \tag{6.40}$$

Therefore the correlation ratio is given by

$$\eta^2 = \frac{ss_b}{ss_t} = \frac{1}{Nn(n-1)^2}\frac{\mathbf{x}'E'E\mathbf{x}}{\mathbf{x}'\mathbf{x}} = \frac{\mathbf{x}'H_n\mathbf{x}}{\mathbf{x}'\mathbf{x}}, \tag{6.41}$$

where

$$H_n = \frac{1}{Nn(n-1)^2}E'E. \tag{6.42}$$

In our example,

$$H_n = \frac{1}{5\times 3\times(3-1)^2}E'E = \begin{bmatrix} 0.1333 & -0.0667 & -0.0667 \\ -0.0667 & 0.2000 & -0.1333 \\ -0.0667 & -0.1333 & 0.2000 \end{bmatrix}.$$

Note that H_n is identical with Guttman's C, and that the row totals and the column totals of H_n are all zero (i.e., H_n is double centred). As before, maximization of (6.41) amounts to solving the eigenequation

$$(H_n - \eta^2 I)\mathbf{x} = \mathbf{0}, \tag{6.43}$$

subject to the condition that $\mathbf{x}'\mathbf{x}$ is constant, say

$$\mathbf{x}'\mathbf{x} = n. \tag{6.44}$$

There are three points one should note about this approach. First, unlike Guttman's formulation, (6.43) does not have the trivial solution due to the marginal constraint, for H_n is double centred. Second, dual scaling of E makes the calculation of optimal weights for subjects a straightforward procedure, an aspect which is not obvious in Guttman's approach. Third, Nishisato's approach extends dual scaling to a data matrix consisting of *negative* elements as well as positive, a significant step for the development of dual scaling. In the present formulation, D and D_N are scalar matrices (i.e., diagonal matrices with constant diagonal elements), which make optimal and locally optimal vectors collinear with the corresponding orthogonal weight vectors. In other words, vectors $\mathbf{x}_j$ and $\mathbf{w}_j$, discussed as distinct in previous chapters, are now identical in their information content. Thus, without loss of information, we can choose orthogonal weight vectors for $\mathbf{x}$ (stimulus scale values) and $\mathbf{y}$ (weights for subjects) to simplify the formulas and computation. The weight vector for subjects is then given by

$$\mathbf{y} = \frac{1}{\eta}D_N{}^{-1/2}ED^{-1/2}\mathbf{x} = \frac{1}{\eta}\left(\frac{1}{Nn(n-1)^2}\right)^{1/2}E\mathbf{x}. \tag{6.45}$$

In multidimensional analysis, the kth solution is given by the set

$$\left.\begin{aligned} \mathbf{x}_k &= \frac{1}{\eta_k}\left(\frac{1}{Nn(n-1)^2}\right)^{1/2}E'\mathbf{y}_k, \\ \mathbf{y}_k &= \frac{1}{\eta_k}\left(\frac{1}{Nn(n-1)^2}\right)^{1/2}E\mathbf{x}_k, \end{aligned}\right\} \tag{6.46}$$

where

$$\mathbf{x}_k{}'\mathbf{x}_t = \mathbf{y}_k{}'\mathbf{y}_t = \begin{cases} n & \text{if } k=t, \\ 0 & \text{if } k\neq t. \end{cases} \tag{6.47}$$

It is also straightforward to see that

$$\mathbf{y}_k{}'D_n{}^{-1/2}ED^{-1/2}\mathbf{x}_k = \eta_k n, \tag{6.48}$$

or

$$\left(\frac{1}{Nn(n-1)^2}\right)^{1/2} \mathbf{y}_k{}'E\mathbf{x}_k = \eta_k n. \tag{6.49}$$

6.2.3 *Slater's method*

Slater (1960a) proposed a method which deals only with F^*, Guttman's (f_{ij}). Since the mean of each row of F^* is $(n-1)/2$, he first subtracts the mean from each element of F^*; the new matrix is indicated by E^*,

$$E^* = F^* - \tfrac{1}{2}(n-1)\mathbf{1}_N\mathbf{1}_n, \tag{6.50}$$

where $\mathbf{1}_N$ and $\mathbf{1}_n$ are respectively the $N \times 1$ and $n \times 1$ vectors of 1's. Slater subjects $E^{*\prime}E^*$ and $E^*E^{*\prime}$ to principal component analysis to obtain the vector of scale values, $\mathbf{x}$, and the vector of weights for subjects, $\mathbf{y}$, respectively. In our example, these matrices are

$$E^* = \begin{bmatrix} 2 & 0 & 1 \\ 1 & 0 & 2 \\ 1 & 1 & 1 \\ 2 & 1 & 0 \\ 1 & 0 & 2 \end{bmatrix} - \frac{3-1}{2}\begin{bmatrix} 1 & 1 & 1 \\ 1 & 1 & 1 \\ 1 & 1 & 1 \\ 1 & 1 & 1 \\ 1 & 1 & 1 \end{bmatrix} = \begin{bmatrix} 1 & -1 & 0 \\ 0 & -1 & 1 \\ 0 & 0 & 0 \\ 1 & 0 & -1 \\ 0 & -1 & 1 \end{bmatrix},$$

$$E^{*\prime}E^* = \begin{bmatrix} 2 & -1 & -1 \\ -1 & 3 & -2 \\ -1 & -2 & 3 \end{bmatrix},$$

and

$$E^*E^{*\prime} = \begin{bmatrix} 2 & 1 & 0 & 1 & 1 \\ 1 & 2 & 0 & -1 & 2 \\ 0 & 0 & 0 & 0 & 0 \\ 1 & -1 & 0 & 2 & -1 \\ 1 & 2 & 0 & -1 & 2 \end{bmatrix}.$$

6.2.4 *Tucker-Carroll's method*

In the 1960s and 1970s, an approach which appears different from the three approaches described above was developed. This is the so-called vector model of preferential choice (Tucker 1960; Carroll and Chang 1964, 1968; Carroll 1972). Carroll (1972) presents a succinct formulation of this model. He defines X, Y, and $\hat{S}$ to be, respectively, the $n \times r$ matrix of stimulus coordinate values, the $N \times r$ matrix of coordinates of the termini of subject vectors, and the $N \times n$ matrix of preference scale values. The model is given by

$$\hat{S} = YX'. \tag{6.51}$$

The procedure for maximizing C_1 (one of Carroll's objective functions) is to define first $S \equiv (s_{ij})$ by

$$s_{ij} = (w_i)^{1/2} \sum_{j \neq k}^{n} ({}_ie_{jk} - {}_ie_{kj}), \tag{6.52}$$

where ${}_ie_{jk}$ and ${}_ie_{kj}$ are defined by (6.1) and w_i are introduced in case it is desired to weight subjects differently. If $w_i = 1$, S is the same as Nishisato's E. Carroll then factors S into a product of the form

$$S = U\beta V', \tag{6.53}$$

where U and V are matrices of eigenvectors of SS' and $S'S$, respectively, and β is the diagonal matrix of the square roots of the eigenvalues of SS', which is the same as that for $S'S$. In terms of the r largest eigenvalues and the corresponding eigenvectors, he defines the rank r solution by (6.51), where

$$Y = U_r\beta_r \quad \text{and} \quad X = V_r, \tag{6.54}$$

subscript r indicating matrices made up of the submatrices corresponding to the r largest eigenvalues.

6.2.5 *Comparisons*

Guttman's approach is a straightforward application of the principle of internal consistency, and Guttman states that his formulation does not assume any relation between $X_j > X_k$ and the difference between the two 'discriminal processes,' a term used in the Thurstonian approach. In contrast, Nishisato's approach employs the interpretation that judgment $X_j > X_k$ is a function of the difference between the corresponding scale values, that is, $x_j - x_k$. Yet these different approaches yield the same scale values.

Let us first compare Guttman's and Nishisato's methods. The former yields quadratic form (6.23) with matrix H_g, and the latter leads to quadratic form (6.41) with matrix H_n. As we have seen, H_g is not double centred and hence needs to be transformed to matrix C by (6.28). Thus, if C is proven to be identical with H_n, we can conclude that the two approaches provide the same scale values.

First, note the following relations:

$$C = H_g - \frac{1}{n}\mathbf{1}_n\mathbf{1}_n'$$

$$= \frac{2}{Nn(n-1)^2}(F^{*\prime}F^* + G^{*\prime}G^*) - \frac{1}{n}\mathbf{1}_n\mathbf{1}_n', \tag{6.55}$$

where $\mathbf{1}_n$ is the $n \times 1$ vector of 1's; and

$$H_n = \frac{1}{Nn(n-1)^2} E'E = \frac{1}{Nn(n-1)^2}(F^* - G^*)'(F^* - G^*)$$
$$= \frac{1}{Nn(n-1)^2}(F^{*\prime}F^* + G^{*\prime}G^* - F^{*\prime}G^* - G^{*\prime}F^*). \tag{6.56}$$

We want to show that

$$C = H_n. \tag{6.57}$$

To prove this, first multiply C and H_n by $Nn(n-1)^2/2$. The right-hand sides of (6.55) and (6.56) are respectively

$$F^{*\prime}F^* + G^{*\prime}G^* - \tfrac{1}{2}N(n-1)^2\mathbf{1}_n\mathbf{1}_n' = C^*, \text{ say,} \tag{6.58}$$

$$\tfrac{1}{2}(F^{*\prime}F^* + G^{*\prime}G^* - F^{*\prime}G^* - G^{*\prime}F^*) = H_n^*, \text{ say,} \tag{6.59}$$

Since $F^* + G^* = (n-1)\mathbf{1}_N\mathbf{1}_n'$ by (6.7),

$$(F^* + G^*)'(F^* + G^*) = (n-1)^2\mathbf{1}_n\mathbf{1}_N'\mathbf{1}_N\mathbf{1}_n' = N(n-1)^2\mathbf{1}_n\mathbf{1}_n'. \tag{6.60}$$

Substituting (6.60) in (6.58) gives

$$C^* = F^{*\prime}F^* + G^{*\prime}G^* - \tfrac{1}{2}(F^* + G^*)'(F^* + G^*)$$
$$= \tfrac{1}{2}(F^{*\prime}F^* + G^{*\prime}G^* - F^{*\prime}G^* - G^{*\prime}F^*), \tag{6.61}$$

which is identical with H_n^* of (6.59), thus proving the equivalence of the two approaches. Recall that our numerical example has already verified this relation.

The equivalence of Slater's approach to the above two can easily be shown. Assume that there is no missing response. Then, using (6.7), we can rewrite (6.50) as

$$E^* = F^* - \tfrac{1}{2}(F^* + G^*) = \tfrac{1}{2}(F^* - G^*) = \tfrac{1}{2}E, \tag{6.62}$$

where E is Nishisato's matrix. From (6.62), it is obvious that Slater's method provides the same results (up to the proportionality) as the other two methods.

The relation of the Tucker-Carroll approach to the above three can also be shown by comparing it with the multidimensional version of Nishisato's approach. Define the $N \times r$ matrix of r orthogonal weight vectors, Y^*, the $n \times r$ matrix of r orthogonal vectors of scale values, X^*, and the $r \times r$ diagonal matrix of the r largest correlation ratios, Λ^2. Then Nishisato's approach defines the multidimensional solution by the relation

$$\hat{E} = Y^*\Lambda X^{*\prime}/n, \tag{6.63}$$

where $\hat{E}$ is the rank r approximation of E. The two decompositions (6.53) and (6.63) are basically the same. The main difference between them stems from the fact that Nishisato's approach retains duality of analysis, as indicated by (6.46), whereas the Tucker-Carroll approach scales $\mathbf{x}_k$ and $\mathbf{y}_k$ non-symmetrically, as seen in (6.54).

The above comparisons are based on Nishisato (1978a, c). All the methods are geared to handling individual differences. Which method one should choose may be a matter of preference, for there do not seem to be substantial differences in terms of computation. However, when the investigator encounters the problem of tied responses, missing responses, or order (or position) effects, Nishisato's method may provide some advantages over the others. These problems will be discussed in Section 6.5.

6.3
GENERAL RANK ORDER DATA

One of the basic differences between paired comparison data and rank order data lies in the possible existence or non-existence of intransitive relations. Paired comparisons may provide such intransitive relations as $X_1 > X_2, X_2 > X_3, X_3 > X_1$, but the rank order judgment puts the stimuli always in a single sequence such as $X_1 > X_2 > X_3$. All the methods discussed in the previous section can easily be extended to handle rank order data. We shall now present some of them and then discuss another method which is not as good, but which seems popular in practice.

6.3.1 *Individual differences scaling*

Although rank order data look statistically different from paired comparison data, it is interesting to know that Guttman's approach, discussed in Section 6.2.1, can handle both types of data in exactly the same way. For rank order data collected from N subjects on n stimuli, all the formulas presented in 6.2.1 remain unchanged. The only difference between the two types of data is that rank order data provide a more orderly distribution of responses than paired comparison data. For example, each row of F^* for rank order data always consists of n distinct values, which is not necessarily true for paired comparison data.

Slater's method, discussed in 6.2.3, is based on F^* and can handle rank order data in exactly the same way as paired comparison data.

Nishisato (1976) noted that one could not specify the design matrix such as A of (6.32) for rank order data, but that one could easily construct matrix E which has properties (6.35) and (6.36). Matrix E for rank order data is the subject-by-stimulus $N \times n$ matrix, of which typical element e_{ij} is given by

$$e_{ij} = n + 1 - 2K_{ij}, \tag{6.64}$$

where K_{ij} is the rank of X_j given by subject i and rank K is defined as the kth choice out of n. This regularity stems from the fact that rank order judgments are always transitive. Once E is specified, the rest of the analysis and the formulas are exactly the same as those described for paired comparison data in 6.2.2.

6.3.2 *Analysis of the stimulus-by-rank table*

One often summarizes rank order data in the form of a stimulus-by-rank contingency table. This table no longer contains individual rank orders from N subjects. Therefore, dual scaling of this table does not maximize the between-subject variation in any direct way. Instead, it maximizes the between-rank and the between-stimulus variations.

Let us indicate by C^* the stimulus-by-rank $(n \times n)$ matrix of response frequencies. If each of N subjects ranks all the n stimuli, every row marginal and every column marginal of C^* are constant and equal to N. To make the formulation general enough to accommodate incomplete rank orders, we indicate by D and D_n the diagonal matrix of the column and the row totals, respectively. Let $\mathbf{x}$ be the vector of weights for the n ranks, with x_j being the weight for rank j, and f_t be the total number of responses in C^*. The correlation ratio of responses weighted by $\mathbf{x}$ can be expressed as

$$\eta^2 = \frac{\mathbf{x}'C^{*\prime}D_n^{-1}C^*\mathbf{x}}{\mathbf{x}'D\mathbf{x}} \tag{6.65}$$

$$= \frac{\mathbf{w}'D^{-1/2}C^{*\prime}D_n^{-1}C^*D^{-1/2}\mathbf{w}}{\mathbf{w}'\mathbf{w}} \tag{6.66}$$

where

$$\mathbf{w} = D^{1/2}\mathbf{x}. \tag{6.67}$$

As we have seen earlier, (6.66) takes the maximum value of 1 when

$$\mathbf{w} = D^{1/2}\mathbf{1},$$

irrespective of what data matrix C^* might be. To eliminate this trivial solution, we calculate residual matrix C from

$$C = D^{-1/2}C^{*\prime}D_n^{-1}C^*D^{-1/2} - D^{1/2}\mathbf{1}\mathbf{1}'D^{1/2}/f_t. \tag{6.68}$$

Noting that the weights for n ranks must be ordered, our problem is to determine $\mathbf{x}$ which maximizes η^2 under the order constraint on $\mathbf{x}$. In other words, we must solve

$$(C - \eta^2 I)\mathbf{w} = 0, \tag{6.69}$$

subject to the two conditions

$$x_1 \geq x_2 \geq \cdots \geq x_n, \tag{6.70}$$

$$\mathbf{w}'\mathbf{w} = f_t. \tag{6.71}$$

This type of constrained maximization can be handled by existing methods (e.g., Bradley, Katti, and Coons 1962; Kruskal 1965; Nishisato and Arri 1975) and will be discussed further in Chapter 8.

Once the solution that maximizes η^2 under constraints (6.70) and (6.71) is obtained, we can calculate the vector of scale values of n stimuli from

$$\mathbf{y} = D^{-1}C^{*}\mathbf{x}/\eta. \tag{6.72}$$

This formulation is analogous to dual scaling of the contingency table, discussed in Chapter 4, except for the order constraint. If one wishes to formulate this approach using the response-pattern representation, one needs data matrix F, which is slightly different from the one used in Chapter 4. Suppose three stimuli were ranked by N subjects. Then F can be expressed, for example, as follows:

	Stimulus			Rank		
Subject	1	2	3	1	2	3

$$\begin{matrix} 1 \left\{ \begin{matrix} \\ \\ \\ \end{matrix} \right. \\ \vdots \\ N \left\{ \begin{matrix} \\ \\ \\ \end{matrix} \right. \end{matrix} \begin{bmatrix} 1 & 0 & 0 & 0 & 1 & 0 \\ 0 & 1 & 0 & 0 & 0 & 1 \\ 0 & 0 & 1 & 1 & 0 & 0 \\ \vdots & \vdots & \vdots & \vdots & \vdots & \vdots \\ 1 & 0 & 0 & 0 & 0 & 1 \\ 0 & 1 & 0 & 1 & 0 & 0 \\ 0 & 0 & 1 & 0 & 1 & 0 \end{bmatrix} = \underset{3N\times 6}{F} \tag{6.73}$$

In general, response pattern matrix F for n stimuli and N subjects is $Nn\times 2n$. The cross-product matrix of F is given as

$$\underset{2n\times 2n}{F'F} = \begin{bmatrix} \underset{n\times n}{D_n} & \underset{n\times n}{C'} \\ \underset{n\times n}{C^{*\prime}} & \underset{n\times n}{D} \end{bmatrix}. \tag{6.74}$$

Thus both the canonical correlational approach and the bivariate correlational approach can be used as before, except that we now must impose the order constraint on $\mathbf{x}$, that is, (6.70). These two approaches lead to the same set of equations as (6.69), (6.70), and (6.71).

6.4 STATISTICAL TESTS

Paired comparison data can be represented as an $N\times n(n-1)/2$ matrix of 1's and 0's, which is a kind of response-pattern table. This matrix, however, has such a structural constraint that only n quantities (i.e., scale values) can vary.

Thus the chi-square statistic, discussed for the response-pattern table, does not seem appropriate without modification. Instead, we suggest the use of the statistic discussed for the response-frequency table with the substitution of N, n, and $Nn(n-1)/2$ for n, m, and f_t, respectively. That is,

$$\chi^2 = -\tfrac{1}{2}[Nn(n-1) - N - n - 1]\log_e(1-\eta_j^2), \tag{6.75}$$

$$\text{df} = N + n - 1 - 2j, \tag{6.76}$$

$$\eta_1^2 > \eta_2^2 > \cdots > \eta_j^2.$$

A statistical test for rank order data is not presented here because further research seems needed.

6.5 TIES, MISSING RESPONSES, AND ORDER EFFECTS

When subjects are allowed to make equality judgments (i.e., *tied responses*), the various techniques discussed in this chapter cannot be used without modification. To deal with the problem, Nishisato (1978a) proposed the following modifications of ${}_if_{jk}$ of (6.30) for paired comparison and e_{ij} of (6.64) for ranking:

$$ {}_if_{jk} = \begin{cases} 1 & \text{if subject } i \text{ judges } X_j > X_k, \\ 0 & \text{if subject } i \text{ judges } X_j = X_k, \\ -1 & \text{if subject } i \text{ judges } X_j < X_k, \end{cases} \tag{6.77}$$

$$e_{ij} = n + 1 - 2\overline{K}_{ij}. \tag{6.78}$$

where $\overline{K}_{ij}$ is the average rank of all the stimuli that subject i judged equal to X_j. When (6.77) and (6.78) are used for tied responses, matrix E satisfies (6.35) and (6.36), and the formulation presented in Section 6.2.2 applies without further modification. This type of treatment of ties is also presented by de Leeuw (1973). The Tucker-Carroll approach can handle tied responses in the same manner. The formulations by Guttman and Slater, however, seem to require further modifications to handle this problem, and these are not immediately obvious.

When some responses are *missing*, it is interesting to note that the relation of E to F^* and G^*, that is, (6.36), still holds, and that matrix S of the Tucker-Carroll method is affected in the same way as E. Thus missing data do not alter the relations among the three methods by Guttman, Nishisato, and Tucker-Carroll. However, it is then necessary to use D and D_N, which may no longer be scalar matrices, and to follow the general formulation of dual scaling. This change also means that optimal and conditionally optimal vectors may no longer be collinear with the corresponding orthogonal weight vectors. The use of D and D_N does not create any difficulty for Nishisato's method, but poses some problems for Guttman's method. More specifically, it is not obvious how D and D_N can be incorporated in his approach. As for Slater's approach, $(F^* + G^*)/2$ of (6.62) would no longer be the matrix of row means, which seems to destroy the relation of E^* to E given by (6.62).

When missing responses are introduced by the experimental design (as in incomplete designs for paired comparison), Nishisato's method appears to be conceptually simpler than the others, for it then uses the design matrix which corresponds to that particular incomplete design. In this case, D is the diagonal matrix with the kth diagonal element being N times the number of non-zero (i.e., 1 and -1) elements in the kth column of the design matrix, and D_N is the diagonal matrix (scalar matrix) with each diagonal element being equal to the total number of non-zero elements in the entire design matrix.

As was mentioned in Section 6.2, there are occasions in which order effects or position effects are suspected to arise in soliciting preference judgments. In the traditional Thurstonian framework, Bock and Jones (1968) presented a formulation of a model to handle such effects. Data are then collected with respect to $n(n-1)$ possible pairs rather than $n(n-1)/2$ pairs. Within the dual scaling framework, Nishisato's method can assess order (or position) effects on judgments by specifying an appropriate design matrix to derive matrix E. More specifically, what is needed here is a subject-by-(stimuli, order-effect parameters) matrix E, and such a matrix can easily be constructed. See Bock and Jones (1968) for an example of a design matrix of the type needed to generate the desired matrix, E.

6.6 APPLICATIONS

The following small examples will illustrate the procedures discussed in the previous sections of this chapter.

Example 6.1: *Paired comparisons* Let us consider the two approaches described in Sections 6.2.1 and 6.2.2 and apply them to the artificial data listed in Table 6.3. Assuming that pairs (X_j, X_k) and (X_k, X_j) are equivalent, 10 subjects were asked to make paired comparison judgments for six pairs, that is, $n(n-1)/2$ with n equal to four.

TABLE 6.3

	(X_1,X_2)		(X_1,X_3)		(X_1,X_4)		(X_2,X_3)		(X_2,X_4)		(X_3,X_4)		
Subjects	1*	2	1	2	1	2	1	2	1	2	1	2	Total
1	1	0	1	0	0	1	1	0	0	1	1	0	6
2	1	0	1	0	1	0	1	0	1	0	0	1	6
3	0	1	0	1	0	1	1	0	0	1	1	0	6
4	1	0	0	1	0	1	1	0	0	1	0	1	6
5	1	0	1	0	0	1	1	0	0	1	0	1	6
6	1	0	1	0	1	0	1	0	0	1	0	1	6
7	1	0	1	0	0	1	1	0	1	0	0	1	6
8	0	1	0	1	0	1	0	1	0	1	1	0	6
9	1	0	1	0	1	0	1	0	0	1	0	1	6
10	1	0	1	0	0	1	1	0	0	1	0	1	6
Subtotal	8	2	7	3	3	7	9	1	2	8	3	7	60
Total	10		10		10		10		10		10		60

*Ranks: 1 indicates judgment $X_j > X_k$ for pair (X_j, X_k) and 2, $X_j < X_k$.

In Guttman's approach, we first construct F^* and G^* from Table 6.3:

$$F^* = \begin{bmatrix} 2 & 1 & 1 & 2 \\ 3 & 2 & 0 & 1 \\ 0 & 2 & 2 & 2 \\ 1 & 1 & 1 & 3 \\ 2 & 1 & 0 & 3 \\ 3 & 1 & 0 & 2 \\ 2 & 2 & 0 & 2 \\ 0 & 1 & 3 & 2 \\ 3 & 1 & 0 & 2 \\ 2 & 1 & 0 & 3 \end{bmatrix}, \qquad G^* = \begin{bmatrix} 1 & 2 & 2 & 1 \\ 0 & 1 & 3 & 2 \\ 3 & 1 & 1 & 1 \\ 2 & 2 & 2 & 0 \\ 1 & 2 & 3 & 0 \\ 0 & 2 & 3 & 1 \\ 1 & 1 & 3 & 1 \\ 3 & 2 & 0 & 1 \\ 0 & 2 & 3 & 1 \\ 1 & 2 & 3 & 0 \end{bmatrix}.$$

Note that the sum of each row of both F^* and G^* is equal to $n(n-1)/2$, and that $f_{ij} + g_{ij} = n-1$, where $n=4$, $i=1,2,3,\ldots,N(=10)$, and $j=1,2,3,4$. These relations can be used to check the correctness of F^* and G^*. We then obtain

$$F^{*\prime}F^* = \begin{bmatrix} 44 & 23 & 3 & 38 \\ 23 & 19 & 9 & 27 \\ 3 & 9 & 15 & 15 \\ 38 & 27 & 15 & 52 \end{bmatrix}, \qquad G^{*\prime}G^* = \begin{bmatrix} 26 & 20 & 18 & 8 \\ 20 & 31 & 39 & 12 \\ 18 & 39 & 63 & 18 \\ 8 & 12 & 18 & 10 \end{bmatrix}.$$

The sums of the elements of these two matrices are equal. More specifically,

$$\mathbf{1}'F^{*\prime}F^{*}\mathbf{1} = \mathbf{1}'G^{*\prime}G^{*}\mathbf{1} = N\left[\tfrac{1}{2}n(n-1)\right]^2,$$

which is in this example equal to 360. The sum of the two matrices is

$$F^{*\prime}F^{*} + G^{*\prime}G^{*} = \begin{bmatrix} 70 & 43 & 21 & 46 \\ 43 & 50 & 48 & 39 \\ 21 & 48 & 78 & 33 \\ 46 & 39 & 33 & 62 \end{bmatrix}.$$

All the marginals are now identical and equal to $Nn(n-1)^2/2$, that is, 180 in our example. Matrix H_g is obtained by dividing each element of the above matrix by $Nn(n-1)^2/2$:

$$H_g = \begin{bmatrix} 0.3889 & & & \text{(symmetric)} \\ 0.2389 & 0.2778 & & \\ 0.1167 & 0.2667 & 0.4333 & \\ 0.2556 & 0.2167 & 0.1833 & 0.3444 \end{bmatrix}.$$

All the marginals of H_g are identical and equal to 1. Residual matrix C, which is free from the trivial solution ($\eta^2 = 1, \mathbf{x} = \mathbf{1}$), can be calculated as

$$C = H_g - \frac{1}{n}\mathbf{11}' = \begin{bmatrix} 0.1389 & & & \text{(symmetric)} \\ -0.0111 & 0.0278 & & \\ -0.1333 & 0.0167 & 0.1833 & \\ 0.0056 & -0.0333 & -0.0667 & 0.0944 \end{bmatrix}.$$

This matrix can now be subjected to the iterative method to find the vector of scale values associated with the largest correlation ratio.

Before calculating the results, let us look at Nishisato's approach. Modified matrix F can be obtained from Table 6.3 by deleting the second column of each pair and replacing 0's in the remaining columns with -1's:

$$F = \begin{bmatrix} 1 & 1 & -1 & 1 & -1 & 1 \\ 1 & 1 & 1 & 1 & 1 & -1 \\ -1 & -1 & -1 & 1 & -1 & 1 \\ 1 & -1 & -1 & 1 & -1 & -1 \\ 1 & 1 & -1 & 1 & -1 & -1 \\ 1 & 1 & 1 & 1 & -1 & -1 \\ 1 & 1 & -1 & 1 & 1 & -1 \\ -1 & -1 & -1 & -1 & -1 & 1 \\ 1 & 1 & 1 & 1 & -1 & -1 \\ 1 & 1 & -1 & 1 & -1 & -1 \end{bmatrix}.$$

The design matrix of the method of paired comparisons is

$$A = \begin{bmatrix} 1 & -1 & 0 & 0 \\ 1 & 0 & -1 & 0 \\ 1 & 0 & 0 & -1 \\ 0 & 1 & -1 & 0 \\ 0 & 1 & 0 & -1 \\ 0 & 0 & 1 & -1 \end{bmatrix}.$$

The first row indicates $x_1 - x_2$, the second row $x_1 - x_3$, the third row $x_1 - x_4$, the fourth row $x_2 - x_3$, the fifth row $x_2 - x_4$, and the last row $x_3 - x_4$. Matrix E is the product of F and A:

$$E = FA = \begin{bmatrix} 1 & -1 & -1 & 1 \\ 3 & 1 & -3 & -1 \\ -3 & 1 & 1 & 1 \\ -1 & -1 & -1 & 3 \\ 1 & -1 & -3 & 3 \\ 3 & -1 & -3 & 1 \\ 1 & 1 & -3 & 1 \\ -3 & -1 & 3 & 1 \\ 3 & -1 & -3 & 1 \\ 1 & -1 & -3 & 3 \end{bmatrix}.$$

Note that all the row marginals are zero, that is,

$$\sum_{j=1}^{n} e_{ij} = 0, \qquad i = 1,2,\dots,N.$$

Note also that $E = F^* - G^*$. The product of E' and E is

$$E'E = \begin{bmatrix} 50 & -4 & -48 & 2 \\ -4 & 10 & 6 & -12 \\ -48 & 6 & 66 & -24 \\ 2 & -12 & -24 & 34 \end{bmatrix}.$$

All the marginals are zero. In other words, $E'E$ is double centred. $E'E$ is symmetric, the marginal of the jth row being equal to $\Sigma_i e_{ij}(\Sigma_{k=1}^{n} e_{ik})$, where $\Sigma_{k=1}^{n} e_{ik} = 0$ for every i. Matrix H_n is given by

$$H_n = \frac{1}{Nn(n-1)^2} E'E = \begin{bmatrix} 0.1389 & & & \text{(symmetric)} \\ -0.0111 & 0.0278 & & \\ -0.1333 & 0.0167 & 0.1833 & \\ 0.0056 & -0.0333 & -0.0667 & 0.0944 \end{bmatrix}.$$

We now see that H_n and C are identical. Thus the two approaches provide the

TABLE 6.4

	Dual scaling solution			Case v solution	Mean e_j
	1	2	3		
η^2	0.3131	0.1074	0.0240		
δ	70.4%	94.6%	100.0%		
x_1	1.1859	0.9610	0.8185	0.8416	0.6000
x_2	−0.1953	0.5805	−1.6203	−0.4016	−0.4000
x_3	−1.5114	0.1101	0.8388	−2.3304	−1.6000
x_4	0.5208	−1.6515	−0.0373	1.8904	1.4000

same eigenequations. Using the iterative method, we obtain the results listed in Table 6.4.

For the purpose of comparison, the traditional method of paired comparisons was applied, with Thurstone's Case v assumptions, to the paired comparison proportions obtained from the column marginals of Table 6.3, that is,

$$[P_{jk}] = [\text{Proportion } X_j > X_k] = \begin{bmatrix} (0.5) & 0.8 & 0.7 & 0.3 \\ 0.2 & (0.5) & 0.9 & 0.2 \\ 0.3 & 0.1 & (0.5) & 0.3 \\ 0.7 & 0.8 & 0.7 & (0.5) \end{bmatrix}.$$

In the normal response model, the scale value of X_j is simply the average of the normal deviates which correspond to the paired comparison proportions in row j of the above matrix. The values of x_j thus obtained, rescaled to (6.44), are listed in Table 6.4 under the name Case v solution. For the purpose of comparing the dual scaling approach with the traditional paired comparison approach, another statistic is listed in the last column of Table 6.4. These values are column means of E.

There are some interesting points in Table 6.4. The first is the multidimensionality of the data. Statistic δ indicates that the total variance is distributed over three dimensions with relative variances of 70.4%, 24.2%, and 5.4%, respectively. Since the first dimension is dominant, one might hope that the remaining two dimensions may be of little importance to the interpretation of the data. As we shall see shortly, this is not the case. Indeed, it does not seem appropriate to regard the data as unidimensional.

The second point is related to the first. If the data can be regarded as almost unidimensional, one may expect that the scale values of the first dimension of dual scaling should reveal the same rank order as the Case v solution. Our results show, however, that $x_1 > x_4 > x_2 > x_3$ and $x_4 > x_1 > x_2 > x_3$

for dual scaling and paired comparison scaling, respectively, that is, the order of x_1 and x_4 is reversed in the two solutions. To interpret the discrepancy, we need an additional piece of information, which is conveniently supplied by the subjects' optimal scores as defined by (6.45). The scores, $\mathbf{y}_1$, associated with the first dimension are listed in Table 6.5. Since the first solution generated the order of the scale values as $x_1>x_4>x_2>x_3$, the predicted paired comparison judgments are $X_1>X_2$, $X_1>X_3$, $X_1>X_4$, $X_2>X_3$, $X_2<X_4$, and $X_3<X_4$. Thus, in terms of the coding used in Table 6.3, the predicted response pattern is

$$(1,0|1,0|1,0|1,0|0,1|0,1).$$

In Table 6.5 large positive scores indicate that judgments are relatively consistent with the pattern predicted by $\mathbf{x}_1$. As the scores approach zero, however, the inconsistency of judgment increases. Negative scores indicate the dominance of reversed (1,0) patterns. Table 6.5 lists how many choices are consistent and inconsistent with the prediction. The most important point in this table is probably the fact that dual scaling assigned a larger value to X_1 than X_4 in spite of the fact that there were seven $X_4>X_1$ responses compared with three $X_1>X_4$ responses. Why is it possible that $x_1>x_4$ under these circumstances? The key lies in the internal consistency. Careful inspection of Tables 6.3 and 6.5 reveals that judgment $X_1>X_4$ comes from subjects 6, 9, and 2, whose responses are highly consistent as we can see from their optimal scores. Thus the above query can be answered by stating that the responses of

TABLE 6.5

Individual differences reflected in the optimal solution

		Number of patterns		
Subject	Score $\mathbf{y}_1$	Consistent	Reversed	Reversed patterns (j,k)*
1	0.3215	4	2	(1,4),(3,4)
2	0.6947	5	1	(2,4)
3	−0.4468	2	4	(1,2),(1,3),(1,4),(3,4)
4	0.1962	4	2	(1,3),(1,4)
5	0.7043	5	1	(1,4)
6	0.8296	6	0	
7	0.5694	4	2	(1,4),(2,4)
8	−0.6947	1	5	(1,2),(1,3),(1,4),(2,3),(3,4)
9	0.8296	6	0	
10	0.7043	5	1	(1,4)

*(j,k) indicates pair (X_j,X_k).

these three subjects had a greater impact on the maximization of the internal consistency than those of the remaining seven subjects. This answer reflects the basic principle of our procedure, that is, duality. As the optimal scores for the subjects are proportional to the mean responses weighted by the scale values for the stimuli, the scale values are proportional to the mean responses weighted by the subjects' optimal scores.

The Case v solution and the column means of E, do not make any differential evaluation of subjects, however, but treat them as though every subject contributes equally to the assessment of scale values. Thus, if 70% of subjects judge $X_4 > X_1, X_4$ receives a larger scale value than X_1.

The equal weighting of the subjects, or rather the ignoring of individual differences, can be regarded as a step backward, for it does not fully utilize the information contained in the data. It is bound to present problems when one deals with a heterogeneous group of subjects. It would then be just like putting garbage in and getting garbage out. The approach of dual scaling, in contrast, provides a multidimensional solution where the group is heterogeneous and a means to categorize the subjects into homogeneous subgroups. Many intransitive relations are resolved into several sets of transitive relations. In this regard, dual scaling can be thought of as a procedure to find sets of paired comparison judgments that produce transitive relations.

Example 6.2: *Rank order data* Let us look at a small example (Table 6.6) and see how our methods can be used to find the scale values of the objects. If one wants to use Guttman's method, the first task is to construct two matrices F^* and G^*. From Table 6.6, we obtain

$$F^* = \begin{bmatrix} 3 & 2 & 0 & 1 \\ 2 & 1 & 0 & 3 \\ 3 & 1 & 0 & 2 \\ 3 & 0 & 1 & 2 \\ 1 & 0 & 2 & 3 \\ 3 & 0 & 1 & 2 \\ 1 & 2 & 0 & 3 \\ 3 & 1 & 0 & 2 \\ 2 & 1 & 0 & 3 \\ 1 & 2 & 0 & 3 \end{bmatrix}, \quad G^* = \begin{bmatrix} 0 & 1 & 3 & 2 \\ 1 & 2 & 3 & 0 \\ 0 & 2 & 3 & 1 \\ 0 & 3 & 2 & 1 \\ 2 & 3 & 1 & 0 \\ 0 & 3 & 2 & 1 \\ 2 & 1 & 3 & 0 \\ 0 & 2 & 3 & 1 \\ 1 & 2 & 3 & 0 \\ 2 & 1 & 3 & 0 \end{bmatrix}.$$

Note that the elements of each row of the two matrices are all distinct, unlike in the corresponding matrices of paired comparison data. One can now calculate H_g by (6.24) and then C by (6.28). Before discussing further, let us look at Nishisato's method. In it, one can construct matrix E from Table 6.6 by formula (6.64):

TABLE 6.6

Rank order data

	Stimuli			
Subjects	1	2	3	4
1	1	2	4	3
2	2	3	4	1
3	1	3	4	2
4	1	4	3	2
5	3	4	2	1
6	1	4	3	2
7	3	2	4	1
8	1	3	4	2
9	2	3	4	1
10	3	2	4	1

$$E = \begin{bmatrix} 3 & 1 & -3 & -1 \\ 1 & -1 & -3 & 3 \\ 3 & -1 & -3 & 1 \\ 3 & -3 & -1 & 1 \\ -1 & -3 & 1 & 3 \\ 3 & -3 & -1 & 1 \\ -1 & 1 & -3 & 3 \\ 3 & -1 & -3 & 1 \\ 1 & -1 & -3 & 3 \\ -1 & 1 & -3 & 3 \end{bmatrix}.$$

One can easily verify by this example that $E = F^* - G^*$. Once E is constructed, one can calculate H_n by (6.42). As discussed earlier, H_n is identical with C, which is used in Guttman's method. In the present example,

$$H_n = C = \begin{bmatrix} 0.1389 & & & \text{(symmetric)} \\ -0.0611 & 0.0944 & & \\ -0.0945 & 0.0167 & 0.1833 & \\ 0.0167 & -0.0500 & -0.1056 & 0.1389 \end{bmatrix}.$$

Solving eigenequation (6.43) under constraint (6.44), we obtain three solutions, with the following correlation ratios, relative variances, and δ's, respectively:

$$\eta_1^2 = 0.3302\ (59.5\%), \qquad \delta_1 = 59.5\%,$$
$$\eta_2^2 = 0.1269\ (22.8\%), \qquad \delta_2 = 82.3\%,$$
$$\eta_3^2 = 0.0984\ (17.7\%), \qquad \delta_3 = 100.0\%.$$

Although the first dimension is clearly dominant, it still leaves about 40% of the total variance unaccounted for. The first solution is

$$\eta_1^2 = 0.3302, \qquad \mathbf{x}_1 = \begin{bmatrix} 0.9350 \\ -0.5467 \\ -1.3671 \\ 0.9788 \end{bmatrix}, \qquad \mathbf{y}_1 = \begin{bmatrix} 0.4935 \\ 0.7814 \\ 0.7734 \\ 0.6229 \\ 0.2086 \\ 0.6229 \\ 0.5096 \\ 0.7734 \\ 0.7814 \\ 0.5096 \end{bmatrix}. \tag{6.79}$$

Let us now look at the stimulus-by-rank table. The elements of the table can be obtained from Table 6.6 by replacing 'subjects' by 'ranks.' In this example, we obtain

$$C^* = \begin{bmatrix} 5 & 2 & 3 & 0 \\ 0 & 3 & 4 & 3 \\ 0 & 1 & 2 & 7 \\ 5 & 4 & 1 & 0 \end{bmatrix}.$$

Note that the diagonal matrices of row totals and column totals, D_n and D, are equal to $10I$, and that $f_t = 40$. After eliminating the trivial solution $(\eta^2 = 1, \mathbf{x} = \mathbf{1})$ the residual matrix can be calculated from (6.68) as

$$C = \begin{bmatrix} 0.2500 & & & \text{(symmetric)} \\ 0.0500 & 0.0500 & & \\ -0.0500 & -0.0100 & 0.0500 & \\ -0.2500 & -0.0900 & 0.0100 & 0.3300 \end{bmatrix}.$$

Our problem is to solve (6.69), (6.70), and (6.71). Using the method of successive data modification, of which the procedure is thoroughly discussed in Chapter 8, we obtain the results

$$\eta^2 = 0.5663\ (83\%), \qquad \mathbf{x} = \begin{bmatrix} 1.2651 \\ 0.3856 \\ -0.1589 \\ -1.4918 \end{bmatrix}, \qquad \mathbf{y} = \begin{bmatrix} 0.8797 \\ -0.5254 \\ -1.3787 \\ 1.0244 \end{bmatrix}. \tag{6.80}$$

Note that $x_1 > x_2 > x_3 > x_4$, where x_j is the value of rank j. $\mathbf{y}$ is the vector of scale values of the four stimuli.

Let us compare the two types of analysis, one by Guttman's and Nishisato's methods and the other based on the stimulus-by-rank table. We note that the two matrices involved in the respective eigenequations have different trace values, which result in different values of the correlation ratio and the percentage of the total variance accounted for by the solution. To make the two sets of results comparable, therefore, we substitute the above solution vector for $\mathbf{x}$ in (6.41) to calculate the adjusted correlation ratio. Then this correlation ratio, η_c^2 say, is used to define the percentage of the variance accounted for by this solution as $100\% \, \eta_c^2/\mathrm{tr}(H_n)$. According to this scheme, the adjusted results are

$$\eta_c^2 = 0.3299 \; (59.4\%), \qquad \mathbf{y} = \begin{bmatrix} 0.8797 \\ -0.5254 \\ -1.3787 \\ 1.0244 \end{bmatrix}. \tag{6.81}$$

Although the scale values, $\mathbf{x}_1$ and $\mathbf{y}$ of (6.79) and (6.81) respectively, are somewhat different, η_c^2 is very close to η^2 of (6.79). One can make a few guesses as to why this is so: the closeness may be due to the particular data used, or it may be peculiar to rank order data, which are so tightly constrained that individual differences may not exert any substantial effect on the scaling outcome. The latter conjecture seems reasonable if we note the difference in the contributions of the first solutions of (6.80) and (6.81) to the total variance. The contribution is 83% in (6.80), where individual differences are ignored, but the figure is reduced to 59% in (6.81), where the *same solution* is evaluated in the context of the subject-by-stimulus matrix. The increase from 59% to 83% seems to reflect the relative contribution of individual differences. If so, the present data contain a substantial amount of variations due to individual differences. The closeness of the two results then seems to lead to the second conjecture mentioned above. Either way, however, the solution by Guttman's and Nishisato's methods provides the maximum discrimination with respect to the original (subject-by-stimulus-by-rank) data, and hence should be preferred to the solution based on the reduced (stimulus-by-rank) data matrix.

Example 6.3: *Ipsative rank scores* We have noted that rank order data provide pairwise transitive relations. This regularity presents a potential pitfall for data analysis when it is combined with the so-called rank scores. As mentioned earlier, the rank score for the kth choice out of n stimuli is defined as $(n+1)-k$, which generates n integer scores for n stimuli. Since tied judgments are not allowed, the n scores are distinct. These appear intuitively reasonable scores for the ranks, and are in practice widely used. One may then wish to use rank scores as a priori weights for the ranks and to apply partially optimal scaling to the subject-by-stimulus matrix of rank scores to obtain scale values of the stimuli. This seemingly reasonable and attractive approach, however, is doomed to yield nonsensical numbers.

The problem stems from the fact that n distinct rank scores constitute the so-called ipsative scores (see, for example, Clemans 1965), namely, every subject receives the same total score, which is the sum of integers from 1 up to n. Under this condition, we observe the tendency that if one stimulus receives a high score then the other stimuli tend to receive low scores. If $n=2$, the product-moment correlation of the two stimuli, calculated from rank scores of N subjects, is always -1, whatever the response distribution may be. Likewise, if $n=3$, correlation coefficients of all the possible pairs are likely to be negative, no matter what their 'true' relationships may be. Since partially optimal scaling uses rank scores as numerals (not frequencies) and correlational measures (covariances) as fundamental quantities, it is bound to be influenced by the same bias, and it is likely that such bias cannot be isolated from the 'true' relationship among items.

As an example, consider the rank order data in Table 6.6. The matrix of rank scores can be obtained by adding 1 to each element of F^*, obtained earlier. Using the procedure described in 5.2, we obtain the matrix

$$D_a^{-1/2}A'D_n^{-1}AD_a^{-1/2} = \begin{bmatrix} 0.2500 & & & (\text{symmetric}) \\ -0.0740 & 0.2500 & & \\ -0.0346 & -0.1946 & 0.2500 & \\ -0.2075 & 0.0000 & 0.0227 & 0.2500 \end{bmatrix}.$$

Using the iterative method, we obtain

$$\eta_1^2 = 0.4712, \qquad \eta_2^2 = 0.4439, \qquad \eta_3^2 = 0.0849.$$

Since the trace of (6.82) is unity, the relative contributions of the three solutions are $100\times\eta_j^2, j=1,2,3$. The optimal weights for the four objects are

$$\mathbf{x} = \begin{bmatrix} -1.5700 \\ 0.8421 \\ -0.3992 \\ 1.8945 \end{bmatrix}.$$

Note that this vector has no resemblance to $\mathbf{x}_1$ of (6.79) or $\mathbf{y}$ of (6.81). Thus partially optimal scaling of rank scores is a case of misapplication.

7 Multidimensional tables

7.1 GENERAL PROCEDURE

As discussed in Chapter 1, a wide variety of complex tables are classified as multidimensional tables. In practice, most of these tables are represented as two-way tables with subcategories in the rows and/or columns. This arrangement is used here as the general format for a multidimensional table, and dual scaling of data in this format will be discussed. Since the general format accommodates a wide variety of tables, it is difficult to describe the scaling procedure for each distinct case. Instead, only a few examples will be given to provide some working knowledge.

A good introduction to our general procedure has already been given in the previous chapter, in our discussion of Nishisato's approach to paired comparison and rank order data. As we recall, subject-by-paired stimulus ($N \times n(n-1)/2$) table F was converted by design matrix A into subject-by-stimulus ($N \times n$) matrix E, and the latter was subjected to dual scaling, resulting in the most discriminative scale values for n stimuli and the most discriminative weights for N subjects. The data matrix was a multidimensional table with structured columns, and hence only the columns of the data matrix were transformed from pairs of stimuli to individual stimuli. The rationale for transforming the data matrix lies in the fact that dual scaling maximizes simultaneously the between-row and the between-column sums of squares of the input matrix, relative to the total sum of squares, and hence that the rows and the columns of the transformed matrix should be represented by variables of interest (Nishisato 1976).

This approach can be expanded to the general format of multidimensional tables. Let us first define the notation:

N = the number of rows of the table,

M = the number of columns of the table,

F = the $N \times M$ multidimensional data matrix,

$\mathbf{z}_r$ = the $N \times 1$ vector of weights (scores) for the rows of F,

$\mathbf{z}_c$ = the $M \times 1$ vector of weights (scores) for the columns of F.

As we have seen in the previous chapter, transformation of the multidimensional table into the matrix of interest amounts to postulating some structures for $\mathbf{z}_r$ and $\mathbf{z}_c$, expressed in terms of quantities of interest. To make our formulation general, let us indicate by $\mathbf{x}$ and $\mathbf{y}$ the vectors of interest associated with the rows and the columns of F, respectively, and suppose that $\mathbf{z}_c$ and $\mathbf{z}_r$ can be expressed as

$$\mathbf{z}_c = A_c\mathbf{x}, \tag{7.1}$$

$$\mathbf{z}_r = A_r\mathbf{y}, \tag{7.2}$$

where A_c and A_r are, respectively, $M \times m$ and $N \times n$ design matrices for the columns and rows of F, and $\mathbf{x}$ and $\mathbf{y}$ are $m \times 1$ and $n \times 1$, respectively. Using (7.1) and (7.2), we obtain

$$\mathbf{z}_r'F\mathbf{z}_c = \mathbf{y}'A_r'FA_c\mathbf{x} = \mathbf{y}'E\mathbf{x}, \tag{7.3}$$

where

$$\underset{n\times m}{E} = A_r'FA_c. \tag{7.4}$$

Thus dual scaling of E results in the most discriminative set of vectors $(\mathbf{x}, \mathbf{y})$. Our general procedure then is to transform F to E by (7.4), and to subject E to dual scaling to obtain optimal vectors $\mathbf{x}$ and $\mathbf{y}$. In the earlier example of paired comparison data, we used the identity matrix for A_r and the familiar paired-comparison design matrix, A, for A_c.

The above approach applies when scale values associated with the rows and the columns of E are of direct interest (e.g., paired comparisons, rank order). In practice, however, there are probably more occasions in which the investigator's interest lies in another set of numbers associated with $\mathbf{x}$ and $\mathbf{y}$ of E. For example, suppose that N subjects answered k multiple-choice questions, and that dual scaling was carried out to determine M option weights of the k items so as to discriminate maximally between the two subgroups of subjects. Then, by our general procedure we would convert the $N \times M$ data matrix to $2 \times M$ matrix E, and would obtain the 2×1 vector of optimal group scores and the $M \times 1$ vector of optimal option weights. Suppose, however, that the investigator wants to obtain N optimal scores for the subjects, rather than two for the groups, which maximize the group difference. Such a situation arises when the investigator wants to use subjects as units for further statistical analysis. Likewise, one may convert the $N \times M$ data matrix to $N \times 3$ matrix E in order to obtain optimal scores for the N rows that discriminate maximally among three sets of columns. The same problem arises here if the investigator wants M weights that maximize the between-set discrimination.

An immediate solution to this type of problem is to calculate the vector of interest from $\mathbf{x}$ and/or $\mathbf{y}$ of E. More specifically, given that $\mathbf{y}'E\mathbf{x}$ is a relative maximum we can calculate the corresponding vectors for the rows and the columns of the data matrix, respectively, by

$$D_N{}^{-1}FA_c\mathbf{x}/\eta = \mathbf{y}_r,\ \text{say}, \tag{7.5}$$

$$D^{-1}F'A_r\mathbf{y}/\eta = \mathbf{x}_c,\ \text{say}, \tag{7.6}$$

where D_N and D are diagonal matrices of row totals and column totals of F, respectively. Note that (7.5) and (7.6) are nothing but the formulas for optimal scores and weights based on F, that is, $\mathbf{z}_r$ and $\mathbf{z}_c$. However, since optimization was carried out in terms of E, $\mathbf{y}$, and $\mathbf{x}$, $\mathbf{y}_r$ and $\mathbf{x}_c$ are used for $\mathbf{z}_r$ and $\mathbf{z}_c$, respectively.

Let us consider a small example. Suppose that two subjects were randomly chosen from each of the four groups 'smoker, alcoholic,' 'smoker, non-alcoholic,' 'non-smoker, alcoholic,' 'non-smoker, non-alcoholic,' and that they answered four dichotomous questions. Data matrix F then is 8×8 (subject-by-option), consisting of 1's and 0's. For example,

$$\begin{array}{c} \text{Items} \\ \begin{array}{cccc} 1 & 2 & 3 & 4 \end{array} \end{array} \qquad \text{Groups}$$

$$F = \begin{bmatrix} 1 & 0 & 0 & 1 & 0 & 1 & 1 & 0 \\ 1 & 0 & 0 & 1 & 1 & 0 & 1 & 0 \\ 0 & 1 & 1 & 0 & 1 & 0 & 0 & 1 \\ 1 & 0 & 1 & 0 & 1 & 0 & 0 & 1 \\ 0 & 1 & 1 & 0 & 1 & 0 & 1 & 0 \\ 0 & 1 & 1 & 0 & 0 & 1 & 0 & 1 \\ 1 & 0 & 0 & 1 & 0 & 1 & 0 & 1 \\ 0 & 1 & 0 & 1 & 1 & 0 & 1 & 0 \end{bmatrix} \begin{matrix} \left.\vphantom{\begin{matrix}1\\1\end{matrix}}\right\} 1 \\ \left.\vphantom{\begin{matrix}1\\1\end{matrix}}\right\} 2 \\ \left.\vphantom{\begin{matrix}1\\1\end{matrix}}\right\} 3 \\ \left.\vphantom{\begin{matrix}1\\1\end{matrix}}\right\} 4 \end{matrix} \tag{7.7}$$

Dual scaling of F maximizes $\mathbf{z}_r{}'F\mathbf{z}_c$, relative to the total sum of squares. Suppose, however, that the investigator is interested in option weights and scores for the subjects that maximize the difference between smokers and non-smokers and the difference between alcoholic and non-alcoholic subjects. What kind of design matrix A_r should he use for this problem?

Let us first consider the complete design of the two-way analysis of variance. Then the decomposition of the score vector, $\mathbf{z}_r$, can be expressed as follows:

$$\mathbf{z}_r = \begin{bmatrix} z_{r1} \\ z_{r2} \\ z_{r3} \\ z_{r4} \\ z_{r5} \\ z_{r6} \\ z_{r7} \\ z_{r8} \end{bmatrix} = \begin{bmatrix} \mu+\alpha_1+\beta_1+\gamma_{11} \\ \mu+\alpha_1+\beta_1+\gamma_{11} \\ \mu+\alpha_1+\beta_2+\gamma_{12} \\ \mu+\alpha_1+\beta_2+\gamma_{12} \\ \mu+\alpha_2+\beta_1+\gamma_{21} \\ \mu+\alpha_2+\beta_1+\gamma_{21} \\ \mu+\alpha_2+\beta_2+\gamma_{22} \\ \mu+\alpha_2+\beta_2+\gamma_{22} \end{bmatrix} + \begin{bmatrix} e_{r1} \\ e_{r2} \\ e_{r3} \\ e_{r4} \\ e_{r5} \\ e_{r6} \\ e_{r7} \\ e_{r8} \end{bmatrix} = A_r\mathbf{y} + \mathbf{e}, \tag{7.8}$$

where

$$A_r = \begin{bmatrix} 1 & 1 & 0 & 1 & 0 & 1 & 0 & 0 & 0 \\ 1 & 1 & 0 & 1 & 0 & 1 & 0 & 0 & 0 \\ 1 & 1 & 0 & 0 & 1 & 0 & 1 & 0 & 0 \\ 1 & 1 & 0 & 0 & 1 & 0 & 1 & 0 & 0 \\ 1 & 0 & 1 & 1 & 0 & 0 & 0 & 1 & 0 \\ 1 & 0 & 1 & 1 & 0 & 0 & 0 & 1 & 0 \\ 1 & 0 & 1 & 0 & 1 & 0 & 0 & 0 & 1 \\ 1 & 0 & 1 & 0 & 1 & 0 & 0 & 0 & 1 \end{bmatrix}, \quad \mathbf{y} = \begin{bmatrix} \mu \\ \alpha_1 \\ \alpha_2 \\ \beta_1 \\ \beta_2 \\ \gamma_{11} \\ \gamma_{12} \\ \gamma_{21} \\ \gamma_{22} \end{bmatrix}, \tag{7.9}$$

and $\mathbf{e}$ is the vector of residuals e_{rj}, $j=1,2,\ldots,8$. μ is the grand mean, α indicates the effects of smokers (α_1) and non-smokers (α_2), β is associated with the alcoholics (β_1) and non-alcoholics (β_2), and all the γ's are the so-called interaction (combination) effects of α and β.

Returning to our problem, we now wish to determine option weights so as to maximize the difference between α_1 and α_2 and the difference between β_1 and β_2. Thus we can eliminate from the complete design matrix those columns associated with the grand mean and the interactions, i.e. the first column and the last four columns of A_r. The design matrix of interest is then

$$\underset{8\times 4}{A_r} = \begin{bmatrix} 1 & 0 & 1 & 0 \\ 1 & 0 & 1 & 0 \\ 1 & 0 & 0 & 1 \\ 1 & 0 & 0 & 1 \\ 0 & 1 & 1 & 0 \\ 0 & 1 & 1 & 0 \\ 0 & 1 & 0 & 1 \\ 0 & 1 & 0 & 1 \end{bmatrix}. \tag{7.10}$$

It is interesting to note that the deletion of μ (the first column) from dual scaling has the effect of setting the mean (hence the sum) of the weighted responses equal to zero (Nishisato 1971a).

With respect to the columns of F, no structure is postulated, and hence we set $A_c = I$. Matrix E for dual scaling then is obtained from (7.7) and (7.10) by (7.4) as

$$E = A_r'F = \begin{bmatrix} 3 & 1 & 2 & 2 & 3 & 1 & 2 & 2 \\ 1 & 3 & 2 & 2 & 2 & 2 & 2 & 2 \\ 2 & 2 & 2 & 2 & 3 & 2 & 3 & 1 \\ 2 & 2 & 2 & 2 & 3 & 1 & 1 & 3 \end{bmatrix}. \tag{7.11}$$

Dual scaling of E results in 4×1 vector $\mathbf{y}$ for the rows and 8×1 vector $\mathbf{x}$ for the columns of E such that $\mathbf{y}'E\mathbf{x}$ is a maximum, relative to the total sum of squares. In this example, weights for the eight options are given by $\mathbf{x}$, but the scores for the eight subjects, $\mathbf{y}_r$, are missing. As mentioned earlier, $\mathbf{y}_r$ may be important for further statistical analysis. If so, one can calculate $\mathbf{y}_r$ from $\mathbf{x}$, A_r, and data matrix F by formula (7.5).

The above problem can also be handled in terms of projection operators in lieu of design matrices A_r and A_c. Since projection operators may be preferred to design matrices in some cases, let us discuss this second approach in detail using the same example. As explained in Appendix A, one can decompose $\mathbf{z}_r$ into orthogonal components due to μ, α, β, γ, and e by means of projection operators. To obtain such decompositions, we first partition the complete design matrix of (7.8) as follows:

$$A_r = (A_\mu, A_\alpha, A_\beta, A_\gamma), \tag{7.12}$$

where $A_\mu, A_\alpha, A_\beta, A_\gamma$ consist of the first column, the next two columns, the next two columns, and the last four columns of A_r, respectively. Define hat matrices H_t by

$$H_t = A_t(A_t'A_t)^{-1}A_t', \tag{7.13}$$

where $t = \mu, \alpha, \beta, \gamma$. Also define orthogonal (disjoint) projection operators P_t by

$$\begin{aligned} P_\mu &= H_\mu = \mathbf{1}(\mathbf{1}'\mathbf{1})^{-1}\mathbf{1}', \\ P_\alpha &= H_\alpha - H_\mu, \\ P_\beta &= H_\beta - H_\mu, \\ P_\gamma &= H_\gamma - H_\alpha - H_\beta + H_\mu, \\ P_e &= I - H_\gamma. \end{aligned} \tag{7.14}$$

In terms of these projection operators, $\mathbf{z}_r$ can be expressed as

$$\mathbf{z}_r = P_\mu \mathbf{z}_r + P_\alpha \mathbf{z}_r + P_\beta \mathbf{z}_r + P_\gamma \mathbf{z}_r + P_e \mathbf{z}_r. \tag{7.15}$$

Note that all the hat matrices, hence all the projection operators, are square and of the same dimension. In our example, they are 8×8. If one wishes to determine weights so as to maximize the difference between α_1 and α_2, and between β_1 and β_2, one simply needs to include the corresponding projection operators in the specification of the design matrix, namely,

$$A_r = P_\alpha + P_\beta. \tag{7.16}$$

In this way A_r can be specified for any combination of effects of interest by simply adding the corresponding projection operators. In this approach, the dimension of E, that is, $A_r'F$, is the same as that of F since P_t are square. Similarly, one can construct the design matrix for the columns, A_c, in terms of projection operators. Thus, E can be $A_r'F$, FA_c, or $A_r'FA_c$, depending on the case. The dimension of E, however, is always the same as that of F, that is $N \times M$.

When A_r and A_c are expressed in terms of projection operators, (7.5) and (7.6) are no longer correct formulas for scores for the rows and the columns of F, associated with solution vectors $\mathbf{x}$ and $\mathbf{y}$ for E. The appropriate formulas for this second approach are

$$\mathbf{y}_r = D_N{}^{-1}F\mathbf{x}/\eta, \tag{7.17}$$

$$\mathbf{x}_c = D^{-1}F'\mathbf{y}/\eta. \tag{7.18}$$

Although E has the same dimension as F, it is used only as a transitional matrix to obtain quantities which optimize a stated criterion.

There are some differences between the two approaches discussed above. First, the dimension of E generated by projection operators is greater than that of E generated by the design matrix. Therefore the use of projection operators means working with a larger matrix than in the first approach, and this may be a disadvantage. Second, when projection operators are used, the dimension of E is the same as that of F, whatever the choice of the effects to be maximized, while the dimension of E is generally affected in the first approach by the choice of the effects. The constancy of the dimension of E may be a computational advantage for the approach using projection operators. Third, the design matrix used in the first approach does not involve negative numbers, but projection operators, except P_μ, have negative elements. As was seen in the discussion of Nishisato's method for paired comparison and rank order data in Chapter 6, negative elements in matrix E create the problem of redefining two diagonal matrices, D_N and D. Fourth, when one wants to maximize the effects of interactions, for instance, one would note that in general the two approaches provide different results. The discrepancy in the results arises from the fact that projection operators are defined in reference to orthogonal subspaces, but the partitioned submatrices of A are not. In most situations, one would be interested in maximizing independent contributions of the chosen effects. Then the solution can be obtained by the approach using projection operators. The next section will present numerical examples to illustrate the two approaches and their relations.

7.2 APPLICATIONS

Two examples will be presented. The first will explore the similarities and differences of the two approaches in terms of the scale values of the chosen effects. The second will illustrate the effects of specific maximization schemes, as reflected in the sum of squares of the chosen effect.

TABLE 7.1

Groups*		Item 1: 1	2	3	Item 2: 1	2	3	Item 3: 1	2	3	4
A_1	B_1	1	0	0	1	0	0	1	0	0	0
		1	0	0	0	0	1	0	1	0	0
		0	1	0	0	1	0	0	0	0	1
	B_2	0	0	1	0	0	1	0	0	0	1
		0	0	1	0	0	1	0	0	1	0
		0	1	0	0	1	0	1	0	0	0
A_2	B_1	0	1	0	0	1	0	0	1	0	0
		0	1	0	0	0	1	0	0	1	0
		0	1	0	0	1	0	0	0	1	0
	B_2	0	0	1	0	1	0	0	0	0	1
		0	0	1	0	1	0	0	1	0	0
		0	0	1	0	0	1	0	1	0	0

(Multiple-choice items 1, 2, 3)

*A_1 = smoker, A_2 = non-smoker; B_1 = alcoholic, B_2 = non-alcoholic.

7.2.1 *Comparisons of scale values*

Consider a small example, in which three subjects were randomly chosen according to the 2×2 crossed design, say (smoker, non-smoker)$\times$(alcoholic, non-alcoholic). Suppose that the subjects answered three multiple-choice questions, and that the responses were summarized as in Table 7.1. Data matrix F is the 12×10 matrix of 1's and 0's, arranged as in Table 7.1, and complete design matrix A_r is given by

$$A_r = \begin{bmatrix} 1 & 1 & 0 & 1 & 0 & 1 & 0 & 0 & 0 \\ 1 & 1 & 0 & 1 & 0 & 1 & 0 & 0 & 0 \\ 1 & 1 & 0 & 1 & 0 & 1 & 0 & 0 & 0 \\ 1 & 1 & 0 & 0 & 1 & 0 & 1 & 0 & 0 \\ 1 & 1 & 0 & 0 & 1 & 0 & 1 & 0 & 0 \\ 1 & 1 & 0 & 0 & 1 & 0 & 1 & 0 & 0 \\ 1 & 0 & 1 & 1 & 0 & 0 & 0 & 1 & 0 \\ 1 & 0 & 1 & 1 & 0 & 0 & 0 & 1 & 0 \\ 1 & 0 & 1 & 1 & 0 & 0 & 0 & 1 & 0 \\ 1 & 0 & 1 & 0 & 1 & 0 & 0 & 0 & 1 \\ 1 & 0 & 1 & 0 & 1 & 0 & 0 & 0 & 1 \\ 1 & 0 & 1 & 0 & 1 & 0 & 0 & 0 & 1 \end{bmatrix}. \tag{7.19}$$

$$\mu \qquad \underbrace{\quad}_{\alpha} \qquad \underbrace{\quad}_{\beta} \qquad \underbrace{\qquad\quad}_{\gamma}$$

For both approaches it is convenient to partition A_r into the form of (7.12), where A_μ is the 12×1 vector of 1's, A_α and A_β are 12×2 matrices, consisting of columns 2,3 and 4,5 of A_r, respectively, and A_γ is the 12×4 matrix consisting of columns 6,7,8,9 of A_r. If we indicate by $\mathbf{1}_p$ the $p\times1$ vector of 1's, the partitioned design matrices are

$$A_\mu = \mathbf{1}_{12},$$

$$A_\alpha = \begin{bmatrix} \mathbf{1}_6 & \mathbf{0} \\ \mathbf{0} & \mathbf{1}_6 \end{bmatrix}, \quad A_\beta = \begin{bmatrix} \mathbf{1}_3 & \mathbf{0} \\ \mathbf{0} & \mathbf{1}_3 \\ \mathbf{1}_3 & \mathbf{0} \\ \mathbf{0} & \mathbf{1}_3 \end{bmatrix}, \quad A_\gamma = \begin{bmatrix} \mathbf{1}_3 & \mathbf{0} & \mathbf{0} & \mathbf{0} \\ \mathbf{0} & \mathbf{1}_3 & \mathbf{0} & \mathbf{0} \\ \mathbf{0} & \mathbf{0} & \mathbf{1}_3 & \mathbf{0} \\ \mathbf{0} & \mathbf{0} & \mathbf{0} & \mathbf{1}_3 \end{bmatrix},$$

where $\mathbf{0}$ is the null vector of the same dimension as $\mathbf{1}_p$. The first approach employs matrices A_t or combinations of them for A_r. The second approach requires H_t of (7.13) and P_t of (7.15). For the present example, matrices H_t are

$$H_\mu = A_\mu(A_\mu' A_\mu)^{-1} A_\mu' = \mathbf{1}_{12}\mathbf{1}_{12}'/12,$$

$$H_\alpha = A_\alpha(A_\alpha' A_\alpha)^{-1} A_\alpha' = \frac{1}{6}\begin{bmatrix} \mathbf{1}_6\mathbf{1}_6' & \mathbf{0}\mathbf{0}' \\ \mathbf{0}\mathbf{0}' & \mathbf{1}_6\mathbf{1}_6' \end{bmatrix},$$

and similarly

$$H_\beta = \frac{1}{6}\begin{bmatrix} \mathbf{1}_3\mathbf{1}_3' & \mathbf{0}\mathbf{0}' & \mathbf{1}_3\mathbf{1}_3' & \mathbf{0}\mathbf{0}' \\ \mathbf{0}\mathbf{0}' & \mathbf{1}_3\mathbf{1}_3' & \mathbf{0}\mathbf{0}' & \mathbf{1}_3\mathbf{1}_3' \\ \mathbf{1}_3\mathbf{1}_3' & \mathbf{0}\mathbf{0}' & \mathbf{1}_3\mathbf{1}_3' & \mathbf{0}\mathbf{0}' \\ \mathbf{0}\mathbf{0}' & \mathbf{1}_3\mathbf{1}_3' & \mathbf{0}\mathbf{0}' & \mathbf{1}_3\mathbf{1}_3' \end{bmatrix},$$

$$H_\gamma = \frac{1}{3}\begin{bmatrix} \mathbf{1}_3\mathbf{1}_3' & \mathbf{0}\mathbf{0}' & \mathbf{0}\mathbf{0}' & \mathbf{0}\mathbf{0}' \\ \mathbf{0}\mathbf{0}' & \mathbf{1}_3\mathbf{1}_3' & \mathbf{0}\mathbf{0}' & \mathbf{0}\mathbf{0}' \\ \mathbf{0}\mathbf{0}' & \mathbf{0}\mathbf{0}' & \mathbf{1}_3\mathbf{1}_3' & \mathbf{0}\mathbf{0}' \\ \mathbf{0}\mathbf{0}' & \mathbf{0}\mathbf{0}' & \mathbf{0}\mathbf{0}' & \mathbf{1}_3\mathbf{1}_3' \end{bmatrix}.$$

To specify projection operators, P_t, it is convenient, from the computational point of view, to set

$$H_\mu \mathbf{z}_r = \mathbf{0}. \tag{7.20}$$

This has the same effect as deleting A_μ in the first approach, which automatically scales the weights so that the sum of the weighted responses is zero. The adoption of (7.20) has also the effect of changing otherwise negative elements of P_t for main (i.e., non-interaction) effects to non-negative. Under this condition, our projection operators are

$$P_\alpha = H_\alpha - H_\mu = H_\alpha, \qquad P_\beta = H_\beta - H_\mu = H_\beta.$$

For the projection operator for the interaction, one can use $-P_\gamma$, instead of P_γ, under (7.20) to avoid negative elements, for

$$-P_\gamma = -H_\gamma + H_\alpha + H_\beta - H_\mu = -H_\gamma + H_\alpha + H_\beta$$

$$= \frac{1}{6}\begin{bmatrix} \mathbf{0}\mathbf{0}' & \mathbf{1}_3\mathbf{1}_3' & \mathbf{1}_3\mathbf{1}_3' & \mathbf{0}\mathbf{0}' \\ \mathbf{1}_3\mathbf{1}_3' & \mathbf{0}\mathbf{0}' & \mathbf{0}\mathbf{0}' & \mathbf{1}_3\mathbf{1}_3' \\ \mathbf{1}_3\mathbf{1}_3' & \mathbf{0}\mathbf{0}' & \mathbf{0}\mathbf{0}' & \mathbf{1}_3\mathbf{1}_3' \\ \mathbf{0}\mathbf{0}' & \mathbf{1}_3\mathbf{1}_3' & \mathbf{1}_3\mathbf{1}_3' & \mathbf{0}\mathbf{0}' \end{bmatrix}.$$

In the present example, we consider substructures only with respect to the rows of F. Therefore the design matrix for the column structure, A_c, is nothing but the identity matrix, which leads to the specification

$$E = A_r'F. \tag{7.21}$$

Let us construct this product matrix for several cases of interest, using the two approaches.

Case A. *Maximization of the effect of* α The design matrix for this case is given by A_α. Thus, for the first approach, we obtain

$$E_1 = A_\alpha'F = \begin{bmatrix} 2 & 2 & 2 & 1 & 2 & 3 & 2 & 1 & 1 & 2 \\ 0 & 3 & 3 & 0 & 4 & 2 & 0 & 3 & 2 & 1 \end{bmatrix}.$$

The product matrix for the second approach is given by

$$E_2 = P_\alpha'F = \frac{1}{6}\begin{bmatrix} 2 & 2 & 2 & 1 & 2 & 3 & 2 & 1 & 1 & 2 \\ 2 & 2 & 2 & 1 & 2 & 3 & 2 & 1 & 1 & 2 \\ 2 & 2 & 2 & 1 & 2 & 3 & 2 & 1 & 1 & 2 \\ 2 & 2 & 2 & 1 & 2 & 3 & 2 & 1 & 1 & 2 \\ 2 & 2 & 2 & 1 & 2 & 3 & 2 & 1 & 1 & 2 \\ 2 & 2 & 2 & 1 & 2 & 3 & 2 & 1 & 1 & 2 \\ 0 & 3 & 3 & 0 & 4 & 2 & 0 & 3 & 2 & 1 \\ 0 & 3 & 3 & 0 & 4 & 2 & 0 & 3 & 2 & 1 \\ 0 & 3 & 3 & 0 & 4 & 2 & 0 & 3 & 2 & 1 \\ 0 & 3 & 3 & 0 & 4 & 2 & 0 & 3 & 2 & 1 \\ 0 & 3 & 3 & 0 & 4 & 2 & 0 & 3 & 2 & 1 \\ 0 & 3 & 3 & 0 & 4 & 2 & 0 & 3 & 2 & 1 \end{bmatrix}.$$

Note that the two rows of E_1, divided by six, are each repeated six times in E_2. Since collinear (proportional) response vectors receive identical scores in dual scaling, we can conclude that E_1 and E_2, hence the two approaches, provide the same optimal weight vector for the 10 options. The squared correlation ratio and the optimal weight vector for E_1 and E_2 are

$$\eta^2 = 0.22037\ (\delta = 100\%), \qquad \mathbf{x} = \begin{bmatrix} 2.1302 \\ -0.4260 \\ -0.4260 \\ 2.1302 \\ -0.7101 \\ 0.4260 \\ 2.1302 \\ -1.0651 \\ -0.7101 \\ 0.7101 \end{bmatrix}. \tag{7.22}$$

Case B. *Maximization of the effect of* β Matrix E_1 is given by

$$E_1 = A_\beta{}'F = \begin{bmatrix} 2 & 4 & 0 & 1 & 3 & 2 & 1 & 2 & 2 & 1 \\ 0 & 1 & 5 & 0 & 3 & 3 & 1 & 2 & 1 & 2 \end{bmatrix}.$$

E_2 is given by $P_\beta{}'F$, the rows of which consist of vectors proportional to the rows of E_1. More specifically, rows 1,2,3 and 7,8,9 of E_2 are all equal to the first row of E_1 divided by six, and rows 4,5,6 and 10,11,12 of E_2 are all equal to the second row of E_1 divided by six. For the same reason as in case A, the two approaches provide identical results. The numerical results are

$$\eta^2 = 0.29630\ (\delta = 100\%), \qquad \mathbf{x} = \begin{bmatrix} 1.8371 \\ 1.1023 \\ -1.8371 \\ 1.8371 \\ 0.0000 \\ -0.3674 \\ 0.0000 \\ 0.0000 \\ 0.6124 \\ -0.6124 \end{bmatrix}. \tag{7.23}$$

Case C. *Maximization of the effects of* α *and* β We have so far seen that the two approaches provide identical results. However, as soon as we consider combinations of effects, some discrepancies between the two emerge. In the first approach, the product matrix is given by

$$E_1 = \begin{bmatrix} A_\alpha{}' \\ A_\beta{}' \end{bmatrix} F = \begin{bmatrix} A_\alpha{}'F \\ A_\beta{}'F \end{bmatrix}$$

$$= \begin{bmatrix} 2 & 2 & 2 & 1 & 2 & 3 & 2 & 1 & 1 & 2 \\ 0 & 3 & 3 & 0 & 4 & 2 & 0 & 3 & 2 & 1 \\ 2 & 4 & 0 & 1 & 3 & 2 & 1 & 2 & 2 & 1 \\ 0 & 1 & 5 & 0 & 3 & 3 & 1 & 2 & 1 & 2 \end{bmatrix}.$$

Note that E_1 is nothing but a partitioned matrix, consisting of the two submatrices used in cases A and B. E_2 is constructed in a way quite different from E_1, for E_2 is a sum of matrices,

$$E_2 = (P_\alpha' + P_\beta')F = P_\alpha' F + P_\beta' F$$

$$= \frac{1}{6}\begin{bmatrix} 4 & 6 & 2 & 2 & 5 & 5 & 3 & 3 & 3 & 3 \\ 4 & 6 & 2 & 2 & 5 & 5 & 3 & 3 & 3 & 3 \\ 4 & 6 & 2 & 2 & 5 & 5 & 3 & 3 & 3 & 3 \\ 2 & 3 & 7 & 1 & 5 & 6 & 3 & 3 & 2 & 4 \\ 2 & 3 & 7 & 1 & 5 & 6 & 3 & 3 & 2 & 4 \\ 2 & 3 & 7 & 1 & 5 & 6 & 3 & 3 & 2 & 4 \\ 2 & 7 & 3 & 1 & 7 & 4 & 1 & 5 & 4 & 2 \\ 2 & 7 & 3 & 1 & 7 & 4 & 1 & 5 & 4 & 2 \\ 2 & 7 & 3 & 1 & 7 & 4 & 1 & 5 & 4 & 2 \\ 0 & 4 & 8 & 0 & 7 & 5 & 1 & 5 & 3 & 3 \\ 0 & 4 & 8 & 0 & 7 & 5 & 1 & 5 & 3 & 3 \\ 0 & 4 & 8 & 0 & 7 & 5 & 1 & 5 & 3 & 3 \end{bmatrix}.$$

Note that the first three rows of E_2 are the sum of rows 1 and 3 of E_1. Likewise, the other three distinct sets of three identical rows of E_2 are the sums of rows 1 and 4, rows 2 and 3, and rows 2 and 4 of E_1. Note that none of the rows of E_2 is proportional to any row of E_1. Despite this apparent difference between E_1 and E_2, however, the two matrices provide the same optimal vector of the 10 option weights. The squared correlation ratios for E_1 and E_2 are different, and η^2 for E_2 is exactly one-half that for E_1. The results are:

for E_1, $\eta^2 = 0.1692$ ($\delta = 100\%$);

for E_2, $\eta^2 = 0.0846$ ($\delta = 100\%$);

$$\text{for } E_1 \text{ and } E_2,\ \mathbf{x} = \begin{bmatrix} 2.3575 \\ 0.7096 \\ -1.6526 \\ 2.3575 \\ -0.2937 \\ -0.1191 \\ 0.8811 \\ -0.4406 \\ 0.1984 \\ -0.1984 \end{bmatrix}. \qquad (7.24)$$

Case D. *Maximization involving* γ It is interesting to know that once γ is involved in maximization the two approaches lead to different results. In the present example, the cases which involve γ are those of γ, $\alpha+\gamma$, $\beta+\gamma$, and

$\alpha+\beta+\gamma$. Instead of demonstrating discrepancies of various comparisons, we shall simply present one case in which the two approaches provide the same results. When combination $\alpha+\beta+\gamma$ is chosen for maximization, the projection operator is given by

$$P_\alpha + P_\beta + P_\gamma = \frac{1}{6}\begin{bmatrix} \mathbf{1}_3\mathbf{1}_3' & \mathbf{0}\mathbf{0}' & \mathbf{0}\mathbf{0}' & \mathbf{0}\mathbf{0}' \\ \mathbf{0}\mathbf{0}' & \mathbf{1}_3\mathbf{1}_3' & \mathbf{0}\mathbf{0}' & \mathbf{0}\mathbf{0}' \\ \mathbf{0}\mathbf{0}' & \mathbf{0}\mathbf{0}' & \mathbf{1}_3\mathbf{1}_3' & \mathbf{0}\mathbf{0}' \\ \mathbf{0}\mathbf{0}' & \mathbf{0}\mathbf{0}' & \mathbf{0}\mathbf{0}' & \mathbf{1}_3\mathbf{1}_3' \end{bmatrix}.$$

This matrix consists of the columns of A_γ, each repeated three times and divided by six. Thus, using the same reasoning as in cases A and B, we can conclude that the design matrices, A_γ and $P_\alpha+P_\beta+P_\gamma$, used in the first and the second approaches, respectively, provide the same results.

The above comparisons are focused only on similarities and differences of the two approaches. Let us now look at the effects of the choice of particular design matrices.

7.2.2 *Comparisons of sums of squares*

Let us use the same example (Table 7.1) and specify the vector of scores for the 12 subjects, $\mathbf{y}$, simply by $F\mathbf{x}$. Consider the following design matrices, A_r, to generate matrix E of (7.21):

(a) A_α or P_α,

(b) A_β or P_β,

(c) (A_α, A_β) or $P_\alpha + P_\beta$,

(d) A_γ or $P_\alpha + P_\beta + P_\gamma$,

(e) $(A_\alpha, A_\beta, A_\gamma)$,

(f) P_γ.

In the first four cases, 'or' indicates that the two matrices lead to the same results. The score vectors obtained under these different maximization schemes are summarized in Table 7.2. The corresponding 2×2 tables of cell totals and marginals are given in Table 7.3. Although Table 7.3 shows some of the effects of maximization, it is not always clear. A further step is thus taken to calculate the sums of squares due to α, β, and γ, using the data in Table 7.1. Results are listed in Table 7.4. Since the total sum of squares is not the same for all the cases, direct comparisons of the different schemes may be difficult. To mitigate this difficulty, the second table is included in Table 7.4, with all the total sums of squares adjusted to 100. Some of the interesting

TABLE 7.2

Scores for 12 subjects, obtained under different maximization schemes

		(a)*	(b)	(c)	(d)	(e)	(f)
A_1	B_1	6.39	3.67	5.60	6.86	5.93	3.89
		1.49	1.47	1.80	2.65	1.97	2.53
		−0.43	0.49	0.22	−0.23	0.15	−0.52
	B_2	0.71	−2.82	−1.97	−0.66	−1.82	0.65
		−0.71	−1.59	−1.57	−2.03	−1.68	−1.94
		0.99	1.10	1.30	0.48	1.21	−1.17
A_2	B_1	−2.20	1.10	−0.02	−0.70	−0.13	−0.19
		−0.71	1.35	0.78	−1.37	0.49	−3.50
		−1.85	1.71	0.61	−1.60	0.30	−3.11
	B_2	−0.42	−2.45	−2.14	−0.89	−2.01	1.04
		−2.20	−1.84	−2.39	−1.36	−2.30	1.36
		−1.07	−2.20	−2.21	−1.13	−2.11	0.97

*Design matrices used for these maximization schemes are: (a) A_α and P_α, (b) A_β and P_β, (c) (A_α, A_β) and $P_\alpha + P_\beta$, (d) A_γ and $P_\alpha + P_\beta + P_\gamma$, (e) $(A_\alpha, A_\beta, A_\gamma)$, (f) P_γ.

points are:

1. A_α and P_α maximize the sum of squares due to α, and A_β and P_β maximize the sum of squares due to β.
2. (A_α, A_β) and $P_\alpha + P_\beta$ maximize the sums of squares due to α and β.
3. P_γ maximizes the sum of squares due to the interaction, that is, γ.
4. A_γ and $P_\alpha + P_\beta + P_\gamma$ maximize the sums of squares due to α, β, and γ.
5. The meaning of $(A_\alpha, A_\beta, A_\gamma)$ is not immediately obvious.

More empirical work is needed for applications of the two approaches. The projection operator provides results which are interpretable in the context of the analysis of variance, but this approach is not always preferred to the other. First, the dimension of P_t may be unnecessarily large in comparison with its rank. Second, related to the first problem is the problem of redundant computation, which in some cases may no longer be negligible. Third, the elements of P_t may be negative, which would require a redefinition of D_N and D for E. Regarding the use of design matrices A_t, it has computational advantages over that of P_t, but the meaning of maximization with A_t is not always clear.

TABLE 7.3

2×2 tables of cell totals

(a)

	A_1	A_2	
B_1	7.46	−4.76	2.70
B_2	0.99	−3.69	−2.70
	8.45	−8.45	

(b)

	A_1	A_2	
B_1	5.63	4.17	9.80
B_2	−3.31	−6.49	−9.80
	2.32	−2.32	

(c)

	A_1	A_2	
B_1	7.61	1.38	8.99
B_2	−2.25	−6.74	−8.99
	5.36	−5.36	

(d)

	A_1	A_2	
B_1	9.28	−3.68	5.60
B_2	−2.21	−3.39	−5.60
	7.07	−7.07	

(e)

	A_1	A_2	
B_1	8.05	0.65	8.70
B_2	−2.28	−6.42	−8.70
	5.77	−5.77	

(f)

	A_1	A_2	
B_1	5.89	−6.80	−0.91
B_2	−2.46	3.37	0.91
	3.43	−3.43	

When the columns of F have substructures, design matrix A_c is required. Such an example was already given in relation to the paired comparison data in Chapter 6. Another interesting example of this type will be fully discussed in Chapter 8. To combine A_r and A_c into the general form of (7.4) is a straightforward matter, at least conceptually. In this case, the two design matrices pick out from F those responses which are jointly related to the combinations of the two sets of structural parameters. Thus, matrix E can be considered as a matrix of joint responses associated with the parameters of interest.

The general strategy adopted here was proposed by Nishisato (1976). His model opened an avenue for dual scaling of data matrices with negative elements. Since then some empirical research has been carried out (e.g., Poon 1977), and a slightly different, theoretical study has also been conducted

TABLE 7.4

Sums of squares due to α, β, and γ

		Maximization schemes					
		(a)	(b)	(c)	(d)	(e)	(f)
Between	α	23.80	1.79	9.58	16.66	11.10	3.92
	β	2.43	32.01	26.94	10.45	25.23	0.28
	γ	4.74	0.25	0.25	11.57	0.88	28.58
Within		29.19	13.74	22.06	29.12	23.55	20.34
Total		60.16	47.79	58.83	67.80	60.76	53.12

Adjusted to $ss_t = 100$, and rounded

		(a)	(b)	(c)	(d)	(e)	(f)
Between	α	40	3	16	25	18	7
	β	4	67	46	15	42	1
	γ	8	1	0	17	1	54
Within		48	29	38	43	39	38
Total		100	100	100	100	100	100

(McDonald, Torii, and Nishisato 1979). However, it seems that much more empirical work is necessary to put this general scheme into wide practical use. Construction of design matrices, construction of projection operators, handling of negative elements in input matrices, and specification of diagonal matrices D_N and D may be major problems when a complex multidimensional table is to be analysed. Statistical evaluation is another topic which has to await further research.

8

Miscellaneous topics

8.1 TREATMENT OF ORDERED CATEGORIES

Dual scaling has so far been applied to a wide variety of cases, without any order constraints on the derived weights (scores). There are occasions, however, when the categories to be weighted are clearly ordered (e.g., always > often > sometimes > rarely > never; excellent > good > mediocre > poor), or when the investigator has prior knowledge of partially ordered categories (e.g., multiple-choice options in a mathematics test). In these cases, our natural expectation would be that the order of 'optimal' weights for the categories should conform to the a priori order.

Let us consider one example. In preparing multiple-choice items for a mathematics test, the item writer generally has some ideas about the order of correctness of the response options. For example, he may know only the correct option, or the best and the second best options, or the complete meritorious order of the options. In general, item writers try to prepare 'good distractors' in relation to the right answers, and distractors undoubtedly have decisive effects on the difficulty level of a particular item. In spite of great efforts on the part of the item writers, however, dual scaling does not guarantee that the scaled weights for the response options will conform to the expected order. The discrepancy between the order of optimally scaled options and that of the investigator's a priori judgments does not necessarily imply that the investigator's judgment is poor, for it is easy to find examples of clear-cut ordered categories which produce disordered optimal weights. What is needed then is a way to incorporate a particular set of order constraints in dual scaling. From a practical point of view, the use of order constraints on ordered categories can be considered as a way to extract more interpretable information, and possibly to increase the validity of the scores. From a technical point of view, it implies that the space becomes 'tighter' by confining the configuration of the variables (options) within the permissible region.

Optimization problems with order constraints have been extensively investigated by those engaged in operations research. A variety of linear and non-linear programming procedures have been proposed for the problems. Other algorithms which handle order constraints have also been investigated, such as least-squares monotone regression or isotonic regression (Ayer, Brunk, Ewing, Reid, and Silverman 1955; Bartholomew 1959, 1961; Barton and Mallows 1961; Barlow, Bartholomew, Bremner, and Brunk 1972; Barlow and Brunk 1972; R.M. Johnson 1975; Kruskal 1971; Miles 1959). These algorithms have been in routine use in non-metric multidimensional scaling (e.g., Kruskal 1964a, b, 1965; Kruskal and Carmone 1969; Kruskal and Carroll 1969; Young and Torgerson 1967). Within the general framework of dual scaling, the constraint of the complete order of categories (options) has been discussed in at least seven papers (Bradley, Katti, and Coons 1962; Kruskal 1965; Nishisato and Inukai 1972; de Leeuw 1973; Nishisato and Arri 1975; de Leeuw, Young, and Takane 1976; Nishisato 1978b) and that of partial order in two papers (Nishisato 1973b; Nishisato and Arri 1975). Here partial order means that order relations are specified only among some of the elements in the weight vector. Of the many methods, we shall discuss only two in detail, which can easily be implemented within the framework of dual scaling as discussed in previous chapters.

8.1.1 *Method of successive data modifications* (SDM)

This method is due to Nishisato (1973b), and its applications are discussed in Nishisato and Arri (1975). Some of the advantages of this method over others are: (a) it is very simple to use; (b) it is computationally fast; (c) it can handle both partial and complete orders of categories, and (d) it always provides a non-trivial solution.

Let F be the $N \times m$ data matrix. Suppose that one wishes to impose k non-redundant pairwise constraints on the $m \times 1$ weight vector, $\mathbf{x}$. For example, one may wish to specify only the best (x_1) and the second best (x_2) of the five options, which we indicate by

$$x_1 \geq x_2 \geq \{x_3, x_4, x_5\}. \tag{8.1}$$

In this case, non-redundant pairwise constraints are

$$x_1 \geq x_2, \qquad x_2 \geq x_3, \qquad x_2 \geq x_4, \qquad x_2 \geq x_5.$$

The other relations such as $x_1 \geq x_j$, $j = 3, 4, 5$, are redundant because these relations are implied by the relation $x_1 \geq x_2$. Define $k \times m$ matrix T such that $T\mathbf{x} \geq \mathbf{0}$ when k pairwise constraints are satisfied. In our example of (8.1), we obtain the four non-redundant pairwise constraints

$$x_1 - x_2 \geq 0, \qquad x_2 - x_3 \geq 0, \qquad x_2 - x_4 \geq 0, \qquad x_2 - x_5 \geq 0.$$

Thus matrix T is 4×5 and is given by

$$T = \begin{bmatrix} 1 & -1 & 0 & 0 & 0 \\ 0 & 1 & -1 & 0 & 0 \\ 0 & 1 & 0 & -1 & 0 \\ 0 & 1 & 0 & 0 & -1 \end{bmatrix}, \tag{8.2}$$

with $\mathbf{x}'$ being $(x_1, x_2, x_3, x_4, x_5)$. The main reason why pairwise relations are restricted to non-redundant ones is simply to keep the number of rows of T relatively small for computational efficiency. However, this has no bearing on the outcome of the computation. In other words, one can include redundant pairwise relations in T without affecting the solution.

The procedure for the method of successive data modifications (hereafter abbreviated as SDM) can be described as follows:

1. Subject data matrix F to the standard procedure of dual scaling, and obtain the optimal weight vector, $\mathbf{x}$, without order constraints. Test to see if

$$T\mathbf{x} \geq \mathbf{0}. \tag{8.3}$$

If (8.3) is satisfied, $\mathbf{x}$ is the solution vector and analysis ends at this point. If (8.3) is not satisfied, follow the procedure described in step 2.

2. Let $T\mathbf{x}=\mathbf{t}=(t_j)$, and t_s be the smallest number (i.e., the negative number with the largest absolute value) in $\mathbf{t}$. Suppose that $t_s = x_p - x_q$. Although the constraint was $t_s \geq 0$, that is, $x_p \geq x_q$, the result was $t_s < 0$, that is, $x_p < x_q$. The smallest possible adjustment necessary for observed x_p and x_q to satisfy the constraint is to set $x_p = x_q$. The SDM attains this equality by modifying the data matrix itself. There are a number of ways to modify F so that x_p becomes equal to x_q, but there seem to be only two distinct classes, the first being to merge columns p and q of F, and the second to make column p proportional to column q. Of these possibilities, the one which seems to be the simplest from the computational (programming) point of view is that referred to as Nishisato's method in Nishisato and Arri (1975). Nishisato's method does not disturb the dimension of F, but simply modifies columns p and q as follows:

$$\textit{Modified } f_{ip} = \textit{Modified } f_{iq} = (f_{ip} + f_{iq})/2, \tag{8.4}$$

$$i = 1, 2, \ldots, N.$$

In other words, columns p and q are now replaced by their averages. The modified data matrix, which we indicate by F^*, is now subjected to the standard procedure of dual scaling.

The above transformation of the data matrix guarantees that the new weights, x_p and x_q, are equal. Once $\mathbf{x}$ is obtained, test to see if (8.3) is satisfied. If the result is affirmative, $\mathbf{x}$ is the SDM solution that satisfies the given order constraints. If one or more negative values are obtained in $T\mathbf{x}$, F^* is further modified by (8.4), and the modified data matrix is subjected to dual

TABLE 8.1

Order categories (Bradley, Katti, and Coons 1962, p. 366, Example 3)

Treatment	1*	2	3	4	5	Total
1	9	5	9	13	4	40
2	7	3	10	20	4	44
3	14	13	6	7	0	40
4	11	15	3	5	8	42
5	0	2	10	30	2	44
Total	41	38	38	75	18	210

*Categories: 1 = excellent, 2 = good, 3 = fair, 4 = poor, 5 = terrible.

scaling. This process is continued until all the constraints are satisfied. To illustrate the SDM procedure, let us look at a numerical example.

Example 8.1 (*Bradley, Katti, and Coons* 1962) The data in Table 8.1 are from the study by Bradley, Katti, and Coons (1962, p. 366, Example 3), and the following SDM analyses are based on the study by Nishisato and Arri (1975). From Table 8.1, we obtain 5×5 data matrix F, $\mathbf{f}' = (41, 38, 38, 75, 18)$, $D_N = \text{diag}(40, 44, 40, 42, 44)$, $N = m = 5$, and $f_t = 210$. The matrix involved in the eigenequation is given by

$$C = D^{-1/2} F' D_N{}^{-1} F D^{-1/2} - D^{1/2} \mathbf{1}\mathbf{1}' D^{1/2} / f_t \tag{8.5}$$

$$= \begin{bmatrix} 0.07109 & & & & \text{(symmetric)} \\ 0.06744 & 0.09543 & & & \\ -0.02324 & -0.04193 & 0.02128 & & \\ -0.08614 & -0.09663 & 0.03979 & 0.11740 & \\ 0.00432 & 0.01773 & -0.01614 & -0.02706 & 0.04642 \end{bmatrix}. \tag{8.6}$$

Subjecting C to the standard procedure, we obtain

$$\eta^2 = 0.2844, \qquad \mathbf{x} = \begin{bmatrix} 1.0432 \\ 1.3047 \\ -0.5500 \\ -1.0658 \\ 0.4714 \end{bmatrix}. \tag{8.7}$$

Although the five categories of the original data are completely ordered, let us just forget this fact and consider the following partial order:

$$x_1 \geq \{x_2, x_3, x_4, x_5\}, \tag{8.8}$$

i.e., we only know that x_1 is the best option. The other weights are free to vary as long as they do not exceed x_1. In this case, matrix is given by

$$T = \begin{bmatrix} 1 & -1 & 0 & 0 & 0 \\ 1 & 0 & -1 & 0 & 0 \\ 1 & 0 & 0 & -1 & 0 \\ 1 & 0 & 0 & 0 & -1 \end{bmatrix}. \tag{8.9}$$

Then, using (8.7) and (8.9), we obtain

$$T\mathbf{x} = \begin{bmatrix} -0.2616 \\ 1.5932 \\ 2.1090 \\ 0.5718 \end{bmatrix}. \tag{8.10}$$

The first element of $T\mathbf{x}$, which is $x_1 - x_2$, is negative. Thus we modify columns 1 and 2 of F by (8.4), giving

$$F^* = \begin{bmatrix} 7 & 7 & 9 & 13 & 4 \\ 5 & 5 & 10 & 20 & 4 \\ 13.5 & 13.5 & 6 & 7 & 0 \\ 13 & 13 & 3 & 5 & 8 \\ 1 & 1 & 10 & 30 & 2 \end{bmatrix}. \tag{8.11}$$

Note that the first two columns of F^* are identical and equal to the average of the two columns of F. The matrix for the eigenequation, C, can be calculated, using F^* for F, by (8.5),

$$C = \begin{bmatrix} 0.07509 & & & & \text{(symmetric)} \\ 0.07509 & 0.07509 & & & \\ -0.03240 & -0.03240 & 0.02128 & & \\ -0.09127 & -0.09127 & 0.03979 & 0.11740 & \\ 0.01089 & 0.01089 & -0.01614 & -0.02706 & 0.04642 \end{bmatrix}. \tag{8.12}$$

Note that the first two columns (rows) are identical and that the submatrix for the last three categories is the same as before. Subjecting this matrix to the iterative procedure, we obtain

$$\eta^2 = 0.2827, \qquad \mathbf{x} = \begin{bmatrix} 1.1733 \\ 1.1733 \\ -0.5454 \\ -1.0711 \\ 0.4649 \end{bmatrix}. \tag{8.13}$$

and hence $T\mathbf{x} \geq \mathbf{0}$. Since all the elements of $T\mathbf{x}$ are non-negative (i.e., x_1 is not smaller than any other x), (8.13) is the SDM solution.

Suppose now that we wish to impose the constraints of the complete order, that is, $x_1 \geq x_2 \geq x_3 \geq x_4 \geq x_5$. In this case, non-redundant pairwise constraints are

$$x_1 - x_2 \geq 0, \qquad x_2 - x_2 \geq 0, \qquad x_3 - x_4 \geq 0, \qquad x_4 - x_5 \geq 0.$$

Therefore matrix T is given by

$$T = \begin{bmatrix} 1 & -1 & 0 & 0 & 0 \\ 0 & 1 & -1 & 0 & 0 \\ 0 & 0 & 1 & -1 & 0 \\ 0 & 0 & 0 & 1 & -1 \end{bmatrix}. \tag{8.14}$$

As before, we first obtain the optimal solution from the original data matrix (Table 8.1), which is in our example given by (8.7). Using this solution and (8.14), we obtain

$$T\mathbf{x} = \begin{bmatrix} -0.2616 \\ 1.8547 \\ 0.5159 \\ -1.5372 \end{bmatrix}. \tag{8.15}$$

There are two negative elements in $T\mathbf{x}$, and the smaller one is -1.5372, which corresponds to $x_4 - x_5$. Thus we modify columns 4 and 5 by (8.4) to obtain

$$F^* = \begin{bmatrix} 9 & 5 & 9 & 8.5 & 8.5 \\ 7 & 3 & 10 & 12 & 12 \\ 14 & 13 & 6 & 3.5 & 3.5 \\ 11 & 15 & 3 & 6.5 & 6.5 \\ 0 & 2 & 10 & 16 & 16 \end{bmatrix}. \tag{8.16}$$

Dual scaling of F^* provides the vectors

$$\mathbf{x} = \begin{bmatrix} 1.1566 \\ 1.4062 \\ -0.5599 \\ -0.8557 \\ -0.8557 \end{bmatrix}, \qquad T\mathbf{x} = \begin{bmatrix} -0.2495 \\ 1.9661 \\ 0.2958 \\ 0.0000 \end{bmatrix}. \tag{8.17}$$

The first element of $T\mathbf{x}$, which is $x_1 - x_2$, is negative, and it is now necessary to modify columns 1 and 2 by (8.4), resulting in the new matrix

$$F^* = \begin{bmatrix} 7 & 7 & 9 & 8.5 & 8.5 \\ 5 & 5 & 10 & 12 & 12 \\ 13.5 & 13.5 & 6 & 3.5 & 3.5 \\ 13 & 13 & 3 & 6.5 & 6.5 \\ 1 & 1 & 10 & 16 & 16 \end{bmatrix}. \tag{8.18}$$

Dual scaling of F^* gives

$$\eta^2 = 0.2436, \qquad \mathbf{x} = \begin{bmatrix} 1.2799 \\ 1.2799 \\ -0.5539 \\ -0.8609 \\ -0.8609 \end{bmatrix}, \tag{8.19}$$

so that

$$T\mathbf{x} = \begin{bmatrix} 0.0000 \\ 1.8338 \\ 0.3070 \\ 0.0000 \end{bmatrix} \geq \mathbf{0}. \tag{8.20}$$

(8.20) satisfies the constraints of the complete order, and is therefore the SDM solution.

There are a few points one should bear in mind in using the SDM procedure:

1. When $\mathbf{x}$ maximizes η^2, the reflected vector of $\mathbf{x}$, that is $-\mathbf{x}$, also maximizes η^2. This problem of reflection of vectors has an important bearing on the use of $T\mathbf{x}$ as the only check for the SDM solution. A recommended strategy then is to use $T\mathbf{x}$ if the number of non-negative elements of $T\mathbf{x}$ is greater than, or equal to, $k/2$, and to use the reflected vector $(-T\mathbf{x})$ otherwise, where k is the number of rows of T (i.e., the number of pairwise constraints).

2. The SDM procedure is normally restricted to only one solution. Under the current formulation, there exists only one solution, which satisfies the constraints of complete order. However, when one considers constraints of partial orders, there are elements which are free to vary. In consequence, one can consider extracting more than one SDM solution satisfying the constraint. However, all the SDM solutions then are likely to be highly correlated, and it does not seem obvious whether they are of great practical use.

3. Once the SDM solution is obtained, one can test the significance of η^2 and calculate the scores (weights) for the rows of F, using F instead of F^*, in the same way as discussed in Chapter 2.

4. When the order constraints are satisfied without any modification of F, the solutions are the stationary interior points of the original non-constrained problem. When F is modified to arrive at the SDM solution, some of the elements of the solution vector are on the boundaries of the constrained problem. The modified eigenequation ensures that the solution is optimum subject to being on the boundary. Empirically, it has been demonstrated (Nishisato and Arri 1975) that the SDM procedure yields the solution which is the same as the one obtained by a much more complex approach of non-linear programming.

The SDM procedure is easy to understand and simple to implement. The fact that it can handle any number and any type of order constraints may be regarded as one of its main advantages over some other techniques. Although the example used here is a contingency/response-frequency type, the SDM procedure can easily be applied to multiple-choice data. The SDM procedure has not been widely applied to practical problems and its empirical evaluation is yet to be made.

8.1.2 *Individual differences scaling of category boundaries and stimuli*

The method to be presented here was developed by Nishisato (1978b), and this is a dual scaling analogue of the method of successive categories, due to Thurstone (see, for example, Guilford 1954; Torgerson 1958; Bock and Jones 1968; Nishisato 1975). Unlike the traditional method of successive categories, however, the current method considers differential weights for responses from different subjects instead of averaging them out, and its procedure was developed in the same fashion as Nishisato's approach to paired comparison and rank order data, discussed in Chapter 6.

The current method differs from the SDM procedure in that it determines the values of category *boundaries* and not the weights for categories. Although the SDM procedure is so general that it can handle order constraints on several kinds of categorical data, the current method is restricted to the case in which all the stimuli are judged by all the subjects in terms of a *common* set of ordered categories.

Let us consider $m+1$ response categories which are successively ordered from category 1 up to category $m+1$. These categories are assumed to divide the judgmental continuum into $m+1$ sections. Let us indicate by τ_k the category boundary between category k and category $k+1$, $k=1,2,\ldots,m$. In total, there are m category boundaries. n stimuli are presented one by one to each of N subjects, and the subjects are asked to classify each stimulus into the most appropriate category. On the basis of their responses, our main task is to determine the optimal values of m category boundaries, τ_k, and scale values of n stimuli, μ_j. The current procedure also enables us to determine the optimal weights for N subjects.

$$
{}_if_{jk} = \begin{cases} 1 & \text{if subject } i \text{ classifies stimulus } j, X_j, \text{ into category } k \text{ or any preceding category,} \\ -1 & \text{if subject } i \text{ classifies } X_j \text{ into category } k+1 \text{ or any higher category,} \end{cases} \tag{8.21}
$$

$i = 1,2,\ldots,N; j = 1,2,\ldots,n;$

$k = 1,2,\ldots,m+1.$

If we indicate by $\hat{\tau}_k$ and $\hat{\mu}_j$ the perceived values of τ_k and μ_j by subject i, respectively, ${}_i f_{jk}$ of (8.21) can be written as

$$ {}_i f_{jk} = \begin{cases} 1 & \text{if } \hat{\tau}_k - \hat{\mu}_j > 0, \\ -1 & \text{if } \hat{\tau}_k - \mu_j < 0. \end{cases} \tag{8.22} $$

To illustrate the procedure, let us consider a small example as given in Table 8.2. Here seven subjects judged three stimuli in terms of three response categories, that is, $N=7$, $n=3$, $m=2$ ($m+1=3$ categories). If we express this table in terms of ${}_i f_{jk}$, we obtain Table 8.3. Let us indicate this data matrix by

$$ \underset{N \times n(m+1)}{F^*} = \left[F_1^*, F_2^*, \ldots, F_j^*, \ldots, F_n^* \right], \tag{8.23} $$

where

$$ \underset{N \times (m+1)}{F_j^*} = \begin{bmatrix} {}_1 f_{j1} & {}_1 f_{j2} & \cdots & {}_1 f_{j(m+1)} \\ {}_2 f_{j1} & {}_2 f_{j2} & \cdots & {}_2 f_{j(m+1)} \\ \vdots & \vdots & \ddots & \vdots \\ {}_N f_{j1} & {}_N f_{j2} & \cdots & {}_N f_{j(m+1)} \end{bmatrix}. \tag{8.24} $$

Since the elements of the last column of F_j^* are all 1's for all values of j, delete the last column of F_j^* and indicate the reduced matrix by F, where

$$ \underset{N \times nm}{F} = \left[F_1, F_2, \ldots, F_j, \ldots, F_n \right]. \tag{8.25} $$

Suppose that matrix F is subjected to dual scaling, and that we obtain the vector of weights for subjects (rows of F), $\mathbf{y}$, and the vector of weights for the

TABLE 8.2

Successive categories data in raw form, indicating choice ('x')

	Stimulus 1			Stimulus 2			Stimulus 3		
Subjects	1*	2	3	1	2	3	1	2	3
1		x		x					x
2	x				x			x	
3	x			x			x		
4			x			x			x
5		x			x		x		
6			x			x		x	
7	x			x				x	

*Categories.

TABLE 8.3

Successive categories data in coded form (${}_if_{jk}$)

Subjects	Stimulus 1			Stimulus 2			Stimulus 3		
	1*	2	3	1	2	3	1	2	3
1	−1	1	1	1	1	1	−1	−1	1
2	1	1	1	−1	1	1	−1	1	1
3	1	1	1	1	1	1	1	1	1
4	−1	−1	1	−1	−1	1	−1	−1	1
5	−1	1	1	−1	1	1	1	1	1
6	−1	−1	1	−1	−1	1	−1	1	1
7	1	1	1	1	1	1	−1	1	1

*Categories.

nm columns, $\mathbf{z}$. We are, however, not interested in $\mathbf{z}$ but in the vector of parameters, say $\mathbf{x}$, where

$$\mathbf{x}' = (\tau_1, \tau_2, \ldots, \tau_m, \mu_1, \mu_2, \ldots, \mu_n). \tag{8.26}$$

We note from (8.22), (8.24), and (8.25) that

$$\mathbf{z} = A\mathbf{x}, \tag{8.27}$$

where A is the familiar design matrix for the method of successive categories (e.g., see Nishisato 1975), that is,

$$\underset{nm\times(n+m)}{A} = \begin{bmatrix} I_m & -\mathbf{1}_m & \mathbf{0} & \cdots & \mathbf{0} \\ I_m & \mathbf{0} & -\mathbf{1}_m & \cdots & \mathbf{0} \\ \vdots & \vdots & \vdots & \ddots & \vdots \\ I_m & \mathbf{0} & \mathbf{0} & \cdots & -\mathbf{1}_m \end{bmatrix} = [\mathbf{1}_n \otimes I_m, -I_n \otimes \mathbf{1}_m], \tag{8.28}$$

where I_m is the identity matrix of order m and $\otimes$ is the operator of direct matrix product (i.e., Kronecker product). For example, if $n=3$ and $m=2$, then $\mathbf{x}'=(\tau_1, \tau_2, \mu_1, \mu_2, \mu_3)$. The design matrix is

$$\underset{6\times 5}{A} = \begin{bmatrix} 1 & 0 & -1 & 0 & 0 \\ 0 & 1 & -1 & 0 & 0 \\ 1 & 0 & 0 & -1 & 0 \\ 0 & 1 & 0 & -1 & 0 \\ 1 & 0 & 0 & 0 & -1 \\ 0 & 1 & 0 & 0 & -1 \end{bmatrix}, \quad \text{so that } A\mathbf{x} = \begin{bmatrix} \tau_1 - \mu_1 \\ \tau_2 - \mu_1 \\ \tau_1 - \mu_2 \\ \tau_2 - \mu_2 \\ \tau_1 - \mu_3 \\ \tau_2 - \mu_3 \end{bmatrix}.$$

Since dual scaling of F maximizes $\mathbf{y}'F\mathbf{z}$, relative to the total sum of squares, we obtain from (8.27) that it maximizes

$$\mathbf{y}'F\mathbf{z} = \mathbf{y}'FA\mathbf{x} = \mathbf{y}'E\mathbf{x}, \tag{8.29}$$

where

$$\underset{N\times(n+m)}{E} = FA. \tag{8.30}$$

Matrix E is the desired matrix of subjects-by-parameters of interest, of which dual scaling provides optimal vectors $\mathbf{y}$ and $\mathbf{x}$. Because of the structure of the design matrix, the row marginals of E are all zero, that is,

$$E\mathbf{1} = \mathbf{0}. \tag{8.31}$$

As was true in Nishisato's formulation for paired comparison and rank order data, (8.33) implies that E has negative elements and that neither its marginals nor the sums of the absolute values of its elements would in general provide the numbers of responses for the rows and the columns, quantities necessary to define diagonal matrices D_N and D, respectively. Since our objective is to estimate τ_k and μ_j, we are interested in the number of comparisons between τ_k and μ_j. In terms of these parameters, elements 1 and -1 of F can be replaced by $\tau_k - \mu_j$ and $\mu_j - \tau_k$, respectively. Then the total number of parameters involved in the $N\times nm$ matrix of 1's and -1's can easily be calculated to be $2nmN$, that is,

$$f_t = 2nmN. \tag{8.32}$$

Matrix E can be regarded as a rearrangement and summary of τ_k and μ_j in the order indicated by (8.26). Suppose, for example, that a subject chose category 2 for X_1, category 3 for X_2, and category 1 for X_3, where $n=3$ and $m=2$. Then the corresponding row of F is

$$(-1,1\,|-1,-1\,|\,1,1),$$

which can be translated as

$$(\mu_1-\tau_1,\tau_2-\mu_1\,|\,\mu_2-\tau_1,\mu_2-\tau_2\,|\,\tau_1-\mu_3,\tau_2-\mu_3).$$

There are three τ_1's $(-\tau_1,-\tau_1,\tau_1)$, three τ_2's $(\tau_2,-\tau_2,\tau_2)$, two μ_1's $(\mu_1,-\mu_1)$, two μ_2's (μ_2,μ_2), and two μ_3's $(-\mu_3,-\mu_3)$. The corresponding row of matrix E represents the vector of summaries within parameters (i.e., the sums of the coefficients of the parameters within parentheses), that is, $(-1,1,0,2,-2)$, or in terms of the parameters $(-\tau_1,\tau_2,0\mu_1,2\mu_2,-2\mu_3)$. Note, however, that these are summaries and that there were three τ_k's and two μ_j's. Since there are N subjects, we can generalize on the basis of this small example that the two diagonal matrices for the data with m category boundaries and n stimuli are

$$\underset{N\times N}{D_N} = \mathrm{diag}[2nm, 2nm, \ldots, 2nm] = 2nmI \tag{8.33}$$

for the N rows, and

$$\underset{(n+m)\times(n+m)}{D} = \text{diag}\big[\underbrace{nN, nN, \ldots, nN,}_{m \text{ terms}} \quad \underbrace{mN, mN, \ldots, mN}_{n \text{ terms}}\big] \tag{8.34}$$

for the $n+m$ columns of matrix E.

Now we can subject E to dual scaling with D_N, D, and f_t as specified above. The solution vectors satisfy the relations

$$\left.\begin{aligned} \mathbf{y} &= D_N^{-1}E\mathbf{x}/\eta, \\ \mathbf{x} &= D^{-1}E'\mathbf{y}/\eta, \end{aligned}\right\} \tag{8.35}$$

$$\mathbf{y}'D_N\mathbf{y} = \mathbf{x}'D\mathbf{x} = 2nmN, \tag{8.36}$$

and

$$\mathbf{y}'E\mathbf{x} = 2nmN\eta. \tag{8.37}$$

Nishisato (1978b) also noted that (8.31) and (8.35) yield the relations

$$\mathbf{y}'E\mathbf{1} = 0, \tag{8.38}$$

$$\mathbf{1}'D\mathbf{x} = 0. \tag{8.39}$$

Let us now look at a small illustrative example.

Example 8.2 (*Nishisato* 1978*b*) The data in Table 8.4 were reported by Nishisato (1978b). Ten ($N=10$) subjects judged three stimuli ($n=3$) in terms

TABLE 8.4

Successive categories data (Nishisato 1978b)

	Stimulus 1			Stimulus 2			Stimulus 3		
Subjects	1*	2	3	1	2	3	1	2	3
1			x		x		x		
2		x			x			x	
3		x			x		x		
4			x		x		x		
5			x	x			x		
6			x			x		x	
7		x			x			x	
8			x			x		x	
9		x			x		x		
10			x	x			x		
Total	0	4	6	2	6	2	7	3	0

*Categories.

of a common set of three successive categories ($m=2$). In Table 8.4, each choice is indicated by an 'x.' The two diagonal matrices are obtained by (8.33) and (8.34) as

$$D_N = \text{diag}[12, 12, \ldots, 12], \qquad D = \text{diag}[30, 30, 20, 20, 20].$$

D_N is 10×10 and D is 5×5. Matrix E can be calculated as

$$E = FA$$

$$= \begin{bmatrix} -1 & -1 & -1 & 1 & 1 & 1 \\ -1 & 1 & -1 & 1 & -1 & 1 \\ -1 & 1 & -1 & 1 & 1 & 1 \\ -1 & -1 & -1 & 1 & 1 & 1 \\ -1 & -1 & 1 & 1 & 1 & 1 \\ -1 & -1 & -1 & -1 & 1 & 1 \\ -1 & 1 & -1 & 1 & -1 & 1 \\ -1 & -1 & -1 & -1 & -1 & 1 \\ -1 & 1 & -1 & 1 & 1 & 1 \\ -1 & -1 & 1 & 1 & 1 & 1 \end{bmatrix} \begin{bmatrix} 1 & 0 & -1 & 0 & 0 \\ 0 & 1 & -1 & 0 & 0 \\ 1 & 0 & 0 & -1 & 0 \\ 0 & 1 & 0 & -1 & 0 \\ 1 & 0 & 0 & 0 & -1 \\ 0 & 1 & 0 & 0 & -1 \end{bmatrix}$$

$$= \begin{bmatrix} -1 & 1 & 2 & 0 & -2 \\ -3 & 3 & 0 & 0 & 0 \\ -1 & 3 & 0 & 0 & -2 \\ -1 & 1 & 2 & 0 & -2 \\ 1 & 1 & 2 & -2 & -2 \\ -1 & -1 & 2 & 2 & -2 \\ -3 & 3 & 0 & 0 & 0 \\ -3 & -1 & 2 & 2 & 0 \\ -1 & 3 & 0 & 0 & -2 \\ 1 & 1 & 2 & -2 & -2 \end{bmatrix}.$$

A simple computational check can be made using (8.31), that is, the relation that all the row marginals of E should be zero. The elements of each row of E indicate the order in which the subject judged the parameters of the judgmental continuum. For example, the first row of E indicates that subject 1 judged the parameters in the order $\mu_1 > \tau_2 > \mu_2 > \tau_1 > \mu_2$. Note also that responses from some subjects do not necessarily contain information to distinguish among the parameters. For example, the responses from subject 2 do not tell us the order among μ_1, μ_2, and μ_3, as seen from the second row of E. In regard to the category boundaries, however, the values of τ_k perceived by each subject always satisfy the weak order

$$\hat{\tau}_1 \leq \hat{\tau}_2 \leq \hat{\tau}_3 \leq \cdots \leq \hat{\tau}_m. \tag{8.40}$$

Because of this built-in constraint, successive categories data as represented by E tend to be more internally consistent than, for example, paired comparison data.

Subjecting E to dual scaling, we obtain the results

$$\eta^2 = 0.2438\ (\delta=49.3\%),\quad \mathbf{y} = \begin{bmatrix} 1.20 \\ 0.87 \\ 1.15 \\ 1.20 \\ 1.01 \\ 0.80 \\ 0.87 \\ 0.52 \\ 1.15 \\ 1.01 \end{bmatrix},\quad \mathbf{x} = \begin{bmatrix} \tau_1 \\ \tau_2 \\ \mu_1 \\ \mu_2 \\ \mu_3 \end{bmatrix} = \begin{bmatrix} -0.69 \\ 1.03 \\ 1.16 \\ -0.14 \\ -1.52 \end{bmatrix}.$$

The optimal vector for the parameters shows that

$$\mu_1 > \tau_2 > \mu_2 > \tau_1 > \mu_3.$$

Subjects 1 and 4, who have the largest weights, perceived the parameters in this order. If we look at the responses of the other subjects as summarized in E, the parameters are more or less disordered. For example,

Subject 3 ($y_3=1.15$): $\tau_2 > (\mu_1, \mu_2) > \tau_1 > \mu_3$,

Subject 8 ($y_8=0.52$): $(\mu_1, \mu_2) > \mu_3 > \tau_2 > \tau_1$.

When these cases are compared with the optimum solution, it seems that the differential weights for the subjects present a clear interpretation.

The traditional Thurstonian approach ignores individual differences, assuming that these are a matter of chance fluctuation. This assumption is the same as assigning a unit weight to every subject in the analysis with a specific probability model (e.g., normal, angular, and logistic models). Although the two approaches are not directly comparable, the tenability of the Thurstonian assumption may be inferred from the extent to which y_i of the dual scaling solution are similar to one another and equal to unity. This inference may be tested in the same manner as Fisher's analysis of variance of non-numerical data (1948). To test the equality of y_i, we can form an asymptotic $\mathbf{F}$ statistic,

$$\mathbf{F} = \frac{\eta^2/(N-1)}{(1-\eta^2)/((N-1)(m+n)-N)}, \tag{8.41}$$

$$\mathrm{df}_1 = N-1, \qquad \mathrm{df}_2 = (N-1)(m+n) - N.$$

This is based on the one-way analysis of variance with df_2 adjusted for the loss of degrees of freedom due to scaling. That is, df_2 is the total degrees of freedom, $N(m+n)-1$, minus the degrees of freedom between subjects, $N-1$, minus the number of parameters estimated, $m+n$. In using this statistic, however, Green (1979) noted some peculiarity of the present formulation, that is, that even when all the subjects respond to stimuli in an identical manner, the value of η^2 varies, depending on the particular way in which their responses are consistent. This leaves the value of η^2 difficult to interpret, for it suggests some sort of normalization problem. Under these circumstances, it may not be meaningful to test the significance of the equality of y_i through η^2.

The present approach per se does not provide multidimensional analysis. Although the optimum solution always assigns the values to category boundaries in the correct order, the remaining solutions associated with the other eigenvalues are most likely not to satisfy even the weak order of (8.40). How to analyse multidimensionality of the stimulus space and of the person space is an interesting problem for further research.

8.1.3 *Other techniques*

Many useful references for several other techniques and discussions on them can be found in the book by Barlow, Bartholomew, Bremner, and Brunk (1972). For example they discuss the up-and-down blocks algorithm developed by Kruskal (1964b), the minimum violation algorithm by Thompson (1962), the pool-adjacent-violator algorithm by Ayer, Brunk, Ewing, Reid, and Silverman (1955), and the minimum lower sets and maximum upper sets algorithms by Miles (1959). These methods are based on amalgamation of means, or block means, which provide a simple monotone order. For example, the pool-adjacent-violators algorithm can be illustrated as follows. Suppose that the average January temperature in an area appeared to be getting warmer over the past 10 years, and that the following data were obtained:

Year	1	2	3	4	5	6	7	8	9	10
Temperature (°C)	2	5	3	6	5	3	6	8	7	8

The algorithm starts with individual points, joins the points into a block, and combines the blocks until the block means attain a non-descending (monotone) order. In the above example, the amalgamation process can be illustrated as follows:

1. Combine disordered adjacent pairs into blocks and calculate block means:

$$2 \quad \underbrace{5 \quad 3} \quad \underbrace{6 \quad 5} \quad 3 \quad 6 \quad \underbrace{8 \quad 7} \quad 8$$

$$2 \quad 8/2 \quad 11/2 \quad 3 \quad 6 \quad 15/2 \quad 8$$

2. Repeat 1 using the newly generated set of numbers:

$$2 \quad 8/2 \quad \underbrace{11/2 \quad 3} \quad 6 \quad 15/2 \quad 8$$

$$2 \quad 4 \quad 14/3 \quad 6 \quad 7.5 \quad 8$$

The last row shows a monotonically increasing order of block means, (2, 4, 4.67, 6, 7.5, 8), obtained from the amalgamation [2, (5, 3), (6, 5, 3), 6, (8, 7), 8]. Gebhardt (1970) presents algorithms for a simple order and for a partial order arising from the requirement that a regression function should be monotonic non-decreasing in each of two independent variables.

In non-metric multidimensional scaling (Shepard 1962a, b; Kruskal 1964a, b), Kruskal's algorithm is in routine use. In this area, another distinct algorithm has been developed by Lingoes and Guttman, the so-called rank-image regression (see, for example, Lingoes 1973). These two algorithms are compared and evaluated by Lingoes and Roskam (1973). See also McDonald (1976).

There are two other distinct algorithms which were developed to solve the order constraint problem of dual scaling. These are the methods of Bradley, Katti, and Coons (1962) and of Nishisato and Arri (1975). The procedure employed by Bradley, Katti, and Coons is an iterative one. First, they transform weight vector **x** to a vector of non-negative contrasts of x_j, **z**, that is, $\mathbf{z} = P\mathbf{x} \geq \mathbf{0}$, say. Once the squared correlation ratio is expressed in terms of **z**, the procedure chooses one element of **z** as a variable, and assigns non-negative arbitrary values to the remaining $k-1$ elements. The variable element then is determined, under the condition that it be non-negative, so as to maximize the squared correlation ratio, η^2. Once the value is determined, another element of **z** is chosen as a variable, and its value is determined in the same way, except that the fixed $k-1$ values now contain the first weight just determined. This process is continued with the revised set of $k-1$ weights as fixed and one weight allowed to vary until η^2 attains its optimum value. The method used by Nishisato and Arri is a non-linear programming approach. Their procedure first transforms η^2 expressed in terms of **x** into a function of non-negative contrasts, **z**, and then expresses this function in the separable programming format (see, for example, Hadley 1964). In other words, all the functions are transformed into single-variable functions by eliminating all the product terms. Then the procedure specifies the so-called mesh points, which provide a set of values for a polygonal approximation to each single-variable function. Once the problem is formulated in terms of mesh points, it is in the format of linear programming, and can be solved by the simplex method with the restricted basis entry (see, for example, Hadley 1962; Abraham and Arri 1973). Theoretically, the solution thus obtained is an approximation, and hence can be improved by the iterative process, called grid refinement. Nishisato and Arri (1975) defined mesh points around the SDM solution. Since the SDM solution is, at least empirically, optimum, there does not seem to be any need for grid refinement. In their study the non-linear programming solution was identical with the SDM solution, and the non-linear programming part was therefore not needed. If mesh points are determined arbitrarily, however, the above procedure requires not only consecutive grid refinements, but also a wider range of mesh points so that the solution points may surely lie within the interval defined by the two extreme mesh points. This strategy, however, often introduces too many variables to work with, and is not very practical.

8.2 ANALYSIS OF VARIANCE OF CATEGORICAL DATA

In Chapter 4, it was noted that contingency tables are often subjected to the analysis of variance (abbreviated hereafter as ANOVA), using the so-called *log linear model*. Some references were also provided. We are now interested in applying dual scaling to ANOVA of categorical data. One natural question is whether or not dual scaling is any better than the log linear approach. Unfortunately, there does not seem to be any comparative study available, and hence our discussion must remain speculative.

The log linear approach postulates an experimental design model with respect to logarithmic transforms of cell frequencies of the contingency table, and its analysis is guarded with all the assumptions entertained in the so-called normal theory. The model is then amenable to, for example, maximum likelihood estimation or weighted least-squares estimation of parameters, and sampling distributions of the estimates can in general be specified. One can obtain so-called minimum-variance estimates, which implies that the analysis is optimum in the least-squares sense.

The dual scaling approach does not employ any assumption about the distribution of responses, which may appear to present some advantage of dual scaling over the log linear approach. However, the same point makes it difficult to make any probabilistic statement about weights or, more specifically, the sampling stability of the optimal weights. In this approach, the unit of analysis is not necessarily a cell frequency, but can be a single response from a single subject. In other words, it can scale a response-pattern matrix of 1's and 0's so as to maximize the contributions of the ANOVA parameters. Individual differences are effectively used in scaling the data. Unlike the dual scaling approach, the log linear approach assumes that subjects are randomly chosen, and that individual differences are nothing but random fluctuations. Thus the log linear approach cannot derive weights for non-numerical responses of individual subjects.

The next difference is in the transformation of data. The logarithmic transformation is variance-stabilizing and in some sense has the effect of reducing the contribution of higher-order moments or interactions to the total variance. The ANOVA decomposition, however, is unique. Dual scaling, in contrast can choose any effect or combination of effects for optimization, as seen in Chapter 7. The individual sums of squares of effects naturally change, depending on the choice of the optimization criterion. This choice, or scaling itself, introduces an additional step to its procedure, namely, adjustment of degrees of freedom for the error term.

Although some degrees of freedom will be lost in the dual scaling approach to ANOVA as just mentioned above, dual scaling is still blessed with a much larger number of degrees of freedom to manipulate, for it can deal with the response-pattern representation of data, rather than just with the contingency table. This aspect of dual scaling suggests the possibility that this approach might almost always provide a more clear-cut picture of effects of the parameters than the log linear approach. This, however, must remain a conjecture.

The comparison of the two approaches deserves scrutiny, and extensive work may be necessary. Comparisons based on a few numerical examples may be misleading, and hence will not be made. This is an important problem from both theoretical and practical points of view, however, and a detailed analysis is much desired.

Now, this section presents dual scaling of categorical data in the ANOVA context. Fisher (1948) did pioneering work, as briefly reviewed in Chapter 1, and his approach was generalized to multi-way designs by Nishisato (1971a, 1972a, b, 1972a).

In Chapter 2, the squared correlation ratio was derived in the context of the one-way ANOVA. To see the relation between dual scaling and ANOVA, we can present a summary table of the one-way ANOVA (Table 8.5). As we recall, the squared correlation ratio, η^2, is defined as $\mathrm{ss}_b/\mathrm{ss}_t$. Statistic $\mathbf{F}$, which is used in the one-way ANOVA, is defined by

$$\mathbf{F} = \mathrm{MS}_b/\mathrm{MS}_w. \tag{8.42}$$

Both η^2 and $\mathbf{F}$ are associated with the magnitude of group differences, and are related simply as follows:

$$\mathbf{F} = \frac{\mathrm{MS}_b}{\mathrm{MS}_w} = \frac{\mathrm{ss}_b/\mathrm{df}_b}{\mathrm{ss}_w/\mathrm{df}_w} = \frac{\mathrm{ss}_b/\mathrm{df}_b}{(\mathrm{ss}_t - \mathrm{ss}_b)/\mathrm{df}_w} = \frac{\eta^2/\mathrm{df}_b}{(1-\eta^2)/\mathrm{df}_w}. \tag{8.43}$$

In the context of dual scaling, therefore, it follows that maximization of η^2 is equivalent to maximization of $\mathbf{F}$. It is assumed here, of course, that $\eta^2 \neq 1$.

To illustrate further the relation between the two analyses, consider the two-way ANOVA with the crossed design. Table 8.6 is the summary table, which corresponds to Table 8.5. The sums of squares and the degrees of freedom are related as follows:

$$\mathrm{ss}_t = \mathrm{ss}_b + \mathrm{ss}_w = (\mathrm{ss}_A + \mathrm{ss}_B + \mathrm{ss}_{AB}) + \mathrm{ss}_w, \tag{8.44}$$

$$\mathrm{df}_t = \mathrm{df}_b + \mathrm{df}_w = (\mathrm{df}_A + \mathrm{df}_B + \mathrm{df}_{AB}) + \mathrm{df}_w. \tag{8.45}$$

TABLE 8.5

A summary table of one-way analysis of variance

Source of variations	df	Sum of squares	Mean squares	F ratio
Between groups	df_b	ss_b	$MS_b = ss_b/df_b$	MS_b/MS_w
Within groups	df_w	ss_w	$MS_w = ss_w/df_w$	
Total	df_t	ss_t		

Note that the between-group sum of squares has three (disjointed) components, ss_A, ss_B, and ss_{AB}. If dual scaling is carried out so as to maximize ss_b/ss_t, it has the effect of maximizing the overall **F** ratios. It is important to recall that maximization of ss_b/ss_t means minimization of ss_w/ss_t, and that these two together lead to maximization of the **F** ratios. Thus, if dual scaling is carried out to maximize the relative contribution of only a subset of ss_b, say ss_A and ss_B, it does not generally maximize the corresponding **F** ratios. It simply maximizes $(ss_A + ss_B)/ss_t$, and not $(ss_A + ss_B)/ss_w$. In other words, ss_w is not necessarily minimized by the maximization of $(ss_A + ss_B)/ss_t$. It may be ss_{AB}, rather than ss_w, which is substantially reduced.

The idea of quantifying categorical data so as to maximize the effects of particular treatments is probably very appealing to researchers in applied areas. One may be interested in the effects of only a few treatments or those

TABLE 8.6

A summary table of two-way analysis of variance (crossed design)

Source of variations	df	Sum of squares	Mean squares	F
Treatment A	df_A	ss_A	$MS_A = ss_A/df_A$	MS_A/MS_w
Treatment B	df_B	ss_B	$MS_B = ss_B/df_B$	MS_B/MS_w
Interaction A×B	df_{AB}	ss_{AB}	$MS_{AB} = ss_{AB}/df_{AB}$	MS_{AB}/MS_w
Between groups	df_b	ss_b	$MS_b = ss_b/df_b$	
Within groups	df_w	ss_w	$MS_w = ss_w/df_w$	
Total	df_t	ss_t		

of particular interactions. One may not wish to include high-order interactions in the maximization simply because they are too difficult to interpret. No matter what one's interests are, the dual scaling approach offers an interesting way of performing data analysis.

When ANOVA is applied to categorical data, one may be concerned about violations of its underlying assumptions (e.g., normality, homoscedasticity). However, the present approach deals with a linear combination of weighted responses as a unit of analysis, and the central limit theorem would at least ensure its asymptotic normality. The ANOVA procedure is also known to be quite robust with respect to violations of the assumptions. To feel truly comfortable, however, it seems as though we need further work by statisticians. Let us look at two of the possible procedures.

8.2.1 *Method I*

Following the procedure discussed in Chapter 7, one can obtain optimal score vector $\mathbf{y}_r$ for subjects by either (7.5) or (7.17), depending on the case. $\mathbf{y}_r$ then is subjected to the ordinary ANOVA procedure. Let us consider a $p \times q$ crossed design with the same number of subjects in each of the pq cells. Suppose that N subjects answered n multiple-choice questions which have in total m response options. Then the ANOVA results can be summarized as in Table 8.7. Recall that $\mathbf{y}_r$ is already optimally scaled, and that the projection operators are used simply to indicate orthogonal decompositions. Unlike the ordinary ANOVA of $\mathbf{y}$ of N scores, we have Nm responses, with some constraints. We estimate m weights with the condition that the sum of the weighted responses

TABLE 8.7

Two-way ANOVA (Method I)

Source of variations	df	Sum of squares	Mean squares & F*
Treatment A	$p-1$	$\mathbf{y}_r'P_\alpha\mathbf{y}_r$	
Treatment B	$q-1$	$\mathbf{y}_r'P_\beta\mathbf{y}_r$	
Interaction A×B	$(p-1)(q-1)$	$\mathbf{y}_r'P_{\alpha\beta}\mathbf{y}_r$	
Between	$pq-1$	$\mathbf{y}_r'(P_\alpha+P_\beta+P_{\alpha\beta})\mathbf{y}_r$	
Within	$(N-1)(m-n)-pq+1$	$\mathbf{y}_r'(I-P_{\alpha\beta})\mathbf{y}_r$	
Total	$(N-1)(m-n)$	$\mathbf{y}_r'\mathbf{y}_r$	

*See Table 8.6 for appropriate entries.

is zero. As mentioned in Chapter 5, this condition implies that the sum of the weighted responses of each item is zero. In consequence, the total degrees of freedom are the number of independent responses, $N(m-n)$, minus the number of independent estimates $m-n$, that is, $df_t=(N-1)(m-n)$. This change in df_t, however, affects only df_w (see Table 8.7).

8.2.2 *Method II*

Method I is probably satisfactory and also easy enough to put to routine use, but Method II presents another interesting approach to the same problem. In 1971, Nishisato considered a simple generalization of the one-way ANOVA to a multi-way ANOVA through dual scaling. To illustrate his approach, let us consider multiple-choice data, obtained from N subjects who were sampled according to a 2×2 crossed design. Define

F = the $N\times m$ response-pattern matrix;
t = the number of groups of subjects, generated by the two-way classification, that is, $t=4$.
n_{ij} = the number of subjects in the group (i,j), $i=1,2$; $j=1,2$;
D^* = the $t\times N$ matrix such that, in our example,

$$D^* = \begin{bmatrix} \underbrace{\begin{matrix} 1\ 1 \cdots 1 \\ 0\ 0 \cdots 0 \\ 0\ 0 \cdots 0 \\ 0\ 0 \cdots 0 \end{matrix}}_{n_{11}} & \underbrace{\begin{matrix} 0\ 0 \cdots 0 \\ 1\ 1 \cdots 1 \\ 0\ 0 \cdots 0 \\ 0\ 0 \cdots 0 \end{matrix}}_{n_{12}} & \underbrace{\begin{matrix} 0\ 0 \cdots 0 \\ 0\ 0 \cdots 0 \\ 1\ 1 \cdots 1 \\ 0\ 0 \cdots 0 \end{matrix}}_{n_{21}} & \underbrace{\begin{matrix} 0\ 0 \cdots 0 \\ 0\ 0 \cdots 0 \\ 0\ 0 \cdots 0 \\ 1\ 1 \cdots 1 \end{matrix}}_{n_{22}} \end{bmatrix};$$

D_N = the $N\times N$ diagonal matrix of the row totals of F;
D = the $m\times m$ diagonal matrix of the column totals of F;
$F_. = D^*F$ = the $t\times N$ matrix with elements being sums of column entries within groups;
D_t = the $t\times t$ diagonal matrix of the row totals of $F_.$;
$\mathbf{x}$ = the $m\times1$ vector of weights for the m options;
$\boldsymbol{\xi}$ = the $h\times1$ vector of ANOVA parameters.

Then the vector of optimal scores of N subjects and the vector of the cell means of optimal scores are, respectively,

$$\mathbf{y} = D_N^{-1}F\mathbf{x}/\eta, \tag{8.46}$$

$$\mathbf{y}_. = D_t^{-1}F_.\mathbf{x}/\eta = D_t^{-1}D^*F\mathbf{x}/\eta. \tag{8.47}$$

$\mathbf{y}_.$ can be expressed in terms of the ANOVA parameters as

$$\mathbf{y}_. = A\boldsymbol{\xi} + \mathbf{e}_. = KL\boldsymbol{\xi} + \mathbf{e}_. = K\boldsymbol{\theta} + \mathbf{e}_., \tag{8.48}$$

where A is the design matrix, $\mathbf{e}_.$ is the vector of residual, $\mathbf{y}_. - A\boldsymbol{\xi}$, called errors, $\boldsymbol{\theta}$ is the vector of q linearly estimable functions of $\boldsymbol{\xi}$, that is, $\boldsymbol{\theta}=L\boldsymbol{\xi}$, and K is a basis matrix. Assuming that $\mathbf{e}_. \sim N(\mathbf{0}, \sigma^2 D_t^{-1})$, the unbiased minimum variance estimate $\boldsymbol{\theta}$ is given by

$$\hat{\boldsymbol{\theta}} = (K'D_tK)^{-1}K'D_t\mathbf{y}_. = (K'D_tK)^{-1}K'F_.\mathbf{x}/\eta. \tag{8.49}$$

The amount of information accounted for by the ANOVA model is proportional to the sum of squares of $\hat{\boldsymbol{\theta}}$, which can be expressed as

$$\begin{aligned} \text{ss}(\hat{\boldsymbol{\theta}}) &= \mathbf{x}'F_.'K(K'D_tK)^{-1}K'F_.\mathbf{x}/\eta^2 \\ &= \mathbf{x}'F'D^{*\prime}K(K'D_tK)^{-1}K'D^*F\mathbf{x}/\eta^2 \\ &= \mathbf{x}'F'GF\mathbf{x}/\eta^2, \text{ say,} \end{aligned} \tag{8.50}$$

where

$$G = D^{*\prime}K(K'D_tK)^{-1}K'D^*. \tag{8.51}$$

The total sum of squares is

$$\text{ss}_t = \mathbf{y}'\mathbf{y} = \mathbf{x}'F'D_N^{-2}F\mathbf{x}/\eta^2 = \mathbf{x}'F'F\mathbf{x}/n^2\eta^2, \tag{8.52}$$

since $D_N = nI$, where n is the number of items. The correlation ratio can be defined as

$$\eta^2 = \frac{\text{ss}(\hat{\boldsymbol{\theta}})}{\text{ss}_t} = \frac{n^2\mathbf{x}'F'GF\mathbf{x}}{\mathbf{x}'F'F\mathbf{x}}. \tag{8.53}$$

Nishisato (1972a, b) maximized $\text{ss}(\hat{\boldsymbol{\theta}})$ subject to the condition that $\mathbf{x}'D\mathbf{x}$ is constant. This condition was used rather than $\mathbf{x}'F'F\mathbf{x}$ being constant because of the computational simplicity. Since $\text{ss}(\hat{\boldsymbol{\theta}})/\mathbf{x}'D\mathbf{x}$ is not equal to η^2, his method does not maximize **F** ratios, but provides an approximation to the optimum solution associated with (8.53). Solving (8.53) directly involves the equation

$$(n^2F'GF - \eta^2F'F)\mathbf{x} = \mathbf{0}, \tag{8.54}$$

where $F'F$ is singular.

There are several ways to handle such an equation as (8.54), which arises from the maximization of the ratio of two singular quadratic forms. In the context of dual scaling, this problem has been discussed by several investigators (e.g., Hayashi, Higuchi, and Komazawa 1970; de Leeuw 1976; Healy and Goldstein 1976; Nishisato 1976; Saporta 1976; McDonald, Torii, and Nishisato 1979). We shall first present two approaches described by Nishisato (1976) and then a more direct method by McDonald, Torii, and Nishisato (1979).

The two approaches employ reparameterization such that

$$\mathbf{x}'F'F\mathbf{x} = \mathbf{x}^{*\prime}F^{*\prime}F^*\mathbf{x}^*, \tag{8.55}$$

where $F^{*\prime}F^*$ is non-singular. There are an infinite number of such reparameterizations, but the relation between $\mathbf{x}$ and $\mathbf{x}^*$ can be expressed in the general form

$$\underset{k\times 1}{\mathbf{x}^*} = \underset{k\times m}{J}\ \underset{m\times 1}{\mathbf{x}}, \tag{8.56}$$

where

$$r\begin{bmatrix} F \\ J \end{bmatrix} = r(F) = r(J) = r(F^*) = k, \text{ say,} \tag{8.57}$$

and the number of rows of F, that is, N, is much larger than m. Then JJ' is non-singular and F^* of (8.55) can be obtained from F and J by

$$F^* = FJ'(JJ')^{-1}. \tag{8.58}$$

As an example of matrix J, Nishisato (1976) suggests the following:

$$J = \begin{bmatrix} J_1 & O & \cdots & O \\ O & J_2 & \cdots & O \\ \vdots & \vdots & \ddots & \vdots \\ O & O & \cdots & J_n \end{bmatrix},$$

where

$$J_j = \begin{bmatrix} 1 & 0 & 0 & 0 & \cdots & 0 & -1 \\ 0 & 1 & 0 & 0 & \cdots & 0 & -1 \\ 0 & 0 & 1 & 0 & \cdots & 0 & -1 \\ 0 & 0 & 0 & 1 & \cdots & 0 & -1 \\ \vdots & \vdots & \vdots & \vdots & \ddots & \vdots & \vdots \\ 0 & 0 & 0 & 0 & \cdots & 1 & -1 \end{bmatrix} = [I, -\mathbf{1}]. \tag{8.59}$$

We wish to maximize η^2, expressed in terms of $\mathbf{x}^*$, which leads to the eigenequation

$$(n^2F^{*\prime}GF^* - \eta^2F^{*\prime}F^*)\mathbf{x}^* = \mathbf{0}, \tag{8.60}$$

where $F^{*\prime}F^*$ is now non-singular. The two approaches are as follows:

Approach (*a*) Using the similarity transformation, decompose $F^{*\prime}F^*$ as

$$F^{*\prime}F^* = P\Lambda P', \tag{8.61}$$

where Λ is the diagonal matrix of the eigenvalues of $F^{*\prime}F^*$ and P consists of the corresponding eigenvectors in the columns. Then

$$\mathbf{x}^{*\prime}F^{*\prime}F^*\mathbf{x}^* = \mathbf{x}^{*\prime}P\Lambda^{1/2}\Lambda^{1/2}P'\mathbf{x}^* = \mathbf{w}'\mathbf{w}, \text{ say,} \tag{8.62}$$

where

$$\mathbf{w} = \Lambda^{1/2}P'\mathbf{x}^* \quad \text{or} \quad \mathbf{x}^* = P\Lambda^{-1/2}\mathbf{w}. \tag{8.63}$$

Thus we obtain the following equation, which is amenable to the standard method:

$$(n^2\Lambda^{-1/2}P'F^{*\prime}GF^*P\Lambda^{-1/2}-\eta^2I)\mathbf{w}=\mathbf{0}. \tag{8.64}$$

Approach (*b*) The second approach considers the Cholesky factorization of $F^{*\prime}F^*$, that is,

$$F^{*\prime}F^* = TT', \text{ say.} \tag{8.65}$$

Then (8.60) can be written as

$$(n^2F^{*\prime}GF^*-\eta^2TT')\mathbf{x}^*=\mathbf{0}, \tag{8.66}$$

which can be transformed into the form we want:

$$\left[n^2T^{-1}F^{*\prime}GF^*(T^{-1})'-\eta^2I\right]\mathbf{w}=\mathbf{0}, \tag{8.67}$$

where

$$\mathbf{w} = T'\mathbf{x}^* \quad \text{or} \quad \mathbf{x}^* = (T')^{-1}\mathbf{w}. \tag{8.68}$$

Of the two approaches, (b) may be preferred to (a) from the computational point of view, for the Cholesky factorization would be much faster to obtain in most cases than a complete solution of the eigenequation, involved in (a).

McDonald, Torii, and Nishisato (1979; see also Torii 1977) present a more direct way than those described above. It is 'more direct,' however, only in theory and not computationally, for it involves complete solutions of three eigenequations. Their method is based on the Rao-Mitra theorems on proper eigenvalues and eigenvectors, and seeks the solution which satisfies the necessary and sufficient conditions for proper eigenvalues and eigenvectors. It is a complex operation and those interested in the computational procedure are referred to Torii (1977) for a FORTRAN program, and to McDonald, Torii, and Nishisato (1979) for theoretical discussion. Their study shows that the reparameterization method suggested by Nishisato (1976), that is, the one using (8.59), provides proper eigenvalues and eigenvectors, and that a commonly used method of deleting one response option per multiple-choice item so as to make $F^{*\prime}F^*$ non-singular does not work. This method is popular, and is suggested in such articles as those by Hayashi, Higuchi, and Komazawa (1970), Takeuchi and Yanai (1972), and Healy and Goldstein (1976).

These recent developments have not been put into practice. Thus the following examples of applications are based on Nishisato's early studies in 1971 and 1972.

8.2.3 *Applications*

Since the basic procedure involved in Method I was illustrated in Chapter 7, we shall concentrate on Method II.

TABLE 8.8

Response-pattern table of two-way classification

		Polychotomous items																
		1			2			3			4				5			
Groups		1	2	3	1	2	3	1	2	3	1	2	3	4	1	2	3	4
B_0	A_0	0	1	0	0	0	1	1	0	0	0	0	0	1	0	0	1	0
		1	0	0	0	1	0	0	1	0	1	0	0	0	1	0	0	0
		1	0	0	1	0	0	0	1	0	0	1	0	0	0	1	0	0
	A_1	0	0	1	1	0	0	0	0	1	0	0	1	0	0	0	0	1
		0	0	1	0	0	1	0	1	0	0	0	0	1	0	0	1	0
		0	1	0	0	0	1	0	0	1	0	0	1	0	0	1	0	0
B_1	A_0	1	0	0	0	0	1	0	1	0	0	0	1	0	1	0	0	0
		0	1	0	1	0	0	1	0	0	0	1	0	0	0	0	1	0
		1	0	0	0	1	0	1	0	0	1	0	0	0	0	1	0	0
	A_1	0	1	0	0	0	1	0	0	1	0	0	0	1	0	0	0	1
		0	1	0	0	0	1	0	0	1	0	0	0	1	0	0	1	0
		0	0	1	0	0	1	0	0	1	0	0	1	0	0	0	0	1

Example 8.3: *Simple case* In 1971, Nishisato used fictitious data, where it was assumed that three subjects were randomly sampled from each of the high achiever (A_0) and boy (B_0) group, the high achiever and girl (B_1) group, the low achiever (A_1) and boy group, and the low achiever and girl group, and that the 12 subjects were asked to answer five multiple-choice questions (Table 8.8). From Table 8.8, we obtain

F = the 12×17 matrix of 1's and 0's arranged in the same way as in Table 8.8,

$$F_{\cdot} = \begin{bmatrix} 2 & 0 & 0 & 1 & 1 & 1 & 1 & 2 & 0 & 1 & 1 & 0 & 1 & 1 & 1 & 1 & 0 \\ 0 & 1 & 2 & 1 & 0 & 2 & 0 & 1 & 2 & 0 & 0 & 2 & 1 & 0 & 1 & 1 & 1 \\ 2 & 1 & 0 & 1 & 1 & 1 & 2 & 1 & 0 & 1 & 1 & 1 & 0 & 1 & 1 & 1 & 0 \\ 0 & 2 & 1 & 0 & 0 & 3 & 0 & 0 & 3 & 0 & 0 & 1 & 3 & 0 & 0 & 1 & 2 \end{bmatrix},$$

D_t = the 4×4 diagonal matrix with all the diagonal elements being $15 = 15I$,

$$A = \begin{bmatrix} 1 & 1 & 0 & 1 & 0 & 1 & 0 & 0 & 0 \\ 1 & 0 & 1 & 1 & 0 & 0 & 1 & 0 & 0 \\ 1 & 1 & 0 & 0 & 1 & 0 & 0 & 1 & 0 \\ 1 & 0 & 1 & 0 & 1 & 0 & 0 & 0 & 1 \end{bmatrix}.$$

A basis matrix can be constructed by Bock's procedure (e.g., see Bock 1963): (i) consider a coefficient matrix of simple contrasts, L_i, for each treatment; (ii) calculate $K_i = [\mathbf{1}, L_i(L_iL_i')^{-1}]$; (iii) the basis matrix, K, for a two-way crossed design is given by $K_\alpha \otimes K_\beta$, where $\otimes$ is the Kronecker product (direct product) operator. In our example, simple contrasts for the two treatment (grouping) effects are $\alpha_0 - \alpha_1$ and $\beta_0 - \beta_1$. Thus

$$L_\alpha = L_\beta = (1, -1).$$

In consequence,

$$K_\alpha = K_\beta = \left[\mathbf{1}, L_i'(L_iL_i')^{-1}\right] = \begin{bmatrix} 1 & 1/2 \\ 1 & -1/2 \end{bmatrix}.$$

Therefore our basis matrix is given by

$$K = K_\alpha \otimes K_\beta = \begin{bmatrix} 1 & 1/2 & 1/2 & 1/4 \\ 1 & -1/2 & 1/2 & -1/4 \\ 1 & 1/2 & -1/2 & -1/4 \\ 1 & -1/2 & -1/2 & 1/4 \end{bmatrix}.$$

The first column represents the grand mean, the second column the effect of treatment (grouping) A, that is, $\alpha_0 - \alpha_1$, the third column the effect of treatment B, and the last column the interaction effect $\alpha\beta$. If we include all the ANOVA parameters, K is square and $\mathrm{ss}(\hat{\boldsymbol{\theta}})$ can be reduced as follows:

$$\mathrm{ss}(\hat{\boldsymbol{\theta}}) = \mathbf{x}'F_.'K(K'D_tK)^{-1}K'F_.\mathbf{x}$$

$$= \mathbf{x}'F_.'D_t^{-1}F_.\mathbf{x} = \mathrm{ss}_b.$$

If we maximize $\mathrm{ss}(\hat{\boldsymbol{\theta}})/\mathbf{x}'D\mathbf{x}$, rather than $\mathrm{ss}(\hat{\boldsymbol{\theta}})/\mathrm{ss}_t$, for computational convenience, the matrix involved in the standard form of the eigenequation is given by

$$C_0 = D^{-1/2}F_.'D_t^{-1}F_.D^{-1/2}.$$

The matrix which is free from the trivial solution due to the marginal constraints can be calculated from

$$C_1 = C_0 - D^{1/2}\mathbf{1}\mathbf{1}'D^{1/2}/f_t.$$

Following the standard procedure of dual scaling, we obtain vector $\mathbf{x}$ (Table 8.9). We can calculate the optimal scores of the 12 subjects, using $\mathbf{x}$, and then submit them to ANOVA. Or we can calculate the sums of squares directly from $\mathbf{x}$ by the relations indicated in Table 8.10, which is the ANOVA table expressed in terms of $\mathbf{x}$. If one wishes to maximize the contribution of a particular effect, first partition the basis matrix columnwise. For example, if $K = [K_\mu, K_\alpha, K_\beta, K_{\alpha\beta}]$, then the sum of squares due to treatment effect α can be expressed as

TABLE 8.9

Optimal weights for ANOVA

Items	Weights	Items	Weights
1	$x_1=0.1754$ $x_2=-0.0425$ $x_3=-0.1630$	4	$x_{10}=0.1754$ $x_{11}=0.1754$ $x_{12}=-0.0768$ $x_{13}=-0.0986$
2	$x_4=0.0709$ $x_5=0.1754$ $x_6=-0.0805$	5	$x_{14}=0.1754$ $x_{15}=0.0709$ $x_{16}=0.0000$ $x_{17}=-0.1878$
3	$x_7=0.1776$ $x_8=0.0954$ $x_9=-0.1829$		$\lambda=0.52927$

TABLE 8.10

Analysis of variance in terms of optimal weight vector **x**

Source of variations	df	Sum of squares*
Treatments	$q=r(K)$	$\mathbf{x}'F_.'K(K'D_tK)^{-1}K'F_.\mathbf{x}$
Residual	$p-q$	$\mathbf{x}'F_.'D_t^{-1}F_.\mathbf{x}$ $-\mathbf{x}'F_.'K(K'D_tK)^{-1}K'F_.\mathbf{x}$
Between groups	p	$\mathbf{x}'F_.'D_t^{-1}F_.\mathbf{x}$
Within groups	$(N-1)(m-n)-p$	$\mathbf{x}'(F'F-F_.'D_t^{-1}F_.)\mathbf{x}$
Total	$(N-1)(m-n)$	$\mathbf{x}'F'F\mathbf{x}$

*The constant multiplier n^2 is omitted from each term.

$$\mathrm{ss}_\alpha = \mathbf{x}'F_.'K_\alpha\left(K_\alpha'D_t^{-1}K_\alpha\right)^{-1}K_\alpha'F_.\mathbf{x},$$

and matrix C_0 is given by

$$C_0 = D^{-1/2}F_.'K_\alpha\left(K_\alpha'D_t^{-1}K_\alpha\right)^{-1}K_\alpha'F_.D^{-1/2}.$$

Similarly, if one wishes to maximize the effects of α and β, specify K to be $[K_\alpha, K_\beta]$. Nishisato (1971a) noted the interesting fact that the deletion of K_μ from optimization has the effect of setting the sum of the weighted responses equal to zero. His results are shown in Table 8.11. In this table, the grand mean is omitted since it is zero (i.e., $\mathbf{f}'\mathbf{x}=\mathbf{0}$). The **F** statistic for the 'between groups' provides a significance test of η^2, which is equal to 0.81365 ($=ss_b/ss_t = 2.6469/3.2531$). In our example, $\mathbf{F}=11.6398$, $df_b=3$, $df_w=173$, and this value is significant at the 0.01 level (refer to a table of critical values of **F**, which can be found in many books on statistics). Where this statistic is significant, it almost guarantees that at least one treatment effect will be significant. In this example, 'achievement' is highly significant since the critical value of **F** at the 0.01 level ($df_1=1$, $df_w=173$) is approximately 6.80.

Let us look at another example, reported by Nishisato (1972a, 1973a). He used fictitious data, where it was assumed that 48 subjects were randomly selected according to the $2\times2\times2\times2$ factorial design, with three subjects in each cell, and that the subjects were asked to answer eight multiple-choice questions with five options per question. The response data are given in Table 8.12. Nishisato (1972a, 1973a) carried out three types of analysis: (i) ANOVA through selective scaling (the hierarchical deletion scheme); (ii) ANOVA through selective scaling (the selective deletion scheme); and (iii) multidimensional ANOVA. These analyses will be described and discussed here.

Example 8.4: ANOVA *through selective scaling* (*hierarchical deletion scheme*)
Matrices F, $F_{\cdot}$, and D_t can be constructed from Table 8.13 in the same way as in the first example. The basis matrix used here can be generated by the

TABLE 8.11

ANOVA through dual scaling

Source of variation	df	Sum of squares	Mean squares	F	p
Achievement (H vs. L) (A)	1	2.5872	2.5872	34.1319	<0.01
Sex (B)	1	0.0196	0.0196	–	
A×B	1	0.0401	0.0401	–	
Between groups	3	2.6469	0.8823	11.6398	<0.01
Within groups	128	0.6062	0.0758		
Total	131	3.2531			

TABLE 8.12

Response table for four-way classification

				Questionnaire items							
				1	2	3	4	5	6	7	8
Factors				12345	12345	12345	12345	12345	12345	12345	12345
D_0	C_0	B_0	A_0	10000	00100	01000	10000	10000	10000	10000	10000
				10000	00100	01000	10000	01000	10000	10000	01000
				10000	00100	01000	01000	01000	00100	10000	00100
			A_1	10000	10000	10000	10000	10000	10000	10000	10000
				01000	10000	10000	10000	01000	00100	10000	01000
				00100	01000	00100	10000	01000	00100	00001	00100
		B_1	A_0	00100	00010	00100	00001	00010	00100	00010	00010
				00010	00001	00100	00001	00010	00010	00001	00010
				00001	00001	00010	00001	00001	00001	00001	00001
			A_1	01000	01000	00100	00100	01000	10000	01000	00010
				01000	00100	00100	00010	01000	10000	01000	00100
				01000	00010	00010	00010	01000	10000	00100	01000
D_0	C_1	B_0	A_0	01000	00100	01000	01000	01000	01000	01000	10000
				01000	00010	01000	00100	01000	01000	01000	10000
				01000	00010	01000	00010	01000	01000	00100	01000
			A_1	00001	00001	00010	01000	01000	01000	01000	01000
				00001	00001	00010	01000	00100	00100	00100	01000
				00010	00001	00010	00100	00100	00100	00100	01000
		B_1	A_0	00001	00010	00010	00100	00010	00100	00010	00100
				00001	00001	00001	00010	00010	00010	00001	00010
				00001	00001	00001	00001	00001	00001	00001	00001
			A_1	00001	00001	00001	00010	00010	00010	00001	00010
				00001	00001	00001	00001	00001	00001	00010	00001
				00010	00001	00010	00001	00001	00001	00010	00001
D_1	C_0	B_0	A_0	00001	00100	00001	00010	00010	00010	00001	00010
				00001	00010	00001	00010	00001	00010	00001	00001
				00001	00001	00001	00001	00001	00010	00001	00001
			A_1	01000	10000	10000	01000	10000	01000	01000	01000
				01000	01000	10000	01000	01000	10000	01000	01000
				01000	01000	01000	01000	00100	10000	00100	01000
		B_1	A_0	10000	10000	10000	10000	10000	10000	10000	01000
				10000	10000	01000	01000	01000	10000	01000	01000
				10000	01000	00100	01000	00100	10000	01000	01000

TABLE 8.12 (continued)

				Questionnaire items							
				1	2	3	4	5	6	7	8
Factors				12345	12345	12345	12345	12345	12345	12345	12345
			A_1	00001	00010	00010	00001	00010	00001	00100	00001
				00010	00010	00010	00010	00001	00001	00010	00010
				00100	00001	00010	00010	00001	00001	00010	00001
D_1	C_1	B_0	A_0	00001	00001	00010	00100	00010	00010	00001	00010
				00001	00001	00010	00010	00001	00010	00001	00010
				00001	00010	00001	00001	00001	00001	00001	00001
			A_1	10000	10000	10000	10000	10000	01000	01000	00100
				10000	10000	01000	01000	01000	01000	01000	01000
				10000	01000	00100	00100	00010	10000	01000	01000
		B_1	A_0	10000	00100	01000	01000	01000	10000	10000	10000
				10000	01000	00010	01000	01000	10000	01000	01000
				01000	10000	00001	00100	00010	01000	00100	01000
			A_1	00100	00100	00010	00010	00100	00100	01000	00100
				00010	00001	00001	00010	00001	00100	00010	00001
				00001	00001	00001	00001	00001	00010	00001	00001

fourfold Kronecker product

$$K = \begin{bmatrix} 1 & 1/2 \\ 1 & -1/2 \end{bmatrix} \otimes \begin{bmatrix} 1 & 1/2 \\ 1 & -1/2 \end{bmatrix} \otimes \begin{bmatrix} 1 & 1/2 \\ 1 & -1/2 \end{bmatrix} \otimes \begin{bmatrix} 1 & 1/2 \\ 1 & -1/2 \end{bmatrix}.$$

The expansion of the product and the treatment effects are summarized in Table 8.13. In the hierarchical deletion scheme, the computer carries out the analysis (scaling of **x** and ANOVA decompositions) first by using all the ANOVA parameters; then in terms of the main effects, two-treatment-interactions and three-treatment interactions, thus deleting the highest-order interaction; then in terms of the main effects and two-treatment interactions; and finally in terms of only the main effects. Note that even when some columns corresponding to the excluded parameters are deleted from K for dual scaling, ANOVA is carried out in terms of the entire basis matrix, K. Thus, ANOVA tables for various deletions of the parameters for scaling remain the same in their format. The results are summarized in Table 8.14. The general mean is not listed in the table since it does not provide any useful information. The stepwise lines differentiate between parameters that are included in the maximization and those that are not. It is interesting to know that once a few

TABLE 8.13

Basis matrix K ($\times 16$) and identification of factors

Mean	A	B	AB	C	AC	BC	ABC	D	AD	BD	ABD	CD	ACD	BCD	ABCD
16	8	8	4	8	4	4	2	8	4	4	2	4	2	2	1
16	−8	8	−4	8	−4	4	−2	8	−4	4	−2	4	−2	2	−1
16	8	−8	−4	8	4	−4	−2	8	4	−4	−2	4	2	−2	−1
16	−8	−8	4	8	−4	−4	2	8	−4	−4	2	4	−2	−2	1
16	8	8	4	−8	−4	−4	−2	8	4	4	2	−4	−2	−2	−1
16	−8	8	−4	−8	4	−4	2	8	−4	4	−2	−4	2	−2	1
16	8	−8	−4	−8	−4	4	2	8	4	−4	−2	−4	−2	2	1
16	−8	−8	4	−8	4	4	−2	8	−4	−4	2	−4	2	2	−1
16	8	8	4	8	4	4	2	−8	−4	−4	−2	−4	−2	−2	−1
16	−8	8	−4	8	−4	4	−2	−8	4	−4	2	−4	2	−2	1
16	8	−8	−4	8	4	−4	−2	−8	−4	4	2	−4	−2	2	1
16	−8	−8	4	8	−4	−4	2	−8	4	4	−2	−4	2	2	−1
16	8	8	4	−8	−4	−4	−2	−8	−4	−4	−2	4	2	2	1
16	−8	8	−4	−8	4	−4	2	−8	4	−4	2	4	−2	2	−1
16	8	−8	−4	−8	−4	4	2	−8	−4	4	2	4	2	−2	−1
16	−8	−8	4	−8	4	4	−2	−8	4	4	−2	4	−2	−2	1

parameters are deleted maximization of $\mathrm{ss}(\hat{\boldsymbol{\theta}})$ does not seem to lead to minimization of the error term for ANOVA. See (7) and (8) in Table 8.14. This relation is reflected in the general reduction of **F** ratios. In spite of the increased value of the error term, however, one can still see the effects of scaling on differentiation between contributions that are included in the optimization and those that are not. For example, see AB, BD, and ABD. Since the choice of parameters to be deleted from scaling is arbitrary, this approach does not always provide useful ANOVA results. One example where this approach may be appropriate would be the situation in which the analysis involves too many high-order interactions to interpret. In this type of situation, one may as well delete high-order interaction terms from scaling.

Example 8.5: ANOVA *through selective scaling* (*selective deletion scheme*) On the basis of the entire analysis, this approach selects those parameters which have comparatively larger contributions than others for scaling. In Table 8.14 the results of the entire analysis using the 15 contrasts are listed in the first **F** column. We look for the parameters with relatively small **F** values, such as A (**F** = 1.85), D (0.03), AD (0.88), BC (0.11), ABC (0.50), and BCD (0.90). We then delete columns of K corresponding to these parameters. $\mathrm{ss}(\hat{\boldsymbol{\theta}})$, expressed in terms of the reduced basis matrix, is now subjected to dual scaling. As before, however, ANOVA is carried out, using the entire basis matrix, thus resulting in

TABLE 8.14

Scaling ANOVA – hierarchical deletion

Source of variation			15* F	14 F	10 F	4 F
(1)	Main	A	1.85	1.61	4.26	0.66
		B	60.58	60.19	42.16	50.20
		C	20.87	21.16	19.07	14.32
		D	0.03	0.04	0.53	0.76
(2)	2-factor int.	AB	78.27	76.86	74.62	23.73
		AC	7.47	7.92	1.78	2.71
		AD	0.88	0.92	0.08	1.13
		BC	0.11	0.20	0.01	0.04
		BD	74.96	74.38	68.67	29.51
		CD	16.47	16.89	15.98	11.02
(3)	3-factor int.	ABC	0.50	0.42	0.00	0.05
		ABD	239.28	239.61	117.76	82.77
		ACD	17.22	17.56	8.08	10.62
		BCD	0.90	0.71	0.23	0.04
(4)	4-factor int.	ABCD	7.98	6.54	5.74	2.45
(5)	$(ss_M/ss_b)\times 100$		15.80	15.81	18.38	28.67
	$(ss_2/ss_b)\times 100$		33.78	33.75	44.92	29.62
	$(ss_3/ss_b)\times 100$		48.91	49.20	35.10	40.65
	$(ss_4/ss_b)\times 100$		1.51	1.24	1.60	1.06
(6)	$\eta^2 = ss_b/ss_t$		0.9445	0.9442	0.9206	0.8812
(7)	ss_e		0.3532	0.3544	0.4632	0.5996
(8)	ss_t		6.3624	6.3558	5.8296	5.0483
(9)	λ†		0.7511	0.7408	0.4246	0.1594
(10)	Deleted terms		None	(4)	(3), (4)	(2), (3), (4)

*Number of contrasts.
†$\lambda = ss_b/\mathbf{x}'D\mathbf{x}$.

TABLE 8.15

Scaling ANOVA – selective deletion

Source of variation			15* F	9 F	7 F
(1)	Main	A	1.85	1.72	1.05
		B	60.58	60.91	62.92
		C	20.87	21.93	22.71
		D	0.03	0.02	0.02
(2)	2-factor int.	AB	78.27	79.12	80.47
		AC	7.47	6.77	5.46
		AD	0.88	0.53	0.32
		BC	0.11	0.08	0.12
		BD	74.96	74.84	76.99
		CD	16.47	17.34	19.34
(3)	3-factor int.	ABC	0.50	0.41	0.38
		ABD	239.28	236.67	241.63
		ACD	17.22	17.56	18.33
		BCD	0.90	0.50	0.18
(4)	4-factor int.	ABCD	7.98	7.84	6.84
(5)	$(ss_M/ss_b)\times 100$		15.80	16.07	16.15
	$(ss_2/ss_b)\times 100$		33.78	33.96	34.04
	$(ss_3/ss_b)\times 100$		48.91	48.48	48.54
	$(ss_4/ss_b)\times 100$		1.51	1.49	1.27
(6)	$\eta^2 = ss_b/ss_t$		0.9445	0.9444	0.9454
(7)	ss_e		0.3532	0.3537	0.3454
(8)	ss_t		6.3624	6.3583	6.3262
(9)	λ†		0.7511	0.7484	0.7276
(10)	Deleted terms		None	A, D, AD, BC, ABC, BCD	A, D, AC, AD, BC, ABC, BCD, ABCD

*Number of contrasts.
†$\lambda = ss_b/\mathbf{x}'D\mathbf{x}$.

the same output format as the original analysis. The results are summarized in Table 8.15. Again, the general mean is not listed there. In this example, after the initial deletion of six parameters, ANOVA results indicate that there are two other parameters which have relatively small **F** values, but that they are statistically significant. Therefore the analysis should stop at this point. In order to see the effects of further deletions of parameters under this scheme, however, we now delete these two parameters for scaling by striking out the two corresponding columns from K. Once **x** from the reduced K is obtained, ANOVA is carried out with respect to all the parameters. The results are given in the last column of Table 8.15. It is interesting to note that in this approach the deletion of a few parameters from scaling does not increase the contribution of errors (ss_e) very much. Thus the effects of scaling here seem more likely to override the opposite effect of parameter deletions than the approach discussed in example 8.4. The approach used for example 8.5 can bring out effects of some parameters to the level of statistical significance which may otherwise be non-significant, and as such it may provide a useful and effective means of information retrieval.

Example 8.6: *Multidimensional ANOVA* Since **x** is obtained from an eigenequation, there may be more than one significant weighting system. If so, ANOVA decompositions of linear composite scores derived from respective weighting systems can be carried out and the analysis may be called multidimensional ANOVA. This is what Nishisato (1971a) called component analysis of variance. To keep the entire analysis simple, one may include all the ANOVA parameters for scaling, rather than scaling the data with respect to specific parameters. In addition, one may prefer orthogonal weight vectors $\mathbf{w}_j$ to optimal and locally optimal weight vectors. Using the same data as in Table 8.12, we obtain three relatively large eigenvalues ($\lambda_1 = 0.7511, \lambda_2 = 0.2520, \lambda_3 = 0.1948$) and the corresponding eigenvectors ($\mathbf{w}_1, \mathbf{w}_2, \mathbf{w}_3$). ANOVA decompositions of the three dimensions are listed in Table 8.16. As is reflected by the values of ss_t, the first dimension accounts for a much greater portion of the variations in the data than the other two dimensions put together. The major contributors to the first dimension are ABD, BD, AB, B, C, ACD, CD, AC, and ABCD. The second and the third dimensions account for roughly the same amount of the variations in the data, which we can see from ss_t and **F** values for 'between.' The major contributors to the second dimensions are BCD, AD, CD, AC, and C, and the major contributors to the third dimension are AB and A.

As in factor analysis and principal component analysis, the investigator may wish to give names to the respective dimensions. The task may be difficult, but one may at least try either one or both of the following. The first is to name the dimensions by abstracting a pattern of major contributors of ANOVA parameters. The second is to name the dimensions by abstracting the common nature of major contributors of items to the respective dimensions.

TABLE 8.16

Scaling multidimensional ANOVA

Source of variation	F for dimension: I	II	III
Mean	0.21	0.71	0.01
A	2.22	4.57*	23.13**
B	52.97**	0.01	1.89
C	21.91**	9.00**	0.00
D	0.00	0.45	6.65*
AB	77.46**	0.14	45.85**
AC	7.87**	14.89**	6.40*
AD	1.86	19.60**	2.12
BC	0.25	0.51	3.58
BD	76.23**	0.34	0.23
CD	15.84**	16.43**	2.35
ABC	0.77	0.30	1.94
ABD	250.55**	4.76*	4.49*
ACD	16.18**	1.00	4.17*
BCD	1.63	36.31**	1.17
ABCD	7.87**	0.53	4.51*
Between	34.3310**	7.0421**	7.0007**
SS_b	65.0297	17.8116	13.5820
SS_w	3.7763	5.0586	3.8802
SS_t	68.8060	22.8702	17.4622

*Significant at $p<0.05$.
**Significant at $p<0.01$.

The first case is the characterization of dimensions in terms of the features of the rows of data matrix F, and the second is in terms of those of the columns. The two alternatives are possible because of the duality of the analysis.

8.3 MISSING RESPONSES

Up to now we have assumed that the data were in a 'perfect' form for data analysis. In practice, this assumption is often false, and the first task for the investigator is likely to be that of reorganizing the input data, in particular, identifying categories into which some unclassified responses (i.e., missing responses) should have presumably been placed. This is the problem of missing responses in dual scaling, and precludes the situation in which an incomplete plan (e.g., incomplete paired-comparison designs) is used as a design of the study, for then analysis of such data can be carried out as rigorously as analysis of complete data.

Omitted responses in multiple-choice testing often present serious practical problems. Owing to the absence of an appropriate method, omitted responses are generally handled in an arbitrary manner. Some investigators assign subjective penalty scores for omissions and some investigators simply remove subjects who omitted responses. Little is known, however, about possible consequences of such arbitrary procedures in data analysis. In the context of dual scaling, Nishisato and Levine (1975) investigated some aspects of omitted responses and Chan (1978) pursued some of their ideas with Monte Carlo computations.

Dual scaling of multiple-choice data is based on response patterns, where responses are specified only as present (1) or absent (0). Therefore, when some responses are omitted, our immediate concern is how to identify, on the basis of the observed responses, the subject's most likely response choice for each unanswered multiple-choice question. This, however, is not as simple a problem as it may appear. For, if one wants to insert optimal responses for the omissions, the approach of dual scaling may lead to a rather unreasonable outcome. For example, consider an extreme case where all the responses were omitted. Then one can always find a set of optimal responses for the omissions which provides perfect internal consistency. More generally, the procedure of inserting optimal responses for omissions leads to an increment in internal consistency which increases as the number of omissions increases. It is also obvious that the same procedure tends to result in a greater increment in internal consistency when the observed score distribution is less consistent. These observations lead to the view that insertion of optimal responses for omissions does not by itself provide a solution to the problem of omitted responses. What alternatives can we then consider? The following are some of the possible approaches, discussed by Nishisato and Levine (1975).

8.3.1 *Extra categories for omitted responses*

When some subjects omit responses to item k, one can create another category in addition to the m_k alternatives and enter 1's in the (m_k+1)th category for those who omitted responses. Not knowing whether there is any systematic pattern in the distribution of omitted responses, this method of extra categories seems to be an excellent possibility. This approach does not make any distinction between responses and omissions. Therefore the optimal weights for the extra categories are also determined so as to maximize η^2. If high scorers (low scorers) tend to omit responses to certain items, the extra categories of these items are expected to receive larger (smaller) weights so that the information regarding omissions may contribute to further differentiation between high scorers and low scorers. If omissions occur in a random fashion, however, the corresponding extra categories are expected to receive values which would not affect to any substantial degree the distribution of subjects' total scores.

This approach has another important feature. It treats incomplete data as though they are complete data, and thus does not encounter any problems associated with incomplete data (e.g., adjustment of degrees of freedom for certain statistics).

In comparison with the traditional (1,0) scoring scheme, the above procedure extracts information not only from responses to the 'incorrect' alternatives, but also from omitted responses in a differential manner.

Thus this approach appears to offer an intuitively reasonable alternative. However, when the number of omissions is relatively small this procedure may assign some extreme values (e.g., $-8, 30, 50$, when the majority of weights lie between -2 and 2) to the extra categories. If so, the extra categories, thus the omitted responses, may become main contributors to the scores of some subjects. In this regard, this approach calls for some caution. Chan (1978) investigated the recovery of information in the complete data from incomplete data scaled by this method, and showed an instance in which the recovery rate deteriorates rapidly as the percentage of omitted responses increases from 10% to 20%. Such a decline in the rate must be related to the distribution of eigenvalues associated with the complete data. More specifically, the rate must drastically decline when the dominance of the optimal dimension recedes into the background as a result of omitted responses. This relation, however, remains a conjecture. Anyway, one cannot always recommend this approach.

8.3.2 *Averages of observed scores*

Dual scaling provides a special built-in constraint on the weighted responses. The widely used practice in dual scaling is to use the condition that the sum of all the weighted scores is zero. Then, if subjects are asked to choose only one response alternative per item, the above condition also implies that the sum of the weighted responses to each item is zero (Guttman 1941). In other words, dual scaling is based on the condition that the average contributions of individual items to the total score of all the subjects are kept equal. When some responses are missing from data matrix F, dual scaling of F still satisfies the above condition. In this case, the diagonal elements of D_N are reduced from n to the actual number of responses, and score vector $\mathbf{y}$ is given by a vector which is collinear with the mean vector of optimally weighted *observed* responses. This is the second alternative.

This approach simply ignores omitted responses, and only the information of observed responses is used in optimization. This is a conservative approach, compared with the first one, in the sense that it does not attempt to extract information from omitted responses. In this regard, it is less vulnerable than the first to freak influences of omitted responses. The approach would work satisfactorily if all the items are relatively homogeneous in the statistical sense and if omissions occur in a random manner. Chan (1978) confirmed the empirical findings of Nishisato and Levine (1975) that this method is most robust against random and non-random omissions, and that it can recover comparatively well the information of the complete data even when 25% of responses are omitted.

8.3.3 *Insertion of the least consistent responses*

The most conservative way to deal with omitted responses is perhaps to insert for omissions those responses that would minimize the internal consistency. Let F_e be data matrix F with a certain number of cells empty. These empty cells correspond to the omitted responses. Now η^2 can be regarded as a function, say g, of the configuration of inserted responses. From the finite set of configurations, choose one which minimizes η^2 and calculate the corresponding weight vector $\mathbf{x}$ and score vector $\mathbf{y}$.

We can then test this minimal η^2 for significance in the usual way. If η^2 is significant, it provides strong support for $\mathbf{x}$ and $\mathbf{y}$ because any other configuration of inserted responses would also have produced significance. The question of which configuration to use presents another problem, however, which will be discussed later. Suppose now that η^2 is not significant. Then $\mathbf{x}$ and $\mathbf{y}$ may not be used. The implications then are that the questionnaire might have failed to show discrimination anyway, or that we might have lost a substantial amount of information through omitted responses.

8.3.4 *Insertion of the most consistent responses*

Given the response matrix F_e with omitted responses, one can insert a configuration of responses which maximizes η^2. This is a well-defined configuration, in which the locations of inserted responses for omissions are determined so as to raise η^2 as high as possible.

Once η^2 is maximized, the corresponding weight vector $\mathbf{x}$ and score vector $\mathbf{y}$ can be calculated, and η^2 can be tested for significance. If η^2 is not significant, one may conclude that, no matter how favourable the omitted responses could have been, the questionnaire would have failed to discriminate among subjects anyway. One must face the possibility that the administration of the questionnaire to another random sample from the same population might fail, even if the percentage of omitted responses were smaller. This remark, of course, remains a conjecture.

8.3.5 *Use of the minimal and maximal correlation ratios*

The minimal and maximal correlation ratios, discussed in 8.3.3 and 8.3.4, can be used together to determine the absolute range of η^2 when η^2 is considered as a function of inserted responses, given the observed responses. There are three distinct outcomes. The first is when the minimal η^2 is significant. The second is when the maximal η^2 is significant, but the minimal η^2 is not. The third is when the maximal η^2 is not significant. The second case is the only one which requires further investigation to determine whether or not the data are worth analysing. To answer this inquiry, one needs to consider the following two approaches. One approach is to identify an 'appropriate' method, whatever it may be, and to see whether the method produces a significant value of η^2. This is just like finding a point estimate, as against an interval estimate, of a parameter in statistics and testing the significance of the parameter under a null hypothesis. The other approach is to determine whether the amount of potential information loss due to omitted responses is substantial or negligible. This is a question regarding the validity of representativeness of sampled information and should therefore be of primary concern for investigators. For the first approach, Nishisato and Levine suggested, on the basis of their empirical research, that the method discussed in 8.3.2 might be a reasonable candidate. For the second approach, they presented the following suggestion.

8.3.6 *Assessment of information loss*

Let us indicate by $\eta^2(F;\mathbf{x})$ the function of the squared correlation ratio of data matrix F weighted by $\mathbf{x}$, and define the optimal function, ψ, by

$$\psi(F;\mathbf{x}) = \max_{\mathbf{x}} \eta^2(F;\mathbf{x}). \tag{8.69}$$

In Section 8.3.3, we defined matrix F_e. We now define another matrix F^*, which is a matrix constructed from F_e by filling the empty cells with arbitrarily assigned responses, 1's and 0's, under the restriction that each subject chooses only one response option per item. The optimal functions for these two matrices are

$$\psi(F_e;\mathbf{x}) = \max_{\mathbf{x}} \eta^2(F_e;\mathbf{x}), \tag{8.70}$$

$$\psi(F^*;\mathbf{x}) = \max_{\mathbf{x}} \eta^2(F^*;\mathbf{x}). \tag{8.71}$$

Of the three functions (8.69), (8.70), and (8.71), the first two are associated with fixed matrices, F and F_e, respectively, but the third varies as a function of arbitrary assignments of responses to the empty cells in F_e. This means that $\psi(F^*;\mathbf{x})$ can be maximized with respect to F^*, that is,

$$\max_{F^*} \psi(F^*;\mathbf{x}) = \max_{F^*}\left(\max_{\mathbf{x}} \eta^2(F^*;\mathbf{x})\right). \tag{8.72}$$

Since there are only a finite number of F^*'s for given F_e, this maximization amounts to computing $\max_{\mathbf{x}}\eta^2(F^*;\mathbf{x})$ for all possible F^*'s and choosing the one that maximizes η^2. It is then obvious that

$$\max_{F^*} \psi(F^*;\mathbf{x}) \geq \psi(F;\mathbf{x}), \tag{8.73}$$

$$\max_{F^*} \psi(F^*;\mathbf{x}) \geq \psi(F_e;\mathbf{x}). \tag{8.74}$$

The remaining possible relation between $\psi(F;\mathbf{x})$ and $\psi(F_e;\mathbf{x})$ depends on whether relatively consistent or inconsistent responses are deleted (omitted) from F to generate F_e. Recall that we are concerned with omited responses, and that the observed data matrix is then given by F_e. Thus (8.74) is the only meaningful relation in this context, and it seems to offer a clear-cut basis for assessment of information loss incurred by omitted responses. If we consider the difference

$$\omega = \max_{F^*} \psi(F^*;\mathbf{x}) - \psi(F_e;\mathbf{x}), \tag{8.75}$$

it presents a monotonically increasing function of the number of omitted responses. Nishisato and Levine then proposed the information-loss function

$$\phi = \log_e \omega. \tag{8.76}$$

This information measure resembles the chi-square statistic, but Nishisato and Levine simply demonstrated the behaviour of ϕ as a function of the number of omitted responses and the unreliability of data, leaving its statistical investigation for the future. It seems that the judgment of 'substantial loss' is a difficult problem to settle. Even with a large number of omitted responses, certain circumstances may not allow the investigator to stop data analysis. If the number of omitted responses is very small, however, the investigator may as well choose to discard those subjects who omitted responses, rather than analyse the entire data set. These two examples are counterexamples of what the function ϕ is intended for, that is, to indicate whether or not the information loss is substantial enough to abort analysis of the data. Anyway, the problem of omitted responses is ubiquitous and real. Further statistical characterizations of ϕ seem to be a worthwhile endeavour for dual scaling methodology.

8.4
CONCLUDING REMARKS

Dual scaling as presented here offers an objective way to look at data. More specifically, it is a way to simplify, reorganize, and evaluate categorical data. Unlike many other statistical procedures, it imposes practically no constraints on the way in which data are obtained. Greenacre (1978b, p. ix) states that 'it [correspondence analysis, a variant of dual scaling] has become a popular technique for analysing almost any matrix of positive numbers' and that 'perhaps the only fully-fledged assumption is that the data elements be nonnegative!' Even this assumption can be relaxed, as was seen in the dual scaling of modified matrix E proposed by Nishisato (1976, 1978a, b) and discussed in Chapters 6, 7, and 8.

Duality of the technique, as discussed in Chapter 3, presents an interesting framework for data analysis. We are so accustomed to handling individual differences by regarding them simply as random fluctuations and hence by using mean responses as basic quantities in data analysis. In dual scaling, however, we have observed that individual subjects are given the same status as stimuli under investigation. Once the data are represented in the subject-by-stimulus matrix, dual scaling determines weights for both subjects and stimuli so as to maximize simultaneously the between-subject and the between-stimulus sums of squares. In a typical data analysis, however, one would perhaps average out individual differences to calculate scale values of the stimuli. In dual scaling, the weights for subjects and those for stimuli are calculated as weighted averages of each other by maximally utilizing both individual differences and stimulus differences. The treatment of variables in the rows and the columns of the data matrix is completely symmetric.

This aspect of dual scaling has also been a target of the criticism that dual scaling capitalizes on chance fluctuations. This criticism, however, seems to require some elaboration. If it is intended to warn that the dual scaling results should not be generalized too far, it is one that is not necessarily restricted to dual scaling. In the normal theory of statistics, the specification of the population, random sampling, estimation of parameters, and sampling distributions of parameter estimates together specify the condition under which one can generalize the sample results to a relevant population with a certain probability. If one is willing to introduce similar assumptions in dual scaling, it may be that dual scaling might produce the results with that much generalizability. If, however, the above criticism is intended to say that we may obtain too good a result by chance to accept, the criticism should be levelled against most current statistical methodology. Mean, standard deviation, product-moment correlation, regression coefficient, multiple correlation, and canonical correlation are calculated from a sample of data and are examples of 'optimum' quantities in statistics. In this regard, they are not different from our optimal vectors. A basic goal of data analysis is to explain data. The process is carried out sometimes by using a statistical model and sometimes by examining internal relationships among the elements of data. No matter which course one may take, the basic strategy should be to utilize as much information in the data as possible. Indeed it sounds even strange that one should not utilize full information in the data to explain the data. 'To capitalize on chance fluctuation' leads to maximizing reproducibility of *individual responses* from scaled quantities.

As has been made clear in this book, dual scaling can be used as a transformation method of categorical data for analysis which requires continuous variates. Nishisato's *transform factor analysis* (1971b) shows a general framework, of which the work by Ishizuka (1976) and McDonald, Ishizuka, and Nishisato (1976) can be regarded as a special case. Since 1976 or so, Young, de Leeuw, and Takane have extended a similar line of work much further in relation to different types of analysis. Their work has presented a new vision to research in scaling, and may be called 'the second breakthrough, after Shepard's (1962a, b), in the history of MDS (multidimensional scaling) methodology' (Nishisato 1978d, p. 11). Dual scaling will undoubtedly be developed much further. To assist its development, however, empirical research is much desired. In spite of the ubiquity of categorical data, the author has found it rather difficult to find some types of real data for dual scaling, more specifically categorical data with 'subjects' as one of the classification variables. Most published data are in summarized forms such as contingency tables and tables of proportions, and are not amenable to analysis using individual differences as a source of information. Summarizing data often reveals a pattern hidden in the data, but almost always at the expense of information in the aggregated variates. Dual scaling, in contrast, makes maximum use of information in individual responses so as to reveal any hidden pattern in the data much more clearly than the simple aggregation of responses. The fact that dual scaling does not waste any information in raw data may be regarded as one of the most remarkable features of the technique.

Dual scaling can be, and should be, used much more than it now is for a wide variety of data analysis. It is a simple technique, but its application is not always so simple. 'Experience' seems to be the key word for further development of dual scaling methodology. Through constant interplay between empirical and theoretical work one can perhaps expand the scope of its applications in effective and useful ways. Inferential aspects of dual scaling deserve immediate attention by researchers.

APPENDICES

A
Matrix calculus

A.1
BASIC CONCEPTS AND OPERATIONS OF MATRIX ALGEBRA

A *matrix* is a rectangular array of numbers, the numbers being called *elements*. It is customary to indicate a matrix by a capital letter such as A, B, C, or D, and the elements by small letters with subscripts such as a_{ij}, b_{ij}, c_{ij}, or d_{ij}, where the first subscript i refers to the row and the second subscript j to the column; for example,

$$A = \begin{bmatrix} a_{11} & a_{12} & a_{13} \\ a_{21} & a_{22} & a_{23} \end{bmatrix} \begin{matrix} 1 \\ 2 \end{matrix} \quad \text{rows}$$

columns 1 2 3

A matrix such as A with two rows and three columns is often referred to as a 2×3 (two-by-three) matrix, and written as $A_{2\times3}$. The expression '2×3' is called the *order* or *dimension* of the matrix.

The *transpose* of A, denoted by A', is the matrix obtained from A by interchanging rows and columns. For example, if

$$A_{3\times2} = \begin{bmatrix} 512 & 23 \\ 3 & 7 \\ 21 & 0 \end{bmatrix},$$

then

$$A'_{2\times3} = \begin{bmatrix} 512 & 3 & 21 \\ 23 & 7 & 0 \end{bmatrix}.$$

A *vector* is a single row or column of elements. A *row vector* can be considered to be a $1\times m$ matrix, and a column vector an $n\times1$ matrix. For example, [2 7 3] is a row vector, and $\begin{bmatrix} 3 \\ 8 \end{bmatrix}$ is a column vector. Column vectors are often indicated by boldface small letters such as **a**, **b**, **c**, or **d**, and row vectors by **a**′, **b**′, **c**′, or **d**′ as if these were transposes of column vectors.

Let us write the matrices A and B in terms of their typical elements a_{ij} and b_{ij} as $A=(a_{ij})$ and $B=(b_{ij})$. Then if two matrices A and B are of the same order, that is, if the dimension of A is the same as the dimension of B, A and B are said to be *conformable for addition and subtraction*, which are defined as follows:

$$A + B = (a_{ij}) + (b_{ij}) = (a_{ij} + b_{ij}),$$

$$A - B = (a_{ij}) - (b_{ij}) = (a_{ij} - b_{ij}).$$

The operations are carried out on the corresponding elements of A and B. For example,

$$\begin{bmatrix} 2 & 3 \\ 1 & 5 \end{bmatrix} + \begin{bmatrix} 5 & 0 \\ 4 & -1 \end{bmatrix} = \begin{bmatrix} 7 & 3 \\ 5 & 4 \end{bmatrix}, \quad \begin{bmatrix} 6 \\ 3 \end{bmatrix} - \begin{bmatrix} 2 \\ 4 \end{bmatrix} = \begin{bmatrix} 4 \\ -1 \end{bmatrix}.$$

If the number of *columns* of A is equal to the number of *rows* of B, A is said to be *conformable to B for multiplication AB*. If A is $m \times r$ and B is $r \times n$, then the product AB is $m \times n$, that is, the number of rows of A times the number of columns of B. *Multiplication AB* is defined by

$$A_{m\times r} B_{r\times n} = C_{m\times n} = (c_{ij}) = (a_{ik})(b_{kj}) = \left(\sum_{k=1}^{r} a_{ik} b_{kj} \right).$$

A typical element c_{ij} of the product AB is the sum of the elementwise product of the ith *row* of A and the jth *column* of B. For example,

$$\begin{bmatrix} a_{11} & a_{12} \\ a_{21} & a_{22} \end{bmatrix} \begin{bmatrix} b_{11} & b_{12} \\ b_{21} & b_{22} \end{bmatrix} = \begin{bmatrix} (a_{11}b_{11} + a_{12}b_{21}) & (a_{11}b_{12} + a_{12}b_{22}) \\ (a_{21}b_{11} + a_{22}b_{21}) & (a_{21}b_{12} + a_{22}b_{22}) \end{bmatrix},$$

$$\begin{bmatrix} 1 & 2 \\ 3 & 4 \end{bmatrix} \begin{bmatrix} 8 & 6 \\ 7 & 5 \end{bmatrix} = \begin{bmatrix} (1\times 8 + 2\times 7) & (1\times 6 + 2\times 5) \\ (3\times 8 + 4\times 7) & (3\times 6 + 4\times 5) \end{bmatrix}$$

$$= \begin{bmatrix} 22 & 16 \\ 52 & 38 \end{bmatrix},$$

$$\begin{bmatrix} 2 \\ 5 \end{bmatrix} \begin{bmatrix} 3 & 1 \end{bmatrix} = \begin{bmatrix} 6 & 2 \\ 15 & 5 \end{bmatrix}, \quad \begin{bmatrix} 2 & 5 \end{bmatrix} \begin{bmatrix} 3 \\ 1 \end{bmatrix} = 11.$$

Note that the multiplication operation is that of 'row by column':

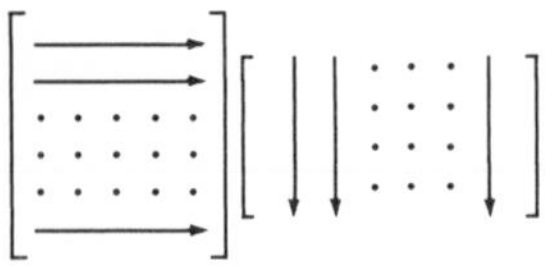

The multiplication of a matrix by a constant k (scalar) is termed scalar multiplication and is defined by

$$kA = (ka_{ij}).$$

For example,

$$10\begin{bmatrix} 1 & 2 \\ 3 & 4 \end{bmatrix} = \begin{bmatrix} 10 & 20 \\ 30 & 40 \end{bmatrix}, \qquad \frac{1}{4}\begin{bmatrix} 12 \\ 4 \end{bmatrix} = \begin{bmatrix} 3 \\ 1 \end{bmatrix}.$$

Elements a_{jj} of an $n \times n$ square matrix A are called *diagonal elements*, and the others, that is, a_{ij}, $i \neq j$, are called *off-diagonal elements*. A square matrix whose off-diagonal elements are all zero is called a *diagonal matrix*, which is often denoted by D. The following are examples of diagonal matrices:

$$D = \begin{bmatrix} d_{11} & 0 & 0 \\ 0 & d_{22} & 0 \\ 0 & 0 & d_{33} \end{bmatrix}, \qquad D = \begin{bmatrix} 1 & 0 & 0 & 0 \\ 0 & 5 & 0 & 0 \\ 0 & 0 & 4 & 0 \\ 0 & 0 & 0 & 7 \end{bmatrix}.$$

A diagonal matrix is often written in terms of the diagonal elements as $D = (d_{jj})$ or $D = \mathrm{diag}(d_{jj})$. If all the diagonal elements of D are identical, the matrix is called a *scalar matrix*. For example, the following are scalar matrices:

$$D = \begin{bmatrix} 3 & 0 \\ 0 & 3 \end{bmatrix}, \qquad D = \begin{bmatrix} 5 & 0 & 0 \\ 0 & 5 & 0 \\ 0 & 0 & 5 \end{bmatrix}.$$

If the diagonal elements of a scalar matrix are equal to one, the matrix is called the *identity matrix*, and indicated by I. For example,

$$I = \begin{bmatrix} 1 & 0 \\ 0 & 1 \end{bmatrix}, \qquad I = \begin{bmatrix} 1 & 0 & 0 \\ 0 & 1 & 0 \\ 0 & 0 & 1 \end{bmatrix}.$$

The identity matrix has the following property: if the matrices I and A are conformable for multiplication IA and AI, then

$$IA = A \quad \text{and} \quad AI = A.$$

For example,

$$\begin{bmatrix} 1 & 0 \\ 0 & 1 \end{bmatrix}\begin{bmatrix} 2 & 3 & 4 \\ 6 & 5 & 1 \end{bmatrix} = \begin{bmatrix} 2 & 3 & 4 \\ 6 & 5 & 1 \end{bmatrix},$$

$$\begin{bmatrix} 2 & 3 & 4 \\ 6 & 5 & 1 \end{bmatrix}\begin{bmatrix} 1 & 0 & 0 \\ 0 & 1 & 0 \\ 0 & 0 & 1 \end{bmatrix} = \begin{bmatrix} 2 & 3 & 4 \\ 6 & 5 & 1 \end{bmatrix}.$$

Associated with any square matrix A, there exists a quantity, called the *determinant*, which is a function of the elements of A, and closely related to the so-called volume function. The determinant of A is indicated by $|A|$. The determinant of a 2×2 matrix A is defined by

$$|A| = \begin{vmatrix} a_{11} & a_{12} \\ a_{21} & a_{22} \end{vmatrix} = a_{11}a_{22} - a_{12}a_{21}.$$

If the elements of the ith row and jth column are deleted from an $n \times n$ square matrix A, the determinant of the remaining $(n-1)\times(n-1)$ matrix is called the *minor* of the element a_{ij}, and indicated by $|M_{ij}|$. For example, if

$$A = \begin{bmatrix} a_{11} & a_{12} & a_{13} \\ a_{21} & a_{22} & a_{23} \\ a_{31} & a_{32} & a_{33} \end{bmatrix},$$

then

$$|M_{11}| = \begin{vmatrix} a_{22} & a_{23} \\ a_{32} & a_{33} \end{vmatrix} = a_{22}a_{33} - a_{23}a_{32},$$

$$|M_{12}| = \begin{vmatrix} a_{21} & a_{23} \\ a_{31} & a_{33} \end{vmatrix} = a_{21}a_{33} - a_{23}a_{31},$$

$$|M_{13}| = \begin{vmatrix} a_{21} & a_{22} \\ a_{31} & a_{32} \end{vmatrix} = a_{21}a_{32} - a_{22}a_{31},$$

$$|M_{21}| = \begin{vmatrix} a_{12} & a_{13} \\ a_{32} & a_{33} \end{vmatrix} = a_{12}a_{33} - a_{13}a_{32}.$$

and so on. The minor with sign $(-1)^{i+j}$ is called the *co-factor* of a_{ij} and indicated by α_{ij}, that is,

$$\alpha_{ij} = (-1)^{i+j}|M_{ij}|.$$

For example, the co-factors of the above example are

$$\alpha_{11} = (-1)^{1+1}|M_{11}| = a_{22}a_{33} - a_{23}a_{32},$$

$$\alpha_{12} = (-1)^{1+2}|M_{12}| = -a_{21}a_{33} + a_{23}a_{31},$$

$$\alpha_{13} = (-1)^{1+3}|M_{13}| = a_{21}a_{32} - a_{22}a_{31},$$

$$\alpha_{21} = (-1)^{2+1}|M_{21}| = -a_{12}a_{33} + a_{13}a_{32}.$$

The determinant of A can be expressed as a sum of the product of each element of a row (or column) of A times its co-factor:

$$|A| = \sum_{i=1}^{n} a_{ij}\alpha_{ij} \quad \text{for any } j \text{ (column expansion)},$$

$$= \sum_{j=1}^{n} a_{ij}\alpha_{ij} \quad \text{for any } i \text{ (row expansion)}.$$

The following is an example of a row expansion:

$$\begin{vmatrix} 2 & 1 & 0 \\ 3 & 5 & 4 \\ 6 & 7 & 8 \end{vmatrix} = 2(-1)^{1+1}\begin{vmatrix} 5 & 4 \\ 7 & 8 \end{vmatrix} + 1(-1)^{1+2}\begin{vmatrix} 3 & 4 \\ 6 & 8 \end{vmatrix} + 0(-1)^{1+3}\begin{vmatrix} 3 & 5 \\ 6 & 7 \end{vmatrix}$$

$$= 2\begin{vmatrix} 5 & 4 \\ 7 & 8 \end{vmatrix} - \begin{vmatrix} 3 & 4 \\ 6 & 8 \end{vmatrix} = 2[(5\times 8)-(4\times 7)] - [(3\times 7)-(4\times 6)] = 24$$

and column expansion:

$$\begin{vmatrix} 2 & 1 & 0 \\ 3 & 5 & 4 \\ 6 & 7 & 8 \end{vmatrix} = 2\begin{vmatrix} 5 & 4 \\ 7 & 8 \end{vmatrix} - 3\begin{vmatrix} 1 & 0 \\ 7 & 8 \end{vmatrix} + 6\begin{vmatrix} 1 & 0 \\ 5 & 4 \end{vmatrix}$$

$$= 2[(5\times 8)-(4\times 7)] - 3[(1\times 8)-(0\times 7)] + [(1\times 4)-(0\times 5)] = 24.$$

When the matrix is of high order, however, one would not usually calculate the determinant in this manner. There are other techniques appropriate for such matrices. A matrix A is said to be *singular* if $|A|=0$ and *non-singular* if $|A|\neq 0$.

The *inverse* of A, denoted by A^{-1}, is defined for a non-singular matrix A by the relation

$$AA^{-1} = A^{-1}A = I.$$

In matrix algebra, multiplying by an inverse is analogous to multiplying by a reciprocal in scalar mathematics. Singular matrices do not have inverses defined as above.

If $D=\mathrm{diag}(d_{jj})$ is a non-singular diagonal matrix, the inverse of D, D^{-1}, is given by

$$D^{-1} = \mathrm{diag}(1/d_{jj}).$$

For example, if

$$D = \begin{bmatrix} 2 & 0 & 0 \\ 0 & 5 & 0 \\ 0 & 0 & 10 \end{bmatrix},$$

then

$$D^{-1} = \begin{bmatrix} 0.5 & 0 & 0 \\ 0 & 0.2 & 0 \\ 0 & 0 & 0.1 \end{bmatrix}.$$

It is obvious that the inverse of the identity matrix is the identity matrix. The inverse of a matrix A can be expressed generally in terms of the matrix of co-factors of a_{ij} and the determinant of A as follows:

$$A^{-1} = (\alpha_{ij})'/|A| = (\alpha_{ji})/|A|.$$

For example, if

$$A = \begin{bmatrix} 4 & 1 \\ 2 & 3 \end{bmatrix},$$

then

$$\alpha_{11} = (-1)^{1+1}|M_{11}| = 3,$$

$$\alpha_{12} = (-1)^{1+2}|M_{12}| = -2,$$

$$\alpha_{21} = (-1)^{2+1}|M_{21}| = -1,$$

$$\alpha_{22} = (-1)^{2+2}|M_{22}| = 4.$$

Thus

$$(\alpha_{ij})' = (\alpha_{ji}) = \begin{bmatrix} 3 & -1 \\ -2 & 4 \end{bmatrix}.$$

The determinant is

$$|A| = (4\times3) - (1\times2) = 10.$$

Therefore

$$A^{-1} = \frac{1}{10}\begin{bmatrix} 3 & -1 \\ -2 & 4 \end{bmatrix}.$$

Similarly, if

$$A = \begin{bmatrix} a_{11} & a_{12} \\ a_{21} & a_{22} \end{bmatrix},$$

then

$$A^{-1} = \frac{1}{|A|}\begin{bmatrix} a_{22} & -a_{12} \\ -a_{21} & a_{11} \end{bmatrix}.$$

Again, when matrices are of high order, there are other techniques of matrix inversion which are more appropriate than the above.

Let A be an $m\times n$ matrix. Select any s rows and t columns, where $s<m$ and $t<n$. The $s\times t$ matrix which consists of these s rows and t columns is called a *minor* matrix of A. If s is equal to t, the determinant of this matrix is called a *minor determinant* of A. If A contains at least one $k\times k$ minor matrix whose determinant is non-zero, and if all the minor determinants of $(k+1)\times(k+1)$ minor matrices are zero, then A is said to be of *rank* k, which is indicated by $r(A)=k$. For example,

$$A = \begin{bmatrix} 2 & 4 & 1 \\ 1 & 2 & 0 \\ 1 & 2 & 1 \end{bmatrix} \text{ is of rank 2.}$$

Because the determinant of the 3×3 matrix is zero,

$$\begin{aligned} A &= 2\begin{vmatrix} 2 & 0 \\ 2 & 1 \end{vmatrix} - \begin{vmatrix} 4 & 1 \\ 2 & 1 \end{vmatrix} + \begin{vmatrix} 4 & 1 \\ 2 & 0 \end{vmatrix} \\ &= 2[(2\times1)-(0\times2)] - [(4\times1)-(1\times2)] + [(4\times0)-(1\times2)] \\ &= 0, \end{aligned}$$

the rank cannot be three. But there is at least one 2×2 matrix whose minor determinant is non-zero. For instance

$$\begin{vmatrix} 2 & 0 \\ 2 & 1 \end{vmatrix} = 2 \neq 0.$$

Therefore the rank of A, $r(A)$, is two.

A set of n $(m\times1)$ vectors $(\mathbf{a}_1, \mathbf{a}_2, \ldots, \mathbf{a}_n)$ is said to be *linearly dependent* if and only if there exists a set of numbers $(c_1, c_2, \ldots, c_n)$, not all of which are zero, such that

$$c_1\mathbf{a}_1 + c_2\mathbf{a}_2 + \cdots + c_n\mathbf{a}_n = \mathbf{0}.$$

For example, vectors

$$\begin{bmatrix} 2 \\ 3 \\ 1 \end{bmatrix}, \quad \begin{bmatrix} 1 \\ 4 \\ 2 \end{bmatrix}, \quad \begin{bmatrix} 3 \\ 2 \\ 0 \end{bmatrix}$$

are linearly dependent, for the set of numbers $(2, -1, -1)$ yields

$$2\begin{bmatrix} 2 \\ 3 \\ 1 \end{bmatrix} - \begin{bmatrix} 1 \\ 4 \\ 2 \end{bmatrix} - \begin{bmatrix} 3 \\ 2 \\ 0 \end{bmatrix} = \begin{bmatrix} 0 \\ 0 \\ 0 \end{bmatrix}.$$

If a set of vectors is not linearly dependent, it is said to be *linearly independent*. In other words, a set of vectors is linearly independent if and only if the equation $c_1\mathbf{a}_1 + c_2\mathbf{a}_2 + \cdots + c_n\mathbf{a}_n = \mathbf{0}$ implies that $c_1 = c_2 = \cdots = c_n = 0$. For example, vectors

$$\begin{bmatrix} 1 \\ 0 \\ 0 \end{bmatrix}, \quad \begin{bmatrix} 0 \\ 1 \\ 0 \end{bmatrix}, \quad \begin{bmatrix} 0 \\ 0 \\ 1 \end{bmatrix}$$

are linearly independent, for the equation

$$c_1\begin{bmatrix} 1 \\ 0 \\ 0 \end{bmatrix} + c_2\begin{bmatrix} 0 \\ 1 \\ 0 \end{bmatrix} + c_3\begin{bmatrix} 0 \\ 0 \\ 1 \end{bmatrix} = \begin{bmatrix} c_1 \\ c_2 \\ c_3 \end{bmatrix} = \begin{bmatrix} 0 \\ 0 \\ 0 \end{bmatrix}$$

indicates that $c_1 = c_2 = c_3 = 0$.

In terms of linear independence of vectors, the rank of $A, r(A)$, can be defined as the maximum number of linearly independent vectors that one can construct from the set $(\mathbf{a}_1, \mathbf{a}_2, \ldots, \mathbf{a}_n)$.

A.2
SYSTEMS OF LINEAR EQUATIONS

Consider a set of m linear equations in n unknowns $x_1, x_2, \ldots, x_n$:

$$\begin{aligned} a_{11}x_1 + a_{12}x_2 + \cdots + a_{1n}x_n &= c_1, \\ a_{21}x_1 + a_{22}x_2 + \cdots + a_{2n}x_n &= c_2, \\ \vdots \qquad\quad \vdots \qquad\quad & \quad \vdots \\ a_{m1}x_1 + a_{m2}x_2 + \cdots + a_{mn}x_n &= c_m. \end{aligned}$$

where a_{ij} and c_i are known constants. This set of equations can be expressed by

$$A\mathbf{x} = \mathbf{c},$$

where

$$A = \begin{bmatrix} a_{11} & a_{12} & \cdots & a_{1n} \\ a_{21} & a_{22} & \cdots & a_{2n} \\ \vdots & \vdots & \ddots & \\ a_{m1} & a_{m2} & \cdots & a_{mn} \end{bmatrix}, \quad \mathbf{x} = \begin{bmatrix} x_1 \\ x_2 \\ \vdots \\ x_n \end{bmatrix}, \quad \mathbf{c} = \begin{bmatrix} c_1 \\ c_2 \\ \vdots \\ c_m \end{bmatrix}.$$

When A is square (i.e., $m=n$) and non-singular (i.e., $|A| \neq 0$), there exists a unique solution for $\mathbf{x}$, which can be obtained by premultiplying both sides of the equation by the inverse of A,

$$A^{-1}A\mathbf{x} = A^{-1}\mathbf{c}.$$

Since $A^{-1}A\mathbf{x} = I\mathbf{x} = \mathbf{x}$, we obtain

$$\mathbf{x} = A^{-1}\mathbf{c}.$$

For example, consider

$$\begin{aligned} 2x_1 + 3x_2 &= 16, \\ 4x_1 + 8x_2 &= 36, \end{aligned} \quad \text{or} \quad \begin{bmatrix} 2 & 3 \\ 4 & 8 \end{bmatrix} \begin{bmatrix} x_1 \\ x_2 \end{bmatrix} = \begin{bmatrix} 16 \\ 36 \end{bmatrix}.$$

The matrix of coefficients is square and non-singular. Therefore

$$\begin{bmatrix} x_1 \\ x_2 \end{bmatrix} = \begin{bmatrix} 2 & 3 \\ 4 & 8 \end{bmatrix}^{-1} \begin{bmatrix} 16 \\ 36 \end{bmatrix} = \frac{1}{4} \begin{bmatrix} 8 & -3 \\ -4 & 2 \end{bmatrix} \begin{bmatrix} 16 \\ 36 \end{bmatrix} = \begin{bmatrix} 5 \\ 2 \end{bmatrix},$$

that is, $x_1 = 5$ and $x_2 = 2$.

When A is not square (i.e., $m \neq n$), the set of linear equations is solvable only if

$$r(A) = r(A, \mathbf{c}),$$

where the matrix $(A, \mathbf{c})$, that is, A flanked by $\mathbf{c}$, is called the *augmented* matrix. In other words, the set of linear equations is solvable only if the vector of constants, $\mathbf{c}$, is linearly dependent on the columns of A, that is, only if $\mathbf{c}$ can be expressed as a linear combination of the columns of A (note that if $n+1$ vectors constitute a set of linearly dependent vectors, any vector in the set can be expressed as a linear combination of the remaining n vectors). This condition becomes understandable once we express the set of linear equations in the form

$$\begin{bmatrix} a_{11} \\ a_{21} \\ \vdots \\ a_{m1} \end{bmatrix} x_1 + \begin{bmatrix} a_{12} \\ a_{22} \\ \vdots \\ a_{m2} \end{bmatrix} x_2 + \cdots + \begin{bmatrix} a_{1n} \\ a_{2n} \\ \vdots \\ a_{mn} \end{bmatrix} x_n = \begin{bmatrix} c_1 \\ c_2 \\ \vdots \\ c_m \end{bmatrix},$$

or

$$\sum \mathbf{a}_j x_j = \mathbf{c}.$$

It is easy to see that when $\mathbf{c}$ is expressed as a linear combination of the column vectors $\mathbf{a}_j$ of A then the coefficients of $\mathbf{a}_j$, that is, x_j, provide the solution. However, if $r(A) \neq r(A,\mathbf{c})$, it is not possible to find values of x_j such that $\Sigma \mathbf{a}_j x_j = \mathbf{c}$, for the last equation implies that $r(A) = r(A,\mathbf{c})$.

Suppose that a solution exists and that $r(A) = k(<n)$. Then, we assign arbitrary values to $n-k$ unknowns and determine the remaining k unknowns. For example, consider

$$\begin{bmatrix} 4 & 1 & 0 \\ 2 & 3 & 1 \end{bmatrix} \begin{bmatrix} x_1 \\ x_2 \\ x_3 \end{bmatrix} = \begin{bmatrix} 3 \\ 1 \end{bmatrix}, \quad \text{or} \quad A\mathbf{x} = \mathbf{c}.$$

The rank of any $m \times n$ matrix cannot be larger than the smaller value of m and n. In the present example, A is 2×3 and $(A,\mathbf{c})$ is 2×4. Therefore their ranks cannot be greater than two. However, since A has a non-vanishing 2×2 submatrix, the ranks of A and $(A,\mathbf{c})$ have to be greater than one. The conclusion is that $r(A) = r(A,\mathbf{c}) = 2$. Since $n = 3$ and $r(A) = 2$, we assign an arbitrary value to one (i.e., $3-2$) unknown. Let $x_1 = 0$. Then we have

$$\begin{bmatrix} 1 & 0 \\ 3 & 1 \end{bmatrix} \begin{bmatrix} x_2 \\ x_3 \end{bmatrix} = \begin{bmatrix} 3 \\ 1 \end{bmatrix}.$$

Thus

$$\begin{bmatrix} x_2 \\ x_3 \end{bmatrix} = \begin{bmatrix} 1 & 0 \\ 3 & 1 \end{bmatrix}^{-1} \begin{bmatrix} 3 \\ 1 \end{bmatrix} = \begin{bmatrix} 3 \\ -8 \end{bmatrix}.$$

That is, $x_2 = 3$ and $x_3 = -8$ (and $x_1 = 0$).

If $r(A) = r(A,\mathbf{c}) = k$ and $k < m$, retain k independent equations, discarding the others, and solve the set of k equations for k unknowns by assigning arbitrary values to the remaining $n-k$ unknowns. For example, consider

$$\begin{bmatrix} 4 & 1 & 0 \\ 2 & 3 & 1 \\ 4 & 6 & 2 \end{bmatrix} \begin{bmatrix} x_1 \\ x_2 \\ x_3 \end{bmatrix} = \begin{bmatrix} 3 \\ 1 \\ 2 \end{bmatrix}, \quad \text{or} \quad A\mathbf{x} = \mathbf{c}.$$

The second and the third rows of A are linearly dependent. Thus the third row of A and the third element of $\mathbf{c}$ can be eliminated. The remaining set of equations is the same as the above example, and the same solution set $(x_1 = 0, x_2 = 3, x_3 = -8)$ satisfies the equation $A\mathbf{x} = \mathbf{c}$.

When $\mathbf{c} = \mathbf{0}$, that is $A\mathbf{x} = \mathbf{0}$, the set is called a set of *homogeneous* equations, and there always exists a solution, $\mathbf{x} = \mathbf{0}$, called the *trivial* solution. If A is square and non-singular, the trivial solution is the only solution. If A is square but singular, there exists a non-trivial solution. Then the set of homogeneous equations can be changed to a set of non-homogeneous equations by assigning arbitrary values of $n-k$ unknowns, where $k = r(A)$. Then solve the equations by retaining k independent equations. For example, consider

$$\begin{bmatrix} 5 & 3 & 2 \\ 4 & 3 & 1 \\ 3 & 0 & 3 \end{bmatrix} \begin{bmatrix} x_1 \\ x_2 \\ x_3 \end{bmatrix} = \begin{bmatrix} 0 \\ 0 \\ 0 \end{bmatrix}, \quad \text{or} \quad A\mathbf{x} = \mathbf{0}.$$

$n=3$ and $r(A)=2$. Therefore we assign an arbitrary value to one of the three unknowns. Let x_3 be equal to -1. Then we transfer the corresponding vector $\mathbf{a}_3 x_3$, that is, $-\mathbf{a}_3$, to the right-hand side of the equation, which results in the equation

$$\begin{bmatrix} 5 & 3 \\ 4 & 3 \\ 3 & 0 \end{bmatrix} \begin{bmatrix} x_1 \\ x_2 \end{bmatrix} = \begin{bmatrix} 2 \\ 1 \\ 3 \end{bmatrix}, \quad \text{or} \quad A^*\mathbf{x}^* = \mathbf{c}^*.$$

Since $r(A^*)=r(A^*,\mathbf{c}^*)=2$, delete the third row of A^*, which is linearly dependent on the first two rows (i.e., $[3 \;\; 0]=3[5 \;\; 3]-3[4 \;\; 3]$). Then

$$\begin{bmatrix} 5 & 3 \\ 4 & 3 \end{bmatrix} \begin{bmatrix} x_1 \\ x_2 \end{bmatrix} = \begin{bmatrix} 2 \\ 1 \end{bmatrix},$$

and we obtain

$$\begin{bmatrix} x_1 \\ x_2 \end{bmatrix} = \begin{bmatrix} 1 \\ -1 \end{bmatrix}, \quad \text{and} \quad x_3 = -1 \text{ (by assignment)}.$$

A.3
ORTHOGONAL TRANSFORMATION AND EIGENVALUES

A square matrix P is said to be *orthogonal* if and only if its inverse is the same as its transpose, that is

$$P^{-1} = P'.$$

This is the same as stating that

$$PP' = I \quad \text{and} \quad P'P = I.$$

The following are examples of orthogonal matrices:

$$\begin{bmatrix} \frac{1}{\sqrt{2}} & -\frac{1}{\sqrt{2}} \\ \frac{1}{\sqrt{2}} & \frac{1}{\sqrt{2}} \end{bmatrix}, \quad \begin{bmatrix} \cos\theta & \sin\theta \\ -\sin\theta & \cos\theta \end{bmatrix}, \quad \begin{bmatrix} 1 & 0 & 0 \\ 0 & 1 & 0 \\ 0 & 0 & 1 \end{bmatrix}.$$

Note that $|P|$ is either 1 or -1.

Let us consider the rectangular coordinates, (x,y), of a point T and rotate axes X and Y by an angle θ to obtain new coordinates, x^*,y^*, of T. This transformation can be described by the relation

$$\begin{bmatrix} x^* \\ y^* \end{bmatrix} = \begin{bmatrix} \cos\theta & \sin\theta \\ -\sin\theta & \cos\theta \end{bmatrix} \begin{bmatrix} x \\ y \end{bmatrix}$$

or

$$\mathbf{u}^* = P\mathbf{u},$$

where $\mathbf{u}^* = \begin{bmatrix} x^* \\ y^* \end{bmatrix}$, $\mathbf{u} = \begin{bmatrix} x \\ y \end{bmatrix}$, and P is the 2×2 orthogonal matrix. This is called an *orthogonal coordinate transformation* of $\mathbf{u}$. If $|P|=1$, the orientation of the axes is preserved, but if $|P|=-1$, it is altered.

We can apply this transformation to another important topic in matrix algebra, a quadratic form. Suppose A is an $n\times n$ matrix such that $A'=A$, namely, a *symmetric matrix*, and $\mathbf{x}$ is an $n\times1$ column vector. Then the scalar $\mathbf{x}'A\mathbf{x}$ is called a *quadratic form* in $\mathbf{x}$. For example, consider a polynomial of the second degree,

$$ax^2 + 2bxy + cy^2 = \begin{bmatrix} x & y \end{bmatrix} \begin{bmatrix} a & b \\ b & c \end{bmatrix} \begin{bmatrix} x \\ y \end{bmatrix} = \mathbf{u}'A\mathbf{u}, \text{ say.}$$

This is quadratic in x and y. By an orthogonal coordinate transformation, it is always possible to express it in terms of only the quadratic terms,

$$\lambda_1 x^{*2} + \lambda_2 y^{*2} = \begin{bmatrix} x^* & y^* \end{bmatrix} \begin{bmatrix} \lambda_1 & 0 \\ 0 & \lambda_2 \end{bmatrix} \begin{bmatrix} x^* \\ y^* \end{bmatrix} = \mathbf{u}^{*\prime}\Lambda\mathbf{u}^*, \text{ say,}$$

where λ_1 and λ_2 are constants and $\Lambda=\mathrm{diag}(\lambda_j)$. It is known that this problem of transformation can be handled algebraically by solving the equation

$$(A-\lambda_j I)\mathbf{p}_j = \mathbf{0},$$

where $\mathbf{p}_j$ is the jth column of P. This is a set of homogeneous equations and has a trivial solution, $\mathbf{p}_j=\mathbf{0}$. As discussed in the previous section, the system has non-trivial solutions if and only if the determinant of the coefficient matrix vanishes, that is, if and only if

$$|A-\lambda_i I| = 0.$$

This is called the *characteristic equation* of the quadratic form $\mathbf{x}'A\mathbf{x}$. λ_i are called the *eigenvalues* (*latent roots, characteristic roots, proper values*) and $\mathbf{p}_j$ are called the *eigenvectors* (*latent vectors, characteristic vectors, proper vectors*) associated with λ_j. In our example, the characteristic equation is

$$|A-\lambda_j I| = \begin{vmatrix} a-\lambda_j & b \\ b & c-\lambda_j \end{vmatrix} = (a-\lambda_j)(c-\lambda_j) - b^2$$

$$= \lambda_j^2 - (a+c)\lambda_j + (ac-b^2) = 0.$$

Once λ_j are obtained, the corresponding $\mathbf{p}_j$ can be obtained by solving $(A-\lambda_j I)\mathbf{p}_j=\mathbf{0}$, and hence $\mathbf{u}^*$ by the relation $\mathbf{u}^*=P\mathbf{u}$.

The form $\mathbf{u}^*\Lambda\mathbf{u}^*$ thus obtained is called the *canonical form* of $\mathbf{u}'A\mathbf{u}$, and the axes, rotated to the canonical form, are called the *principal axes*.

Let us find the canonical form of

$$5x^2+8xy+5y^2=9,$$

that is,

$$\begin{bmatrix} x & y \end{bmatrix}\begin{bmatrix} 5 & 4 \\ 4 & 5 \end{bmatrix}\begin{bmatrix} x \\ y \end{bmatrix}=9.$$

The characteristic equation is

$$\left|\begin{bmatrix} 5 & 4 \\ 4 & 5 \end{bmatrix}-\lambda\begin{bmatrix} 1 & 0 \\ 0 & 1 \end{bmatrix}\right|=\begin{vmatrix} 5-\lambda & 4 \\ 4 & 5-\lambda \end{vmatrix}=(5-\lambda)^2-16=(\lambda-9)(\lambda-1)=0.$$

The two eigenvalues are $\lambda_1=9$ and $\lambda_2=1$. Hence the canonical form is given by

$$9x^{*2}+y^{*2}=9.$$

The relation between the original equation and its canonical form is illustrated in Figure A.1. Note that the principal axis Y^* cuts through the points where the distance of the ellipse from the origin is largest, and that the principal axis X^* cuts through those where the distance is smallest.

Now, solving

$$\begin{bmatrix} 5-9 & 4 \\ 4 & 5-9 \end{bmatrix}\begin{bmatrix} p_{11} \\ p_{21} \end{bmatrix}=\mathbf{0}, \qquad \begin{bmatrix} 5-1 & 4 \\ 4 & 5-1 \end{bmatrix}\begin{bmatrix} p_{12} \\ p_{22} \end{bmatrix}=\mathbf{0},$$

we obtain $p_{11}=p_{21}$ and $p_{12}=-p_{22}$. Thus we arbitrarily set $p_{11}=p_{21}=1, p_{12}=1$, and $p_{22}=-1$, and then *normalize* them by using the relation

$$p_{ij}^*=p_{ij}/\Sigma p_{ij}^2.$$

The matrix P with elements p_{ij}^* is now orthogonal,

$$P=\frac{1}{\sqrt{2}}\begin{bmatrix} 1 & 1 \\ 1 & -1 \end{bmatrix}.$$

The columns of P are eigenvectors. The orthogonal coordinate transformation for this canonical reduction is given by

$$\begin{bmatrix} X^* \\ Y^* \end{bmatrix}=\frac{1}{\sqrt{2}}\begin{bmatrix} 1 & 1 \\ 1 & -1 \end{bmatrix}\begin{bmatrix} X \\ Y \end{bmatrix}.$$

Since $\cos\theta=1/\sqrt{2}$ and $\sin\theta=1/\sqrt{2}$, we obtain the angle of rotation $\theta=45°$.

The topic of eigenvalues and eigenvectors is extremely important in statistics. Principal component analysis, multiple discriminant analysis, and canonical analysis, to name a few, are all based on finding eigenvalues and

Figure A.1
A quadratic form, $5x^2 + 8xy + 5y^2 = 9$, and its conical form, $9x^{*2} + y^{*2} = 9$

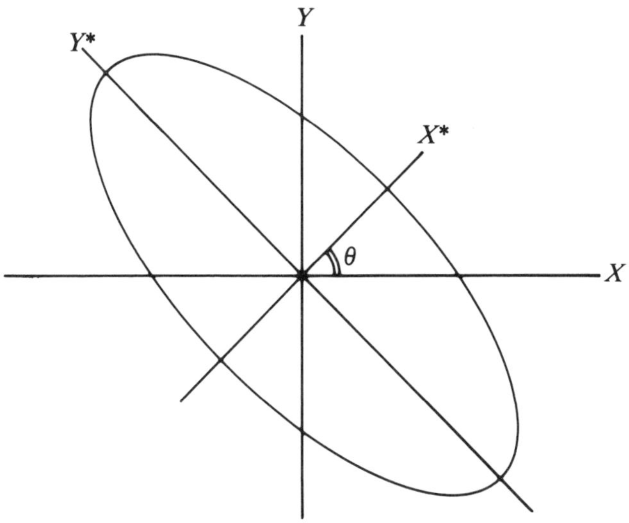

eigenvectors of certain matrices. Dual scaling is also nothing but a procedure to find eigenvalues and eigenvectors of a special matrix. To make the foregoing discussion more general, let us consider the characteristic equation of an $n \times n$ matrix A. This can be expressed as a polynomial of the nth degree in λ,

$$|A - \lambda I| = \begin{vmatrix} a_{11}-\lambda & a_{12} & a_{13} & \cdots & a_{1n} \\ a_{21} & a_{22}-\lambda & a_{23} & \cdots & a_{2n} \\ a_{31} & a_{32} & a_{33}-\lambda & \cdots & a_{3n} \\ \vdots & \vdots & \vdots & \ddots & \vdots \\ a_{n1} & a_{n2} & a_{n3} & \cdots & a_{nn}-\lambda \end{vmatrix}$$

$$= c_0(-\lambda)^n + c_1(-\lambda)^{n-1} + \cdots + c_{n-1}(-\lambda) + c_n$$

$$= \sum_{m=0}^{n} c_m(-\lambda)^{n-m} = 0,$$

where $c_0 = 1$, c_1 is the sum of the diagonal elements of A, which is called the *trace* of A and indicated by $\mathrm{tr}(A)$, c_m is the sum of all the m-rowed principal minor determinants, and $c_n = |A|$. For example,

$$\begin{vmatrix} a_{11}-\lambda & a_{12} & a_{13} \\ a_{21} & a_{22}-\lambda & a_{23} \\ a_{31} & a_{32} & a_{33}-\lambda \end{vmatrix} = c_0(-\lambda)^3 + c_1(-\lambda)^2 + c_2(-\lambda)^1 + c_3(-\lambda)^0$$

$$= -\lambda^3 + c_1\lambda^2 - c_2\lambda + c_3,$$

where

$$c_1 = a_{11} + a_{22} + a_{33},$$

$$c_2 = \begin{vmatrix} a_{11} & a_{12} \\ a_{21} & a_{22} \end{vmatrix} + \begin{vmatrix} a_{11} & a_{13} \\ a_{31} & a_{33} \end{vmatrix} + \begin{vmatrix} a_{22} & a_{23} \\ a_{32} & a_{33} \end{vmatrix}$$

$$= a_{11}a_{22} - a_{12}a_{21} + a_{11}a_{33} - a_{13}a_{31} + a_{22}a_{33} - a_{23}a_{32},$$

and

$$c_3 = |A| = a_{11}\begin{vmatrix} a_{22} & a_{23} \\ a_{32} & a_{33} \end{vmatrix} - a_{21}\begin{vmatrix} a_{12} & a_{13} \\ a_{32} & a_{33} \end{vmatrix} + a_{31}\begin{vmatrix} a_{12} & a_{13} \\ a_{22} & a_{23} \end{vmatrix}$$

$$= a_{11}(a_{22}a_{33} - a_{23}a_{32}) - a_{21}(a_{12}a_{33} - a_{13}a_{32}) + a_{31}(a_{12}a_{23} - a_{13}a_{22}).$$

Let A be an $n \times n$ symmetric matrix (i.e., $A = A'$), Λ be the $n \times n$ diagonal matrix with the eigenvalues of A in the main diagonal, and P be the $n \times n$ matrix with the eigenvectors in its columns,

$$\Lambda = \text{diag}(\lambda_1, \lambda_2, \ldots, \lambda_n) \quad \text{and} \quad P = (\mathbf{p}_1, \mathbf{p}_2, \ldots, \mathbf{p}_n).$$

Then the relation

$$(A - \lambda_j I)\mathbf{p}_j = \mathbf{0} \quad \text{or} \quad A\mathbf{p}_j = \lambda_j \mathbf{p}_j$$

yields the expression

$$AP = P\Lambda.$$

Since P is orthogonal, post-multiplication of both sides of the above equation by P' results in the expression

$$A = P\Lambda P',$$

and premultiplication by P' leads to the equation

$$\Lambda = P'AP.$$

The first indicates that A can be decomposed into the product of Λ and P, and the second represents the so-called *similarity transformation* of A. Suppose k is a positive integer and consider the kth power of A,

$$A^k = A \times A \times \cdots \times A$$

$$= (P\Lambda P') \times (P\Lambda P') \times \cdots \times (P\Lambda P')$$

$$= P\Lambda^k P', \quad \text{since } P'P = I.$$

Thus the eigenvalue of A^k is equal to the eigenvalue of A raised to the power

k. This relation is used in the iterative scheme in dual scaling and is one of the important properties of eigenvalues and eigenvectors.

A.4
NOTATION FOR MATRIX DIFFERENTIATION

In Section A.3, it was mentioned that dual scaling and some other statistical techniques are based on finding eigenvalues and eigenvectors of certain matrices. Such a formula as $A\mathbf{p}=\lambda\mathbf{p}$, however, is usually an outcome of maximization or minimization of a certain function, and an essential tool for such optimization is differential calculus. A typical problem is to set the partial derivatives of a certain function with respect to certain variables equal to zero and to solve them for the variables. This section introduces some symbols for matrix partial differentiation.

Let $y=f(x)$ be a continuous function in a given interval. Choose two points x_0 and x_1. The corresponding values of the function at these points are $y_0=f(x_0)$ and $y_1=f(x_1)$. Then the quotient

$$\frac{y_1-y_0}{x_1-x_0}=\frac{\Delta y}{\Delta x}$$

is known as the difference quotient, and indicates the slope of the straight line passing through the points (x_0,y_0) and (x_1,y_1). If the quotient has a limit as $x_1 \to x_0$, that limit is called the *derivative* of $f(x)$ at x_0, and is indicated by $f'(x_0)$, or $dy/dx|_{x=x_0}$. That is,

$$f'(x_0) = \lim_{x_1\to x_0}\left[\frac{y_1-y_0}{x_1-x_0}\right] = \lim_{\Delta x\to 0}\frac{\Delta y}{\Delta x}$$

$$= \lim_{\Delta x\to 0}\left[\frac{f(x_0+\Delta x)-f(x_0)}{\Delta x}\right] = \left.\frac{dy}{dx}\right|_{x=x_0}.$$

The derivative is the slope of the tangent line to $f(x)$ at the point (x_0,y_0).

Let $u=f(x,y)$ be a continuous function containing a point (x_0,y_0) and consider u as a function of x alone in the neighbourhood of (x_0,y_0), say $u=f(x_0+h,y_0)$. Then, if the limit

$$\lim_{h\to 0}\left[\frac{f(x_0+h,y_0)-f(x_0,y_0)}{h}\right]$$

exists, it is called the *partial derivative* of $f(x,y)$ with respect to x at the point (x_0,y_0). The partial derivative is indicated by '∂' in the following way:

$\partial f/\partial x, \quad \partial f/\partial y, \quad$ etc.

Consider the function of three variables x_1, x_2, and x_3 given by

$$f(x_1, x_2, x_3) = 2x_1^2 + 3x_1x_2 + 5x_3^2.$$

Then the partial derivatives of f with respect to x_1, x_2, and x_3 are

$$\partial f/\partial x_1 = 4x_1 + 3x_2, \qquad \partial f/\partial x_2 = 3x_1, \qquad \partial f/\partial x_3 = 10x_3.$$

In this case, if we regard f as a function of the vector $\mathbf{x}$, where

$$\mathbf{x} = \begin{bmatrix} x_1 \\ x_2 \\ x_3 \end{bmatrix},$$

then we can introduce the expression

$$\partial f/\partial \mathbf{x} = \begin{bmatrix} \partial f/\partial x_1 \\ \partial f/\partial x_2 \\ \partial f/\partial x_3 \end{bmatrix} = \begin{bmatrix} 4x_1 + 3x_2 \\ 3x_1 \\ 10x_3 \end{bmatrix}.$$

To generalize, let $\mathbf{x}$ be an $n \times 1$ vector of variables and f be a function of the elements of $\mathbf{x}$. Then we define

$$\frac{\partial f}{\partial \mathbf{x}} = \begin{bmatrix} \dfrac{\partial f}{\partial x_1} \\ \dfrac{\partial f}{\partial x_2} \\ \vdots \\ \dfrac{\partial f}{\partial x_n} \end{bmatrix}, \quad \text{and} \quad \frac{\partial f}{\partial \mathbf{x}'} = \begin{bmatrix} \dfrac{\partial f}{\partial x_1} & \dfrac{\partial f}{\partial x_2} & \cdots & \dfrac{\partial f}{\partial x_n} \end{bmatrix}.$$

Consider a general quadratic form

$$y = \mathbf{x}'A\mathbf{x} + 2\mathbf{b}'\mathbf{x} + c,$$

where $\mathbf{x}$ is an $n \times 1$ vector of variables, A an $n \times n$ *symmetric* matrix of constants, $\mathbf{b}$ an $n \times 1$ vector of constants, and c a scalar. Then

$$\frac{\partial y}{\partial \mathbf{x}} = \frac{\partial}{\partial \mathbf{x}}(\mathbf{x}'A\mathbf{x} + 2\mathbf{b}'\mathbf{x} + c)$$

$$= \frac{\partial(\mathbf{x}'A\mathbf{x})}{\partial \mathbf{x}} + \frac{\partial(2\mathbf{b}'\mathbf{x})}{\partial \mathbf{x}} + \frac{\partial c}{\partial \mathbf{x}} = 2A\mathbf{x} + 2\mathbf{b}.$$

Similarly,

$$\frac{\partial y}{\partial \mathbf{x}'} = 2\mathbf{x}'A + 2\mathbf{b}'.$$

Let $\mathbf{f}$ be an $m \times 1$ vector of functions f_i of the $n \times 1$ vector variable $\mathbf{x}$. Then

$$\frac{\partial \mathbf{f}}{\partial \mathbf{x}} = \begin{bmatrix} \partial f_1/\partial x_1 & \partial f_2/\partial x_1 & \cdots & \partial f_m/\partial x_1 \\ \partial f_1/\partial x_2 & \partial f_2/\partial x_2 & \cdots & \partial f_m/\partial x_2 \\ \vdots & \vdots & \ddots & \vdots \\ \partial f_1/\partial x_n & \partial f_2/\partial x_n & \cdots & \partial f_m/\partial x_n \end{bmatrix}.$$

Consider a set of linear equations

$$\mathbf{y} = A\mathbf{x} + \mathbf{b},$$

where $\mathbf{y}$ is an $m \times 1$ vector, A an $m \times n$ matrix, $\mathbf{x}$ an $n \times 1$ vector, and $\mathbf{b}$ an $m \times 1$ vector. Then

$$\partial \mathbf{y}/\partial \mathbf{x} = A' \quad \text{and} \quad \partial \mathbf{y}/\partial \mathbf{x}' = A.$$

The matrix differentiations used in the main text are simple applications of these basic forms.

A.5
EXAMPLES OF MATRIX REPRESENTATION

In practice, one often starts with expressing a set of formulas in matrix notation. This may, or may not, be a difficult task, depending on one's experience in applications of matrix algebra. This section presents a few examples of matrix representation which are relevant to the topics discussed in the main text. Let us define

$$\mathbf{1}_n' = [1, 1, \ldots, 1],$$

which is the $1 \times n$ row vectors of 1's. Then

$$\mathbf{1}_n'\mathbf{1}_n = n \quad \text{(the number of elements in } \mathbf{1}_n\text{)}, \tag{1}$$

$$\mathbf{1}_n\mathbf{1}_n' = \begin{bmatrix} 1 & 1 & \cdots & 1 \\ 1 & 1 & \cdots & 1 \\ \vdots & \vdots & \ddots & \vdots \\ 1 & 1 & \cdots & 1 \end{bmatrix} \quad \text{(the } n \times n \text{ matrix of 1's)}. \tag{2}$$

Define

$$\mathbf{x}' = [x_1, x_2, \ldots, x_n],$$

where $\mathbf{x}$ is a vector of n observed scores. Then

$$\sum_{i=1}^{n} x_i = \mathbf{1}_n' x \quad \text{(the sum of the } n \text{ scores)}. \tag{3}$$

The mean of x, m, can be expressed by (3) and (1) as

$$m = \sum x_i/n = \mathbf{1}_n'\mathbf{x}/\mathbf{1}_n'\mathbf{1}_n. \tag{4}$$

Define the $1\times n$ vector of constants,

$$\mathbf{w}' = [w_1, w_2, \ldots, w_n].$$

Then the linear combination of x_i in terms of w_i is

$$\sum w_i x_i = \mathbf{w}'\mathbf{x}. \tag{5}$$

In dual scaling, one often deals with quadratic forms. Some of them are:

$$\sum x_i^2 = \mathbf{x}'\mathbf{x} \quad \text{(the sum of squares of } x_i\text{)}; \tag{6}$$

$$\sum w_i x_i^2 = \mathbf{x}'D\mathbf{x} \quad \left(\text{the sum of squares of } x_i\text{, weighted by } w_i\right), \tag{7}$$

where

$$D = \text{diag}(w_i) = \begin{bmatrix} w_1 & 0 & \cdots & 0 \\ 0 & w_2 & \cdots & 0 \\ \vdots & \vdots & \ddots & \vdots \\ 0 & 0 & \cdots & w_n \end{bmatrix};$$

$$w_1x_1^2 + 2w_2x_1x_2 + w_3x_2^2 = \begin{bmatrix} x_1 & x_2 \end{bmatrix} \begin{bmatrix} w_1 & w_2 \\ w_2 & w_3 \end{bmatrix} \begin{bmatrix} x_1 \\ x_2 \end{bmatrix} = \mathbf{x}'W\mathbf{x}, \text{ say}; \tag{8}$$

$$(w_1x_1 + w_2x_2)^2 = \begin{bmatrix} x_1 & x_2 \end{bmatrix} \begin{bmatrix} w_1 \\ w_2 \end{bmatrix} \begin{bmatrix} w_1 & w_2 \end{bmatrix} \begin{bmatrix} x_1 \\ x_2 \end{bmatrix} = \mathbf{x}'\mathbf{w}\mathbf{w}'\mathbf{x}, \text{ say}; \tag{9}$$

$$(w_1x_1 + w_2x_2)^2 + (w_3x_1 + w_4x_2)^2 = \begin{bmatrix} x_1 & x_2 \end{bmatrix} \begin{bmatrix} w_1 & w_3 \\ w_2 & w_4 \end{bmatrix} \begin{bmatrix} w_1 & w_2 \\ w_3 & w_4 \end{bmatrix} \begin{bmatrix} x_1 \\ x_2 \end{bmatrix}$$
$$= \mathbf{x}'WW'\mathbf{x}, \text{ say}; \tag{10}$$

$$\frac{(w_1x_1 + w_2x_2)^2}{w_5} + \frac{(w_3x_1 + w_4x_2)^2}{w_6}$$
$$= \begin{bmatrix} x_1 & x_2 \end{bmatrix} \begin{bmatrix} w_1 & w_3 \\ w_2 & w_4 \end{bmatrix} \begin{bmatrix} 1/w_5 & 0 \\ 0 & 1/w_6 \end{bmatrix} \begin{bmatrix} w_1 & w_2 \\ w_3 & w_4 \end{bmatrix} \begin{bmatrix} x_1 \\ x_2 \end{bmatrix}$$
$$= \mathbf{x}'WD^{-1}W'\mathbf{x}, \text{ say}. \tag{11}$$

Let us indicate the $n\times 1$ vector consisting of the mean, m, by $\mathbf{x}_m$,

$$\mathbf{x}_m = m\mathbf{1}_n = \begin{bmatrix} m \\ m \\ \vdots \\ m \end{bmatrix}.$$

Then the vector of deviation scores, $\mathbf{x}_d$, can be expressed as

$$\mathbf{x}_d = \begin{bmatrix} x_1 - m \\ x_2 - m \\ \vdots \\ x_n - m \end{bmatrix} = \mathbf{x} - \mathbf{x}_m = \mathbf{x} - m\mathbf{1}_n = \mathbf{x} - \left[\frac{\mathbf{1}_n'\mathbf{x}}{\mathbf{1}_n'\mathbf{1}_n}\right]\mathbf{1}_n$$

$$= \left[I - \frac{\mathbf{1}_n\mathbf{1}_n'}{\mathbf{1}_n'\mathbf{1}_n}\right]\mathbf{x} = P\mathbf{x}, \text{ say.} \tag{12}$$

Note that $P^2 = P$. This property defines the so-called *idempotent matrix*. Furthermore, P is symmetric, that is, $P' = P$. When a matrix is symmetric and idempotent, it is called a *projection operator*. Using two projection operators, we can express the observed score vector in terms of two orthogonal components,

$$\mathbf{x} = \mathbf{x}_m + \mathbf{x}_d = \left[\frac{\mathbf{1}_n\mathbf{1}_n'}{\mathbf{1}_n'\mathbf{1}_n}\right]\mathbf{x} + \left[I - \frac{\mathbf{1}_n\mathbf{1}_n'}{\mathbf{1}_n'\mathbf{1}_n}\right]\mathbf{x}$$

$$= P_m\mathbf{x} + P\mathbf{x}, \text{ say,} \tag{13}$$

where P_m and P are both projection operators. The orthogonality of the decomposition is demonstrated by the relation $P_mP = PP_m = \mathbf{0}$ (null matrix).

As a slightly more complicated case, consider a 2×2 cross design of the analysis of variance with three subjects in each cell. Then the observed scores of the 12 subjects $y_{ijk}(i=1,2;\ j=1,2;\ k=1,2,3)$, can be expressed as follows:

$$\begin{bmatrix} y_{111} \\ y_{112} \\ y_{113} \\ y_{121} \\ y_{122} \\ y_{123} \\ y_{211} \\ y_{212} \\ y_{213} \\ y_{221} \\ y_{222} \\ y_{223} \end{bmatrix} = \begin{bmatrix} \mu+\alpha_1+\beta_1+\gamma_{11} \\ \mu+\alpha_1+\beta_1+\gamma_{11} \\ \mu+\alpha_1+\beta_1+\gamma_{11} \\ \mu+\alpha_1+\beta_2+\gamma_{12} \\ \mu+\alpha_1+\beta_2+\gamma_{12} \\ \mu+\alpha_1+\beta_2+\gamma_{12} \\ \mu+\alpha_2+\beta_1+\gamma_{21} \\ \mu+\alpha_2+\beta_1+\gamma_{21} \\ \mu+\alpha_2+\beta_1+\gamma_{21} \\ \mu+\alpha_2+\beta_2+\gamma_{22} \\ \mu+\alpha_2+\beta_2+\gamma_{22} \\ \mu+\alpha_2+\beta_2+\gamma_{22} \end{bmatrix} + \begin{bmatrix} e_{111} \\ e_{112} \\ e_{113} \\ e_{121} \\ e_{122} \\ e_{123} \\ e_{211} \\ e_{212} \\ e_{213} \\ e_{221} \\ e_{222} \\ e_{223} \end{bmatrix},$$

where μ is the general mean, α_i and β_j are the parameters associated with the effect of the ith level of factor α and the jth level of factor β, respectively, γ_{ij}

are their interaction effects, and e_{ijk} are the residuals. Let us indicate the vector of the observed scores by $\mathbf{y}$ and the vector of the residuals by $\mathbf{e}$ and define the vector of parameters by

$$\boldsymbol{\theta}' = [\mu, \alpha_1, \alpha_2, \beta_1, \beta_2, \gamma_{11}, \gamma_{12}, \gamma_{21}, \gamma_{22}].$$

Then the above formula can be expressed in the form

$$\mathbf{y} = A\boldsymbol{\theta} + \mathbf{e}, \tag{14}$$

where

$$A = \begin{bmatrix} 1 & 1 & 0 & 1 & 0 & 1 & 0 & 0 & 0 \\ 1 & 1 & 0 & 1 & 0 & 1 & 0 & 0 & 0 \\ 1 & 1 & 0 & 1 & 0 & 1 & 0 & 0 & 0 \\ 1 & 1 & 0 & 0 & 1 & 0 & 1 & 0 & 0 \\ 1 & 1 & 0 & 0 & 1 & 0 & 1 & 0 & 0 \\ 1 & 1 & 0 & 0 & 1 & 0 & 1 & 0 & 0 \\ 1 & 0 & 1 & 1 & 0 & 0 & 0 & 1 & 0 \\ 1 & 0 & 1 & 1 & 0 & 0 & 0 & 1 & 0 \\ 1 & 0 & 1 & 1 & 0 & 0 & 0 & 1 & 0 \\ 1 & 0 & 1 & 0 & 1 & 0 & 0 & 0 & 1 \\ 1 & 0 & 1 & 0 & 1 & 0 & 0 & 0 & 1 \\ 1 & 0 & 1 & 0 & 1 & 0 & 0 & 0 & 1 \end{bmatrix}.$$

A is called the *design matrix*. The first column of A indicates μ, the second column indicates the presence (if 1) or the absence (if 0) of α_1, the third column is associated with α_2, and so on. There are nine columns in A corresponding to the nine parameters in $\boldsymbol{\theta}$. Suppose that one wishes to decompose $\mathbf{y}$ into orthogonal components associated with μ, α, β, γ, and e, respectively, as in (13). This problem amounts to that of deriving projection operators for the components. Let us define the following submatrices of A:

A_μ is the 12×1 vector consisting of the first column of A;

A_α is the 12×2 matrix consisting of columns 2 and 3 of A;

A_β consists of columns 4 and 5 of A;

A_γ consists of the last four columns of A.

Define 'hat matrices' H_t (e.g., Hoaglin and Welsch 1978) by

$$H_t = A_t(A_t'A_t)^{-1}A_t',$$

where $t = \mu$, α, β, and γ. In terms of H_t, $\mathbf{y}$ can be expressed in the form

$$\begin{aligned} \mathbf{y} &= H_\mu\mathbf{y} + (H_\alpha - H_\mu)\mathbf{y} + (H_\beta - H_\mu)\mathbf{y} + (H_\gamma - H_\alpha - H_\beta + H_\mu)\mathbf{y} + (I - H_\gamma)\mathbf{y} \\ &= P_\mu\mathbf{y} + P_\alpha\mathbf{y} + P_\beta\mathbf{y} + P_\gamma\mathbf{y} + P_e\mathbf{y}, \text{ say,} \end{aligned} \tag{15}$$

where

$$P_\mu = H_\mu = \mathbf{11}'/\mathbf{1}'\mathbf{1}, \qquad P_\alpha = H_\alpha - H_\mu, \qquad P_\beta = H_\beta - H_\mu,$$
$$P_\gamma = H_\gamma - H_\alpha - H_\beta + H_\mu, \qquad P_e = I - H_\gamma.$$

When cell frequencies f_{ij} are equal or proportional (i.e., $f_{ij}=f_{i.}f_{.j}/f_t$), there exist the following relations:

$$H_\alpha H_\mu = H_\beta H_\mu = H_\alpha H_\beta = H_\mu, \qquad H_\gamma H_\alpha = H_\alpha, \qquad H_\gamma H_\beta = H_\beta.$$

When these relations hold, one can easily prove that P_t specified as above are all disjoint, and that **y** can be decomposed into additive components. For instance, $P_\alpha P_\beta = (H_\alpha - H_\mu)(H_\beta - H_\mu) = H_\alpha H_\beta - H_\alpha H_\mu - H_\mu H_\alpha + H_\mu H_\mu = H_\mu - H_\mu - H_\mu + H_\mu = 0$. Noting that $P_t P_s = P_t$ if $t=s$ and that $P_t P_s = 0$ if $t \neq s$, we obtain the expression

$$\mathbf{y}'\mathbf{y} = \mathbf{y}'P_\mu\mathbf{y} + \mathbf{y}'P_\alpha\mathbf{y} + \mathbf{y}'P_\beta\mathbf{y} + \mathbf{y}'P_\gamma\mathbf{y} + \mathbf{y}'P_e\mathbf{y}.$$

If **y** is distributed as multivariate normal, each sum of squares on the right-hand side of the above decomposition is independently distributed as χ^2, central or non-central, depending on the centrality of **y**, with degrees of freedom equal to the rank of that projection operator (e.g., see Graybill 1961). It is convenient to know that the rank of an idempotent matrix (e.g., hat matrices, project operators) is equal to the trace of the matrix. Note that unequal or non-proportional cell frequencies provide non-additive sums of squares, and that orthogonalization of the decomposition becomes necessary.

One drawback in working with projection operators is that the order of each matrix is often too large to handle. As an alternative, one can consider a general method called 'reparametrization.' First note that a parameter (or a function of parameters) is said to be estimable if there exists a linear combination of the observations, **y**, whose expected value is equal to the parameter (or the function of parameters) (Graybill 1961). Bock (1963) presented a practical way to reparameterize $\boldsymbol{\theta}$. Suppose A is $n \times m$. Then A can be expressed as the matrix product

$$A_{n\times m} = K_{n\times r}L_{r\times m} \tag{16}$$

if and only if

$$\text{rank}\begin{bmatrix} A \\ L \end{bmatrix} = r[A, K] = r(A) = r(K) = r(L) = r. \tag{17}$$

Then vector $\boldsymbol{\zeta} = L\boldsymbol{\theta}$ is a vector of estimable linear functions of $\boldsymbol{\theta}$. In terms of this reparameterized vector, (14) can be rewritten as

$$\mathbf{y} = A\boldsymbol{\theta} + \mathbf{e} = KL\boldsymbol{\theta} + \mathbf{e} = K\boldsymbol{\zeta} + \mathbf{e}, \tag{18}$$

where, once L is given, $K = AL'(LL')^{-1}$ and K is called a *basis matrix*. To choose L, one can use the explicit relation in the first term of (17) that the rows of L should be linearly dependent on, or linear combinations of, the rows of A. If one can choose a set of r linearly independent rows as (17) indicates, then the r reparameterized parameters are estimable. No matter what r-rowed matrix L one may choose, the analytical results of $\mathbf{y}$ remain invariant.

These are some examples of matrix representations. In the main text matrix formulations are presented step by step with the assistance of illustrative examples.

B
Computer program

B.1
LISTING OF DUAL1

```
*****************************************************************************
*                                                                           *
*                   DUAL1  : VERSION 1, FEBRUARY, 1979                      *
*                                                                           *
*****************************************************************************


** AUTHOR   SHIZUHIKO NISHISATO


  THIS IS A SMALL-SCALE PROGRAM FOR DUAL SCALING, BASED ON THE METHODS
  DESCRIBED IN THE BOOK:

      NISHISATO, S.  ANALYSIS OF CATEGORICAL DATA: DUAL SCALING AND
                     ITS APPLICATIONS (MATHEMATICAL EXPOSITIONS NO.24).
                     UNIVERSITY OF TORONTO PRESS, 1980.

  DUAL1 HANDLES THE FOLLOWING TYPES OF DATA;  CONTINGENCY/FREQUENCY
  TABLES (ITYPE=0); MULTIPLE-CHOICE DATA (ITYPE=1); PAIRED COMPARISON
  DATA (ITYPE=2); RANK ORDER DATA (ITYPE=3); SUCCESSIVE CATEGORIES
  DATA (ITYPE=4).

** ORGANIZATION OF DATA **

  (1)  CONTINGENCY/FREQUENCY TABLE:  PREPARE A TWO-WAY TABLE OF FREQUENCIES.
  (2)  MULTIPLE-CHOICE DATA :  PREPARE THE DATA IN EITHER OF THE FOLLOWING
    WAYS; (A) THE SUBJECT-BY-OPTION MATRIX OF (1,0) RESPONSE PATTERNS,
          (B)  THE SUBJECT-BY-ITEM MATRIX WITH ELEMENTS INDICATING THE
            CHOSEN OPTION NUMBERS.
  (3)  PAIRED COMPARISON DATA:  THE SUBJECT-BY-(N(N-1)/2 PAIRS) MATRIX
    WITH ELEMENTS 1 IF THE SUBJECT JUDGES XJ TO BE GREATER THAN XK AND
    -1 IF THE SUBJECT JUDGES XJ TO BE SMALLER THAN XK.  THE N(N-1)/2
    COLUMNS OF THE MATRIX CORRESPOND TO PAIRS IN ORDER:  (X1,X2),
    (X1,X3),...,(X1,XN),(X2,X3),...,(X2,XN),...,(X(N-1),XN).
  (4)  RANK ORDER DATA:  THE SUBJECT-BY-STIMULUS MATRIX OF RANK NUMBERS.
    FOR TIED RANKS, ENTER THEIR AVERAGE RANKS.
  (5)  SUCCESSIVE CATEGORIES DATA:  THE SUBJECT-BY-STIMULUS MATRIX WITH
    ELEMENTS BEING THE CHOSEN CATEGORY NUMBERS FOR THE STIMULI.
    CATEGORY NUMBERS 1,2,3,... CORRESPOND TO THE SUCCESSIVE CATEGORIES
    STARTING FROM THE LOWEST CATEGORY.

  ***PREPARATION OF INPUT DATA ***

  CARD 1:  PUNCH THE TITLE OF YOUR STUDY IN 80 SPACES OR LESS

  CARD 2:  NR, NC, ITYPE, NCAT  IN FORMAT 4I4
```

```
C       NR=THE NUMBER OF ROWS OF THE INPUT MATRIX.
C       NC= THE NUMBER OF COLUMNS OF THE INPUT MATRIX.
C       ITYPE = 0  FOR CONTINGENCY/FREQUENCY TABLES(SEE CHAPTER 4)
C             = 1  FOR MULTIPLE-CHOICE DATA (SEE CHAPTER 5)
C             = 2  FOR PAIRED COMPARISON DATA (SEE CHAPTER 6, NISHISATO'S METHO
C             = 3  FOR RANK ORDER DATA (SEE CHAPTER 6, NISHISATO'S METHOD)
C             = 4  FOR SUCCESSIVE CATEGORIES DATA (SEE CHAPTER 8)
C       NCAT  = 0  FOR ITYPE = 0,1,2,3
C             =  THE NUMBER OF SUCCESSIVE CATEGORIES FOR ITYPE = 4.
C
C     CARD 3:  THE FORMAT IN WHICH THE INPUT DATA ARE TO BE READ IN (USE THE
C       F  FORMAT).  FOR EXAMPLE, TYPE  (10F4.0).
C
C     CARD 4 TO CARD (4+NR-1):  THE DATA PUNCHED IN THE FORMAT SPECIFIED BY
C       CARD 3.  THE DATA ARE READ IN ROW BY ROW
C
C     FOR ITYPE =0,2,3,4, THE ABOVE CARDS CONSTITUTE A SET.  IF MORE THAN
C       ONE SET ARE TO BE ANALYZED, PREPARE THE SECOND, THE THIRD, ...., SETS OF
C       DATA IN THE SAME WAY.  AFTER THE LAST DATA SET, INSERT A BALANK CARD TO
C       TERMINATE THE RUN.
C
C     FOR ITYPE =1, PREPARE THE FOLLOWING ADDITIONAL CARD(S).
C
C     CARD X1:  IFORM, NIT  IN FORMAT 2I4.
C       IFORM = 0  IF THE DATA ARE IN THE SUBJECT-BY-OPTION FORM OF (1,0)
C                  RESPONSE PATTERNS
C             = NON ZERO NUMBER IF THE DATA ARE IN THE SUBJECT-BY-ITEM FORM
C               WITH ELEMENTS INDICATING THE CHOSEN OPTION NUMBERS.
C       NIT   = THE NUMBER OF MULTIPLE-CHOICE ITEMS.
C
C     CARD X2 :  THIS IS REQUIRED ONLY IF IFORM = NON ZERO NUMBER.
C       PUNCH  OPTION(I), I= 1,2,3,..., NIT  IN FORMAT  40I2.
C        OPTION(1) = THE NUMBER OF MULTIPLE-CHOICE OPTIONS FOR ITEM 1
C        OPTION(2) = THE NUMBER OF MULTIPLE-CHOICE OPTIONS FOR ITEM 2
C        AND SO ON TO  OPTION(NIT)
C
C
C     **  RUN IS TERMINATED WHEN ONE OF THE FOLLOWING OCCURS:
C
C     (1)  THE NUMBER OF ROWS OF THE INPUT MATRIX IS LESS THAN ONE
C     (2)  THE NUMBER OF COLUMNS OF THE INPUT MATRIX IS LESS THAN TWO
C     (3)  THE NUMBER OF ITERATION EXCEEDS 1000
C     (4)  THE CORRELATION RATIO IS GREATER THAN ONE OR LESS THAN ZERO
C     (5)  CRONBACH'S ALPHA (A RELIABILITY COEFFICIENT) IS NEGATIVE
C     (6)  THE CHI-SQUARE VALUE IS LESS THAN ITS DEGREES OF FREEDOM
C     (7)  MORE THAN 99.5% OF THE TOTAL VARIANCE IS ACCOUNTED FOR
C     (8)  THE NUMBER OF SOLUTION EXCEEDS 5, NC-1, OR NR-1
C     (9)  THE OPTIMAL SOLUTION IS OBTAINED FOR SUCCESSIVE CATEGORIES DATA
C
C
C     ***  DISCLAIMER :  THIS PROGRAM HAS BEEN THOROUGHLY TESTED.  HOWEVER,
C            NO WARRANTY OF ANY KIND IS MADE BY THE AUTHOR AS TO THE ACCURACY
C            AND FUNCTIONING OF THE PROGRAM.
C
C
C     ***  FOR A LARGE DATA SET, EXPAND THE DIMENSIONS OF THE PROGRAM.  THE LI
C       IS INDICATED BY THE DIMENSION OF F(I,J), WHERE
C       I =  THE NUMBER OF ROWS FOR ALL TYPES OF DATA;
C       J =  THE NUMBER OF COLUMNS (ITYPE= 0,3), THE TOTAL NUMBER OF OPTIONS
C            OF ALL THE ITEMS (ITYPE= 1), THE TOTAL NUMBER OF POSSIBLE PAIRS
C            (ITYPE= 2), THE NUMBER OF STIMULI TIMES THE NUMBER OF CATEGORY
C            BOUNDARIES (ITYPE= 4).
C
C     *** A LARGE-SCALE PRODUCTION PROGRAM, CALLED OPSCALE AND WRITTEN BY
C       K. LEONG, S. NISHISATO AND R. G. WOLFE, IS ALSO AVAILABLE FROM
```

```
INTERNATIONAL EDUCATIONAL SERVICES IN CHICAGO.*
   IN ADDITION TO WHAT DUAL1 HANDLES, OPSCALE CARRIES OUT PARTIALLY
OPTIMAL SCALING (SEE CHAPTER 5) AND ANALYSIS OF ITYPE=0 DATA WITH
ORDER CONSTRAINTS ON COLUMNS BY THE SDM PROCEDURE (SEE CHAPTER 8).
   OPSCALE IS ALSO BASED ON THE METHODS DESCRIBED IN NISHISATO (1980).


***  COPYRIGHT BY SHIZUHIKO NISHISATO, 1979



**********     MAIN PROGRAM      **********



    DIMENSION D2(30),R(30),Y(30),U(30),B(30),G(30),C(30,30),FMT(18) DUAL0010
    DIMENSION E(30,30),W(30),TITLE(18),WT(4,30)                      DUAL0020
    COMMON F(30,30),DN(30),D(30)                                     DUAL0030
IN EXPANDING DIMENSIONS OF THE PROGRAM, INCREASE ALL THE NUMBERS EXCEPT
FMT(18), TITLE(18) AND 4 OF WT(4,J).
    INTEGER CD, P                                                    DUAL0040
    CD=5                                                             DUAL0050
    P= 6                                                             DUAL0060
 99 READ(CD,93) TITLE                                                DUAL0070
READ IN THE PARAMETERS NR(THE NUMBER OF ROWS OF THE DATA MATRIX), NC (THE
NUMBER OF COLUMNS), ITYPE (TYPE OF THE DATA), NCAT (THE NUMBER OF
SUCCESSIVE CATEGORIES FOR ITYPE = 4)
 93 FORMAT(20A4)                                                     DUAL0080
    READ(CD,4) NR, NC, ITYPE, NCAT                                   DUAL0090
    IF(NR.LT.1) GO TO 98                                             DUAL0100
    WRITE(P,1)                                                       DUAL0110
  1 FORMAT(1H1)                                                      DUAL0120
    WRITE(P,2)                                                       DUAL0130
  2 FORMAT(/15X,33('*'))                                             DUAL0140
    WRITE(P,3)                                                       DUAL0150
  3 FORMAT(/15X,'*       DUAL SCALING OF        *')                  DUAL0160
  4 FORMAT(4I4)                                                      DUAL0170
    IF(ITYPE.EQ.0) WRITE(P,5)                                        DUAL0180
    IF(ITYPE.EQ.1) WRITE(P,6)                                        DUAL0190
    IF(ITYPE.EQ.2) WRITE(P,7)                                        DUAL0200
    IF(ITYPE.EQ.3) WRITE(P,8)                                        DUAL0210
    IF(ITYPE.EQ.4) WRITE(P,9)                                        DUAL0220
  5 FORMAT(/15X,'*   CONTINGENCY/FREQUENCY TABLE   *')               DUAL0230
  6 FORMAT(/15X,'      MULTIPLE-CHOICE DATA       *')                DUAL0240
  7 FORMAT(/15X,'*     PAIRED COMPARISON DATA     *')                DUAL0250
  8 FORMAT(/15X,'*        RANK ORDER DATA         *')                DUAL0260
  9 FORMAT(/15X,'*   SUCCESIVE CATEGORIES DATA   *')                 DUAL0270
    WRITE(P,2)                                                       DUAL0280
    WRITE(P,94) TITLE                                                DUAL0290
 94 FORMAT(///5X,'TITLE OF STUDY : ',20A4)                           DUAL0300
    WRITE(P,10)                                                      DUAL0310
 10 FORMAT(//5X,'***  INPUT MATRIX  ***'//)                          DUAL0320
    READ(CD,71) FMT                                                  DUAL0330
 71 FORMAT(18A4)                                                     DUAL0340
    DO 72 I=1,NR                                                     DUAL0350
 72 READ(CD,FMT) (F(I,J),J=1,NC)                                     DUAL0360
    DO 11 I=1,NR                                                     DUAL0370
    WRITE(P,12) (F(I,J), J=1,NC)                                     DUAL0380
 11 CONTINUE                                                         DUAL0390
 12 FORMAT(2X,20F6.0)                                                DUAL0400
PREPARE THE INPUT DATA IN THE FORM AMENABLE TO DUAL SCALING
    IF(ITYPE.EQ.0) CALL FRQNCY(NR,NC,FT)                             DUAL0410
    IF (ITYPE.EQ.1) CALL MULT(NR,NC,FT,NIT)                          DUAL0420
    IF(ITYPE.EQ.2) CALL PAIR(NR,NC,FT)                               DUAL0430
    IF(ITYPE.EQ.3) CALL RANK(NR,NC,FT)                               DUAL0440
```

*In preparation.

```
      IF (ITYPE.EQ.4) CALL SUCCES(NR,NC,FT,NCAT)                        DUAL045
      IF(NC.GT.1) GO TO 96                                              DUAL046
      WRITE(P,97)                                                       DUAL047
      GO TO 99                                                          DUAL048
   96 CONTINUE                                                          DUAL049
      WRITE(P,13) (DN(I),I=1,NR)                                        DUAL050
   13 FORMAT(/2X,'** NUMBER OF RESPONSES FOR ROWS'///3X,7(16F5.0))      DUAL051
      WRITE(P,14) (D(I),I=1,NC)                                         DUAL052
   14 FORMAT(/2X,'** NUMBER OF RESPONSES FOR COLUMNS'///3X,7(16F5.0))   DUAL053
   15 FORMAT(//5X,'***  MATRIX C FOR EIGENEQUATION  ***'///)            DUAL054
      CHK1 =0.0                                                         DUAL055
      CHK2 =0.0                                                         DUAL056
      DO 111 I=1,NR                                                     DUAL057
  111 CHK1=CHK1+DN(I)                                                   DUAL058
      DO 112 I=1,NC                                                     DUAL059
  112 CHK2=CHK2+D(I)                                                    DUAL060
      DCHK=ABS(CHK1-CHK2)                                               DUAL061
      IF(DCHK.GT.0.5) GO TO 113                                         DUAL062
      IF(ITYPE.LT.2) GO TO 19                                           DUAL063
C   PRINT TRANSFORMED DATA MATRIX E FOR PAIRED COMPARISON, RANK ORDER, AND
C   SUCCESSIVE CATEGORIES DATA
      WRITE(P,16)                                                       DUAL064
   16 FORMAT(//5X,'***  TRANSFORMED DATA MATRIX E  ***'//)              DUAL065
      DO 17 I=1,NR                                                      DUAL066
   17 WRITE(P,18) (F(I,J), J=1,NC)                                      DUAL067
   18 FORMAT(2X,30F4.0)                                                 DUAL068
   19 CONTINUE                                                          DUAL069
      WRITE(P,15)                                                       DUAL070
C   CONSTRUCTION OF MATRIX C FOR EIGENEQUATION
      DO 39 I=1,NC                                                      DUAL071
      IF(D(I).LT.0.5) D2(I)=0.0                                         DUAL072
      IF(D(I).GE.0.5) D2(I)=1.0/SQRT(D(I))                              DUAL073
   39 CONTINUE                                                          DUAL074
      DO 21 I=1,NC                                                      DUAL075
      DO 21 J=I,NC                                                      DUAL076
      C(I,J)= 0.0                                                       DUAL077
      DO 20 K=1,NR                                                      DUAL078
      IF(DN(K).LT.0.1) DR=0.0                                           DUAL079
      IF (DN(K).GT.0.1) DR=1.0/DN(K)                                    DUAL080
   20 C(I,J)=C(I,J)+D2(I)*F(K,I)*F(K,J)*D2(J)*DR                        DUAL081
   21 C(J,I)= C(I,J)                                                    DUAL082
      IF(ITYPE.GE.2) GO TO 24                                           DUAL083
      DO 23 I=1,NC                                                      DUAL084
      DO 23 J=I,NC                                                      DUAL085
      C(I,J)=C(I,J)-SQRT(D(I)*D(J))/FT                                  DUAL086
   23 C(J,I)= C(I,J)                                                    DUAL087
   24 CONTINUE                                                          DUAL088
      DO 25 I=1,NC                                                      DUAL089
   25 WRITE(P,26) (C(I,J),J=1,I)                                        DUAL090
   26 FORMAT(2X,18F7.4)                                                 DUAL091
      TRACE =0.0                                                        DUAL092
      DO 27 I=1,NC                                                      DUAL093
   27 TRACE=TRACE+C(I,I)                                                DUAL094
      WRITE(P,28) TRACE                                                 DUAL095
   28 FORMAT(//2X,'** TRACE OF C  =',F8.4)                              DUAL096
C   ITERATIVE POWER METHOD TO FIND THE MAXIMUM EIGENVALUE AND THE CORRESPOND
C   EIGENVECTOR.  FIRST, GENERATE A TRIAL VECTOR FOR THE INITIAL RUN
      NROOT=0                                                           DUAL097
      SSP=0.0                                                           DUAL098
      T=NC-1                                                            DUAL099
      STEP=2.0/T                                                        DUAL100
   31 ITER=1                                                            DUAL101
      G(1)=1.0                                                          DUAL102
      B(1)=1.0                                                          DUAL103
      DO 30 I=2,NC                                                      DUAL104
```

```
      KK=I-1                                                            DUAL1050
      G(I)=G(KK)-STEP                                                   DUAL1060
   30 B(I)=G(I)                                                         DUAL1070
      NROOT=NROOT+1                                                     DUAL1080
      WRITE(P,29)                                                       DUAL1090
   29 FORMAT(/////)                                                     DUAL1100
      WRITE(P,32)                                                       DUAL1110
   32 FORMAT(1X,20('*'))                                                DUAL1120
      WRITE(P,33) NROOT                                                 DUAL1130
   33 FORMAT(1X,'** SOLUTION',I5,'  **')                                DUAL1140
      WRITE(P,32)                                                       DUAL1150
   34 CONTINUE                                                          DUAL1160
      DO 35 I=1,NC                                                      DUAL1170
      R(I)=0.0                                                          DUAL1180
      DO 35 J=1,NC                                                      DUAL1190
   35 R(I)=R(I)+C(I,J)*G(J)                                             DUAL1200
      S=ABS(R(1))                                                       DUAL1210
      DO 36 I=2,NC                                                      DUAL1220
      Z=ABS(R(I))                                                       DUAL1230
      IF(S.GT.Z) GO TO 36                                               DUAL1240
      S=Z                                                               DUAL1250
   36 CONTINUE                                                          DUAL1260
      DO 37 I=1,NC                                                      DUAL1270
   37 G(I)=R(I)/S                                                       DUAL1280
      CR=0.0                                                            DUAL1290
      DO 38 I=1,NC                                                      DUAL1300
      DFR=ABS(B(I)-G(I))                                                DUAL1310
      IF(DFR.LT.0.00005) GO TO 38                                       DUAL1320
      GO TO 88                                                          DUAL1330
   38 CONTINUE                                                          DUAL1340
      GO TO 42                                                          DUAL1350
   88 DO 40 I=1,NC                                                      DUAL1360
   40 B(I)=G(I)                                                         DUAL1370
      ITER=ITER+1                                                       DUAL1380
C    A STOPPING CRITERION
      IF(ITER.LT.1000) GO TO 34                                         DUAL1390
      WRITE(P,41)                                                       DUAL1400
   41 FORMAT(////2X,'***** NO CONVERGENCE -- RUN ABORTED *****')        DUAL1410
      GO TO 99                                                          DUAL1420
C    CALCULATION OF (LOCALLY) OPTIMAL AND ORTHOGONAL WEIGHT VECTORS
C    FOR THE ROWS AND THE COLUMNS OF THE DATA MATRIX
C    A STOPPING CRITERION                                               DUAL1430
   42 IF(S.LE.1.00.AND.S.GE.0.0) GO TO 92                               DUAL1440
  113 WRITE(P,97)                                                       DUAL1450
   97 FORMAT(////2X,'***** ERROR -- CHECK INPUT CARDS **')              DUAL1460
      GO TO 99                                                          DUAL1470
   92 S2=0.0                                                            DUAL1480
      DO 43 I=1,NC                                                      DUAL1490
   43 S2=S2+G(I)**2                                                     DUAL1500
      DO 44 I=1,NC                                                      DUAL1510
   44 G(I)=G(I)/SQRT(S2)                                                DUAL1520
      CRT=SQRT(FT)                                                      DUAL1530
      DO 45 I=1,NC                                                      DUAL1540
   45 G(I)=G(I)*CRT                                                     DUAL1550
      SROOT= SQRT(S)                                                    DUAL1560
      WRITE(P,46) S, SROOT                                              DUAL1570
   46 FORMAT(/3X,'SQUARED CORRELATION RATIO =',F8.4,'    ;  MAXIMUM',   DUAL1580
     1' CORRELATION =',F8.4)                                            DUAL1590
      SSP=SSP+100.0*S/TRACE                                             DUAL1600
      WRITE(P,47) NROOT, SSP                                            DUAL1610
   47 FORMAT(/2X,'DELTA(TOTAL VARIANCE ACCOUNTED FOR BY', I3,2X,        DUAL1620
     1'SOLUTION(S))  =',F7.1,'  (PER CENT)'/)                           DUAL1630
      WRITE(P,48) ITER                                                  DUAL1640
   48 FORMAT(//2X,'NUMBER OF ITERATIONS  =', I4)                        DUAL1650
      IF (ITYPE.NE.1) GO TO 75                                          DUAL1660
```

```
C     CALCULATION OF THE GENERALIZED KUDER-RICHARDSON RELIABILITY COEFFICIENT
C     (CRONBACH'S ALPHA) FOR MULTIPLE-CHOICE DATA
      T=NIT-1                                                           DUAL1670
      TT=1.0/S                                                          DUAL1680
      ALPHA=1.0-(TT-1.0)/T                                              DUAL1690
      WRITE(P,73) ALPHA                                                 DUAL1700
   73 FORMAT(//2X,'RELIABILITY COEFFICIENT (ALPHA)  =',F8.4)           DUAL1710
C     A STOPPING CRITERION
      IF(ALPHA.GT.0.0) GO TO 75                                         DUAL1720
      WRITE(P,74)                                                       DUAL1730
   74 FORMAT(///2X,'NEGATIVE ALPHA -- FURTHER ANALYSIS IS ABORTED')    DUAL1740
      GO TO 99                                                          DUAL1750
   75 DO 49 I=1,NC                                                      DUAL1760
      B(I)=G(I)*D2(I)                                                   DUAL1770
      WT(3,I)=B(I)*SROOT                                                DUAL1780
   49 WT(4,I)=G(I)*SROOT                                                DUAL1790
      TR=1.0/SROOT                                                      DUAL1800
      DO 50 I=1,NR                                                      DUAL1810
      Y(I)=0.0                                                          DUAL1820
      IF(DN(I).LT.0.5) TT=0.0                                           DUAL1830
      IF(DN(I).GE.0.5) TT=1.0/DN(I)                                     DUAL1840
      DO 50 J=1,NC                                                      DUAL1850
   50 Y(I)=Y(I)+TR*F(I,J)*B(J)*TT                                       DUAL1860
      DO 51 I=1,NR                                                      DUAL1870
   51 WT(1,I)=Y(I)*SROOT                                                DUAL1880
      WRITE(P,52)                                                       DUAL1890
   52 FORMAT(/2X,'OPTIMAL/LOCALLY OPTIMAL WEIGHT VECTORS (UNWEIGHTED,', DUAL1900
     1'WEIGHTED BY ETA) FOR'/14X,'** ROWS **',20X,'** COLUMNS **'/      DUAL1910
     2 2(8X,'UNWEIGHTED     WEIGHTED')/)                                DUAL1920
C     UNWEIGHTED VECTORS ARE DIRECT OUTPUTS OF DUAL SCALING, WHILE VECTORS
C     WEIGHTED BY ETA ARE VECTORS ADJUSTED IN PROPORTION TO THE STANDARD DEVIATION
C     OF EACH DIMENSION (SEE CHAPTER 2, REGARDING THE DISTINCTION BETWEEN THE
C     TWO TYPES OF VECTORS)
      IF(NR.LT.NC) GO TO 57                                             DUAL1930
      DO 53 I=1,NC                                                      DUAL1940
   53 WRITE(P,54) I,Y(I),WT(1,I),I,B(I),WT(3,I)                         DUAL1950
   54 FORMAT(1X,I3,2X,2(F12.4),3X,I3,2(F12.4))                          DUAL1960
      IF(NR.EQ.NC) GO TO 61                                             DUAL1970
      K=NC+1                                                            DUAL1980
      DO 55 I=K,NR                                                      DUAL1990
   55 WRITE(P,56) I,Y(I),WT(1,I)                                        DUAL2000
   56 FORMAT(1X,I3,2X,2(F12.4))                                         DUAL2010
      GO TO 61                                                          DUAL2020
   57 DO 58 I=1,NR                                                      DUAL2030
   58 WRITE(P,54) I,Y(I),WT(1,I),I,B(I),WT(3,I)                         DUAL2040
      K=NR+1                                                            DUAL2050
      DO 59 I=K,NC                                                      DUAL2060
   59 WRITE(P,60) I,B(I),WT(3,I)                                        DUAL2070
   60 FORMAT(33X,I3,2(F12.4))                                           DUAL2080
   61 IF(ITYPE.NE.0) GO TO 66                                           DUAL2090
C     CALCULATION OF ORDER K APPROXIMATION TO CONTINGENCY/FREQUENCY DATA
      T=SQRT(S)                                                         DUAL2100
   63 IF(NROOT.GT.1) GO TO 65                                           DUAL2110
      DO 64 I=1,NR                                                      DUAL2120
      DO 64 J=1,NC                                                      DUAL2130
   64 E(I,J)=(DN(I)*D(J)/FT)*(1.0+T*Y(I)*B(J))                          DUAL2140
      GO TO 66                                                          DUAL2150
   65 DO 62 I=1,NR                                                      DUAL2160
      DO 62 J=1,NC                                                      DUAL2170
   62 E(I,J)=E(I,J)+(DN(I)*D(J)/FT)*(T*Y(I)*B(J))                       DUAL2180
C     A STOPPING CRITERION
   66 IF(ITYPE.EQ.4) GO TO 99                                           DUAL2190
      JS=NC-1                                                           DUAL2200
      DO 80 I=1,NR                                                      DUAL2210
      U(I)=Y(I)*SQRT(DN(I))                                             DUAL222
```

```
80 WT(2,I)=U(I)*SROOT                                                   DUAL2230
   WRITE(P,81)                                                          DUAL2240
81 FORMAT(/2X,'ORTHOGONAL WEIGHT VECTORS (UNWEIGHTED, WEIGHTED BY',DUAL2250
  1'ETA) FOR'/14X,'** ROWS **',20X,'** COLUMNS **'/                     DUAL2260
  22(8X,'UNWEIGHTED    WEIGHTED')/)                                     DUAL2270
   IF(NR.LT.NC) GO TO 84                                                DUAL2280
   DO 82 I=1,NC                                                         DUAL2290
82 WRITE(P,54) I,U(I),WT(2,I),I,G(I),WT(4,I)                            DUAL2300
   IF(NR.EQ.NC) GO TO 87                                                DUAL2310
   K=NC+1                                                               DUAL2320
   DO 83 I=K,NR                                                         DUAL2330
83 WRITE(P,56) I,U(I),WT(2,I)                                           DUAL2340
   GO TO 87                                                             DUAL2350
84 DO 85 I=1,NR                                                         DUAL2360
85 WRITE(P,54) I,U(I),WT(2,I),I,G(I),WT(4,I)                            DUAL2370
   K=NR+1                                                               DUAL2380
   DO 86 I=K,NC                                                         DUAL2390
86 WRITE(P,60) I,G(I)                                                   DUAL2400
87 CONTINUE                                                             DUAL2410
   IF(ITYPE.EQ.0.OR.ITYPE.EQ.2) T=NR+NC-1                               DUAL2420
   IF (ITYPE.EQ.1) T=NR+NC-NIT-1                                        DUAL2430
   IF(ITYPE.LE.1) TT=FT                                                 DUAL2440
   IF (ITYPE.EQ.2) TT=(NR*NC*JS)/2                                      DUAL2450
   IF(ITYPE.EQ.3) GO TO 77                                              DUAL2460
C  COMPUTAION OF CHI-SQUARE
   CHI2=(T/2.0-TT+1.0)*ALOG(1.0-S)                                      DUAL2470
   ROOT= NROOT                                                          DUAL2480
   DF=T-2.0*ROOT                                                        DUAL2490
   NDF=DF                                                               DUAL2500
   WRITE(P,76) CHI2,NDF                                                 DUAL2510
76 FORMAT(//2X,'***  CHI-SQUARE  =',F8.2,'   WITH',I5,3X,               DUAL2520
  1 'DEGREES OF FREEDOM')                                               DUAL2530
   IF(ITYPE.NE.0) GO TO 110                                             DUAL2540
   WRITE(P,67) NROOT                                                    DUAL2550
67 FORMAT(/2X,'ORDER',I3,'  APPROXIMATION OF DATA MATRIX'//)            DUAL2560
   DO 68 I=1,NR                                                         DUAL2570
68 WRITE(P,95) (E(I,J),J=1,NC)                                          DUAL2580
95 FORMAT(5X,16F8.1)                                                    DUAL2590
C  THREE MORE STOPPING CRITERIA
110 IF(CHI2.LT.DF) GO TO 99                                             DUAL2600
77 IF(SSP.GT.99.5) GO TO 99                                             DUAL2610
   JT=NR-1                                                              DUAL2620
   IF (NROOT.GE.JS.OR.NROOT.GE.JT.OR.NROOT.GE.5) GO TO 99               DUAL2630
C  CALCULATION OF THE RESIDUAL EIGEN MATRIX
   SSS=0.0                                                              DUAL2640
   DO 90 I=1,NC                                                         DUAL2650
90 SSS=SSS+G(I)**2                                                      DUAL2266
   DO 91 I=1,NC                                                         DUAL2670
   DO 91 J=I,NC                                                         DUAL2680
   C(I,J)=C(I,J)-S*G(I)*G(J)/SSS                                        DUAL2690
91 C(J,I)=C(I,J)                                                        DUAL2700
   GO TO 31                                                             DUAL2710
98 STOP                                                                 DUAL2720
   END                                                                  DUAL2730
C  FOR CONTINGENCY/FREQUENCY TABLES, THIS SUBROUTINE CALCULATES TWO DIAGONAL
C  MATRICES DN AND D, AND THE TOTAL NUMBER OF RESPONSES, FT, INVOLVED IN
C  CONTINGENCY/FREQUENCY DATA
   SUBROUTINE FRQNCY(NR,NC,FT)                                          DUAL2740
   COMMON F(30,30),DN(30),D(30)                                         DUAL2750
   DO 100 J=1,NC                                                        DUAL2760
   D(J)=0.0                                                             DUAL2770
   DO 100 I=1,NR                                                        DUAL2780
100 D(J)=D(J)+F(I,J)                                                    DUAL2790
   FT=0.0                                                               DUAL2800
   DO 101 I=1,NC                                                        DUAL2810
```

```
    101 FT=FT+D(I)                                                    DUAL2820
        DO 102 I=1,NR                                                 DUAL2830
        DN(I)=0.0                                                     DUAL2840
        DO 102 J=1,NC                                                 DUAL2850
    102 DN(I)=DN(I)+F(I,J)                                            DUAL2860
        RETURN                                                        DUAL2870
        END                                                          DUAL2880
C   FOR MULTIPLE-CHOICE DATA, (1) IF RESPONSE-PATTERNS ARE INPUT, OBTAIN
C   DN,D AND FT, (2) IF DATA INDICATE THE CHOSEN OPTION NUMBERS, RATHER THAN
C   (1,0) PATTERNS, CONVERT THE DATA MATRIX TO THAT OF RESPONSE PATTERNS,
C   AND THEN OBTAIN DN,D AND FT
        SUBROUTINE MULT(NR,NC,FT,NIT)                                 DUAL2890
        COMMON F(30,30),DN(30),D(30)                                  DUAL2900
        DIMENSION OPTION(15)                                          DUAL2910
        INTEGER CD,P,OPTION                                           DUAL2920
        CD=5                                                          DUAL2930
        P=6                                                           DUAL2940
        READ(CD,210) IFORM, NIT                                       DUAL2950
    210 FORMAT(2I4)                                                   DUAL2960
        IF (IFORM.NE.0) GO TO 200                                     DUAL2970
        CALL FRQNCY(NR,NC,FT)                                         DUAL2980
        RETURN                                                        DUAL2990
    200 READ(CD,201) (OPTION(I), I=1,NC)                              DUAL3000
    201 FORMAT(40I2)                                                  DUAL3010
        NTOT=0                                                        DUAL3020
        DO 202 I=1,NC                                                 DUAL3030
    202 NTOT=NTOT+OPTION(I)                                           DUAL3040
        NK=NC                                                         DUAL3050
        NC=NTOT                                                       DUAL3060
        DO 205 I=1,NR                                                 DUAL3070
        DO 203 J=1,NK                                                 DUAL3080
    203 D(J)=F(I,J)                                                   DUAL3090
        DO 204 K=1,NC                                                 DUAL3100
    204 F(I,K)=0.0                                                    DUAL3110
        KG=0                                                          DUAL3120
        DO 205 L=1,NK                                                 DUAL3130
        LM=D(L)                                                       DUAL3140
        KL=KG+LM                                                      DUAL3150
        F(I,KL)=1.0                                                   DUAL3160
        KK= OPTION(L)                                                 DUAL3170
        KG=KG+KK                                                      DUAL3180
    205 CONTINUE                                                      DUAL3190
        WRITE(P,206)                                                  DUAL3200
    206 FORMAT(//5X,'CORRESPONDING RESPONSE PATTERN MATRIX'//)        DUAL3210
        DO 207 I=1,NR                                                 DUAL3220
    207 WRITE(P,208) (F(I,J), J=1,NC)                                 DUAL3230
    208 FORMAT(2X,40F3.0)                                             DUAL3240
        CALL FRQNCY(NR,NC,FT)                                         DUAL3250
        RETURN                                                        DUAL3260
        END                                                          DUAL3270
C   FOR PAIRED COMPARISON DATA, TRANSFORM INPUT MATRIX F OF 1'S AND -1'S TO
C   MATRIX E BY POSTMULTIPLYING IT BY THE DESIGN MATRIX A.  DEFINE DN, D, FT
        SUBROUTINE PAIR(NR,NC,FT)                                     DUAL3280
        COMMON F(30,30),DN(30),D(30)                                  DUAL3290
        DIMENSION E(30,30),A(45,10)                                   DUAL3300
C   CONSTRUCTION OF DESIGN MATRIX
        K=2*NC                                                        DUAL3310
        I=1                                                           DUAL3320
    306 I=I+1                                                         DUAL3330
        J=I-1                                                         DUAL3340
        L=I*J                                                         DUAL3350
        IF(L.EQ.K) GO TO 307                                          DUAL3360
        IF(L.LT.K) GO TO 306                                          DUAL3370
        NC=1                                                          DUAL3380
    307 NP=NC                                                         DUAL3390
```

```
      NC=I                                                              DUAL3400
      DO 300 I=1,NP                                                     DUAL3410
      DO 300 J=1,NC                                                     DUAL3420
  300 A(I,J)=0.0                                                        DUAL3430
      JS=NC-1                                                           DUAL3440
      JJ=1                                                              DUAL3450
      DO 301 I=1,JS                                                     DUAL3460
      J=I+1                                                             DUAL3470
      DO 301 K=J,NC                                                     DUAL3480
      A(JJ,I)=1.0                                                       DUAL3490
      A(JJ,K)=-1.0                                                      DUAL3500
  301 JJ=JJ+1                                                           DUAL3510
C      CONSTRUCTION OF MATRIX E
      DO 302 I=1,NR                                                     DUAL3520
      DO 302 J=1,NC                                                     DUAL3530
      E(I,J)=0.0                                                        DUAL3540
      DO 302 K=1,NP                                                     DUAL3550
  302 E(I,J)=E(I,J)+F(I,K)*A(K,J)                                       DUAL3560
      T=NC*JS                                                           DUAL3570
      S=NR*JS                                                           DUAL3580
      DO 303 I=1,NR                                                     DUAL3590
      DO 303 J=1,NC                                                     DUAL3600
  303 F(I,J)=E(I,J)                                                     DUAL3610
      DO 304 I=1,NR                                                     DUAL3620
  304 DN(I)=T                                                           DUAL3630
      DO 305 I=1,NC                                                     DUAL3640
  305 D(I)=S                                                            DUAL3650
      FT=NC                                                             DUAL3660
      RETURN                                                            DUAL3670
      END                                                               DUAL3680
C    FOR RANK ORDER DATA, TRANSFORM MATRIX F TO MATRIX E BY DESIGN MATRIX A
C    DEFINE DN, D AND FT.
      SUBROUTINE RANK(NR,NC,FT)                                         DUAL3690
      COMMON F(30,30),DN(30),D(30)                                      DUAL3700
C      CALCULATION OF MATRIX E
      JS=NC-1                                                           DUAL3710
      T1=NC+1                                                           DUAL3720
      DO 400 I=1,NR                                                     DUAL3730
      DO 400 J=1,NC                                                     DUAL3740
  400 F(I,J)=T1-2.0*F(I,J)                                              DUAL3750
      T=NC*JS                                                           DUAL3760
      S=NR*JS                                                           DUAL3770
      DO 402 I=1,NR                                                     DUAL3780
  402 DN(I)=T                                                           DUAL3790
      DO 403 I=1,NC                                                     DUAL3800
  403 D(I)=S                                                            DUAL3810
      FT=NC                                                             DUAL3820
      RETURN                                                            DUAL3830
      END                                                               DUAL3840
C    FOR SUCCESSIVE CATEGORIES DATA, TRANSFORM MATRIX F TO MATRIX E BY DESIGN
C    MATRIX A,  DEFINE DN, D AND FT
      SUBROUTINE SUCCES(NR,NC,FT,NCAT)                                  DUAL3850
      COMMON F(30,30),DN(30),D(30)                                      DUAL3860
      DIMENSION E(30,30),A(30,20)                                       DUAL3870
      INTEGER P                                                         DUAL3880
      P=6                                                               DUAL3890
      JS=NCAT-1                                                         DUAL3900
      DO 502 I=1,NR                                                     DUAL3910
      M=0                                                               DUAL3920
      DO 516 J=1,NC                                                     DUAL3930
      K=F(I,J)                                                          DUAL3940
      L=K-1                                                             DUAL3950
      IF(L.EQ.0) GO TO 501                                              DUAL3960
      DO 500 KK=1,L                                                     DUAL3970
      M=M+1                                                             DUAL3980
```

```
500 E(I,M)=-1.0                                                    DUAL3990
    IF(L.EQ.JS) GO TO 516                                          DUAL4000
501 DO 510 KKK=K,JS                                                DUAL4010
    M=M+1                                                          DUAL4020
510 E(I,M)=1.0                                                     DUAL4030
516 CONTINUE                                                       DUAL4040
502 CONTINUE                                                       DUAL4050
    KL=JS*NC                                                       DUAL4060
    WRITE(P,511)                                                   DUAL4070
511 FORMAT(///5X,'***  SUCCESSIVE CATEGORIES DATA  ***'//)        DUAL4080
    DO 512 I=1,NR                                                  DUAL4090
512 WRITE(P,513)(E(I,J),J=1,KL)                                    DUAL4100
513 FORMAT(2X,25F5.0)                                              DUAL4110
    T=2*JS*NC                                                      DUAL4120
    DO 503 I=1,NR                                                  DUAL4130
503 DN(I)=T                                                        DUAL4140
    S=NC*NR                                                        DUAL4150
    DO 504 I=1,JS                                                  DUAL4160
504 D(I)=S                                                         DUAL4170
    S=JS*NR                                                        DUAL4180
    L=JS+NC                                                        DUAL4190
    DO 505 I=NCAT,L                                                DUAL4200
505 D(I)=S                                                         DUAL4210
    DO 506 I=1,KL                                                  DUAL4220
    DO 506 J=1,L                                                   DUAL4230
506 A(I,J)=0.0                                                     DUAL4240
    I=1                                                            DUAL4250
    N=JS+NC                                                        DUAL4260
    DO 507 K=NCAT,N                                                DUAL4270
    DO 507 J=1,JS                                                  DUAL4280
    A(I,J)=1.0                                                     DUAL4290
    A(I,K)=-1.0                                                    DUAL4300
507 I=I+1                                                          DUAL4310
    WRITE(P,514)                                                   DUAL4320
514 FORMAT(///5X,'***  DESIGN MATRIX  ***'//)                     DUAL4330
    DO 515 I=1,KL                                                  DUAL4340
515 WRITE(P,513)(A(I,J),J=1,L)                                     DUAL4350
    DO 509 I=1,NR                                                  DUAL4360
    DO 509 J=1,L                                                   DUAL4370
    F(I,J)=0.0                                                     DUAL4380
    DO 509 K=1,KL                                                  DUAL4390
509 F(I,J)=F(I,J)+E(I,K)*A(K,J)                                    DUAL4400
    FT=2.0*NR*JS*NC                                                DUAL4410
    NC=L                                                          DUAL4420
    RETURN                                                         DUAL4430
    END                                                           DUAL4440
```

B.2

EXAMPLES

```
$DATA
 EXAMPLE 8.1 (GUTTMAN, 1971)
    8    5    0    0
(5F4.0)
   61  104    8   22    5
   70  117    9   24    7
   97  218   12   28   14
   32  118    6   28    7
    4   11    1    2    1
  104   48   14   16    9
   81  128   14   52   12
   20   42    2    6    0
```

```
 EXAMPLE B.2 (SAME AS EXAMPLE 5.1)
  10   6   1   0
(6F2.0)
 1 0 1 0 0 1
 0 1 0 1 0 1
 1 0 1 0 1 0
 0 1 0 1 1 0
 0 1 1 0 0 1
 0 1 1 0 1 0
 1 0 1 0 0 1
 1 0 0 1 1 0
 1 0 1 0 1 0
 0 1 0 1 0 1
   0   3
 EXAMPLE B.3 (SAME AS EXAMPLE 5.1)
  10   3   1   0
(3F2.0)
 1 1 2
 2 2 2
 1 1 1
 2 2 1
 2 1 2
 2 1 1
 1 1 2
 1 2 1
 1 1 1
 2 2 2
   1   3
 2 2 2
 EXAMPLE B.4 (SAME AS EXAMPLE 6.1)
  10   6   2   0
(6F3.0)
  1  1 -1  1 -1  1
  1  1  1  1  1 -1
 -1 -1 -1  1 -1  1
  1 -1 -1  1 -1 -1
  1  1 -1  1 -1 -1
  1  1  1  1 -1 -1
  1  1 -1  1  1 -1
 -1 -1 -1 -1 -1  1
  1  1  1  1 -1 -1
  1  1 -1  1 -1 -1
 EXAMPLE B.5 (JOHNSON AND LEONE, 1964, P.276)
   6  12   3   0
(12F3.0)
 11  4  3  6  1  5  7  8 10  2 12  9
  8  7  4  5  3  2  6  9 12  1 11 10
  9  8  5  6  1  7  2 11 12  3 10  4
 11  3  2  8  1  4  5 10  9  6 12  7
  8  7  1  9  4  6  2 11 12  3 10  5
  9  6  4 11  3  7  1  5 10  2 12  8
 EXAMPLE B.6 (SAME AS EXAMPLE 8.2)
  10   3   4   3
(3F2.0)
 3 2 1
 2 2 2
 2 2 1
 3 2 1
 3 1 1
 3 3 1
 2 2 2
 3 3 2
 2 2 1
 3 1 1
```

This section presents the computer printouts of analysis, using the six data sets listed above. Some comments are given after the printout of each example.

Example B.1 (Guttman 1971)

```
***********************************
*          DUAL SCALING OF        *
*  CONTINGENCY/FREQUENCY TABLE    *
***********************************


TITLE OF STUDY :  EXAMPLE B.1 (GUTTMAN, 1971)

***  INPUT MATRIX  ***

 61.  104.    8.   22.    5.
 70.  117.    9.   24.    7.
 97.  218.   12.   28.   14.
 32.  118.    6.   28.    7.
  4.   11.    1.    2.    1.
104.   48.   14.   16.    9.
 81.  128.   14.   52.   12.
 20.   42.    2.    6.    0.

** NUMBER OF RESPONSES FOR ROWS

  200. 227. 369. 191.  19. 191. 287.  70.

** NUMBER OF RESPONSES FOR COLUMNS

  469. 786.  66. 178.  55.

   ***  MATRIX C FOR EIGENEQUATION  ***

 0.0332
-0.0260 0.0245
 0.0103-0.0099 0.0041
-0.0070-0.0011 0.0004 0.0128
 0.0027-0.0040 0.0019 0.0011 0.0029

** TRACE OF C  =  0.0775


********************
** SOLUTION    1 **
********************

 SQUARED CORRELATION RATIO =  0.0597   ;  MAXIMUM CORRELATION =  0.2443

DELTA(TOTAL VARIANCE ACCOUNTED FOR BY  1  SOLUTION(S))  =   77.0  (PER CEN


NUMBER OF ITERATIONS  =   9
```

```
OPTIMAL/LOCALLY OPTIMAL WEIGHT VECTORS (UNWEIGHTED,WEIGHTED BY ETA) FOR
               ** ROWS **                          ** COLUMNS **
      UNWEIGHTED     WEIGHTED          UNWEIGHTED     WEIGHTED

 1      -0.0594      -0.0145        1       1.3380       0.3268
 2      -0.0108      -0.0026        2      -0.8697      -0.2124
 3      -0.5173      -0.1264        3       1.2044       0.2942
 4      -1.2219      -0.2985        4      -0.2709      -0.0662
 5      -0.6682      -0.1632        5       0.4501       0.1100
 6       2.4430       0.5968
 7       0.0747       0.0183
 8      -0.5253      -0.1283

ORTHOGONAL WEIGHT VECTORS (UNWEIGHTED, WEIGHTED BYETA) FOR
               ** ROWS **                          ** COLUMNS **
      UNWEIGHTED     WEIGHTED          UNWEIGHTED     WEIGHTED

 1      -0.8394      -0.2051        1      28.9760       7.0783
 2      -0.1632      -0.0399        2     -24.3813      -5.9559
 3      -9.9370      -2.4274        3       9.7847       2.3902
 4     -16.8869      -4.1251        4      -3.6139      -0.8828
 5      -2.9127      -0.7115        5       3.3381       0.8154
 6      33.7634       8.2477
 7       1.2660       0.3093
 8      -4.3948      -1.0736

*** CHI-SQUARE =   95.18   WITH   10   DEGREES OF FREEDOM

ORDER  1  APPROXIMATION OF DATA MATRIX

       59.2   102.4     8.3    23.0     7.0
       68.3   115.1     9.6    26.0     8.0
       92.5   207.1    13.3    43.7    12.3
       34.6   121.7     5.2    23.6     5.9
        4.5    11.0     0.6     2.3     0.6
      103.7    46.5    13.9    18.3     8.6
       88.7   142.9    12.5    32.7    10.2
       17.5    39.4     2.5     8.3     2.3


********************
** SOLUTION    2  **
********************

 SQUARED CORRELATION RATIO =  0.0153   ;  MAXIMUM CORRELATION =  0.1238


DELTA(TOTAL VARIANCE ACCOUNTED FOR BY  2  SOLUTION(S))  =    96.8  (PER CENT)


NUMBER OF ITERATIONS  =    7

OPTIMAL/LOCALLY OPTIMAL WEIGHT VECTORS (UNWEIGHTED,WEIGHTED BY ETA) FOR
               ** ROWS **                          ** COLUMNS **
      UNWEIGHTED     WEIGHTED          UNWEIGHTED     WEIGHTED

 1       0.2442       0.0302        1       0.4190       0.0519
 2       0.2920       0.0361        2       0.4618       0.0572
 3       1.0453       0.1294        3      -0.6749      -0.0836
 4      -0.6668      -0.0826        4      -2.6347      -0.3262
 5      -0.0101      -0.0013        5      -0.8352      -0.1034
 6       0.2799       0.0346
 7      -1.7851      -0.2210
 8       1.2252       0.1517
```

ORTHOGONAL WEIGHT VECTORS (UNWEIGHTED, WEIGHTED BYETA) FOR

	** ROWS ** UNWEIGHTED	WEIGHTED		** COLUMNS ** UNWEIGHTED	WEIGHTED
1	3,4534	0,4275	1	9,0733	1,1233
2	4,3993	0,5446	2	12,9472	1,6029
3	20,0790	2,4858	3	-5,4832	-0,6788
4	-9,2153	-1,1409	4	-35,1513	-4,3518
5	-0,0440	-0,0055	5	-6,1941	-0,7668
6	3,8681	0,4789			
7	-30,2423	-3,7440			
8	10,2504	1,2690			

*** CHI-SQUARE = 23,89 WITH 8 DEGREES OF FREEDOM

ORDER 2 APPROXIMATION OF DATA MATRIX

60,0	103,8	8,2	21,2	6,9
69,3	117,0	9,4	23,5	7,8
98,6	218,3	11,9	29,3	10,9
32,6	118,0	5,6	28,4	6,3
4,5	11,0	0,6	2,3	0,6
104,5	48,0	13,8	16,3	8,4
80,7	128,0	14,3	51,9	12,1
18,8	41,8	2,2	5,1	2,0

****
** SOLUTION 3 **
****

SQUARED CORRELATION RATIO = 0,0024 ; MAXIMUM CORRELATION = 0,0490

DELTA(TOTAL VARIANCE ACCOUNTED FOR BY 3 SOLUTION(S)) = 99,9 (PER CEN

NUMBER OF ITERATIONS = 4

OPTIMAL/LOCALLY OPTIMAL WEIGHT VECTORS (UNWEIGHTED,WEIGHTED BY ETA) FOR

	** ROWS ** UNWEIGHTED	WEIGHTED		** COLUMNS ** UNWEIGHTED	WEIGHTED
1	1,0542	0,0517	1	0,3688	0,0181
2	0,4267	0,0209	2	0,0235	0,0012
3	-0,9394	-0,0460	3	-0,7186	-0,0352
4	-0,4458	-0,0219	4	0,7283	0,0357
5	-2,6881	-0,1318	5	-4,9750	-0,2439
6	-0,3952	-0,0194			
7	0,0702	0,0034			
8	3,2913	0,1613			

ORTHOGONAL WEIGHT VECTORS (UNWEIGHTED, WEIGHTED BYETA) FOR

	** ROWS ** UNWEIGHTED	WEIGHTED		** COLUMNS ** UNWEIGHTED	WEIGHTED
1	14,9090	0,7309	1	7,9871	0,3915
2	6,4288	0,3151	2	0,6578	0,0322
3	-18,0446	-0,8846	3	-5,8383	-0,2862
4	-6,1611	-0,3020	4	9,7167	0,4763
5	-11,7172	-0,5744	5	-36,8955	-1,8087
6	-5,4624	-0,2678			
7	1,1888	0,0583			
8	27,5371	1,3499			

*** CHI-SQUARE = 3,72 WITH 6 DEGREES OF FREEDOM

ORDER 3 APPROXIMATION OF DATA MATRIX

61.1	104.0	7.9	22.0	5.0
69.8	117.1	9.2	23.9	6.9
96.7	218.1	12.4	27.9	13.9
32.2	118.0	5.8	28.1	7.1
4.2	10.9	0.7	2.1	1.1
104.1	48.0	13.9	16.0	9.0
80.8	128.1	14.2	51.9	11.9
20.1	42.0	1.9	6.0	0.0

Comments The data were obtained from 1,554 Israeli adults with respect to two categorical variables, 'principal worry' and 'country of origin.' The eight rows are, respectively, 'enlisted relative,' 'sabotage,' 'military situation,' 'political situation,' 'economic situation,' 'personal economics,' 'other,' and 'more than one,' and the five columns are, respectively, 'Asia-Africa,' 'Europe-America,' 'Israel; father Asia-Africa,' 'Israel; father Europe-America,' and 'Israel; father Israel.' The optimum solution accounts for 77% of the total variance, and we can see that the order one approximation is already very close to the input matrix. The maximum correlation between the two categorical variables is 0.2443, and the discriminability of the weighting scheme is significant at the 0.01 level. From $\mathbf{y}_1$ and $\mathbf{x}_1$, it is easy to see that dominant variables are rows 6 and 4 and columns 1 and 2. Since more than 99.5% of the total variance is accounted for by three solutions and the chi-square value is less than the number of degrees of freedom, computation is terminated after the third solution. Had it been continued, order four approximation would have reproduced the input matrix completely, except for a possible rounding error. It is interesting to note that the second solution, despite its 19.8% contribution to the total variance, does not seem to contribute very much to the recovery of the input matrix. It should be recalled, however, that the order one approximation is really a rank two approximation, consisting of that of the trivial solution (the outcome expected when the two variables are statistically independent) plus that of the optimum solution. δ_1 of 77%, in contrast, is the contribution of the optimum solution to the variance generated by only non-trivial solutions, and hence provides more straightforward information than the order one approximation. The detailed examination and interpretation of the results are left for the readers.

DUAL1 computes order k approximation only for contingency and response-frequency tables. In other types of data, -1's, 0's, and 1's are typical elements in the data matrix, and approximating them in terms of a few solutions appears neither interesting nor useful.

Example B.2 (*same as example* 5.1 *in the text*)

```
                    **********************************
                    *         DUAL SCALING OF        *
                              MULTIPLE-CHOICE DATA   *
                    **********************************

   TITLE OF STUDY :  EXAMPLE B.2 (SAME AS EXAMPLE 5.1)
   ***  INPUT MATRIX  ***

    1.    0.    1.    0.    0.    1.
    0.    1.    0.    1.    0.    1.
    1.    0.    1.    0.    1.    0.
    0.    1.    0.    1.    1.    0.
    0.    1.    1.    0.    0.    1.
    0.    1.    1.    0.    1.    0.
    1.    0.    1.    0.    0.    1.
    1.    0.    0.    1.    1.    0.
    1.    0.    1.    0.    1.    0.
    0.    1.    0.    1.    0.    1.
** NUMBER OF RESPONSES FOR ROWS
    3.   3.   3.   3.   3.   3.   3.   3.   3.   3.
** NUMBER OF RESPONSES FOR COLUMNS
    5.   5.   6.   4.   5.   5.

   ***  MATRIX C FOR EIGENEQUATION  ***

 0.1667
-0.1667 0.1667
 0.0609-0.0609 0.1333
-0.0745 0.0745-0.1633 0.2000
 0.0333-0.0333-0.0000-0.0000 0.1667
-0.0333 0.0333-0.0000-0.0000-0.1667 0.1667

** TRACE OF C  =  1.0000


********************
** SOLUTION    1  **
********************

  SQUARED CORRELATION RATIO =  0.4849    ;  MAXIMUM CORRELATION =  0.6963

 DELTA(TOTAL VARIANCE ACCOUNTED FOR BY  1  SOLUTION(S))  =   48.5 (PER CEN

 NUMBER OF ITERATIONS  =  22

 RELIABILITY COEFFICIENT (ALPHA)  =  0.4688
```

```
OPTIMAL/LOCALLY OPTIMAL WEIGHT VECTORS (UNWEIGHTED,WEIGHTED BY ETA) FOR
           ** ROWS **                                ** COLUMNS **
      UNWEIGHTED     WEIGHTED           UNWEIGHTED     WEIGHTED

 1        0.7582       0.5279       1        1.2247       0.8528
 2       -1.4891      -1.0369       2       -1.2247      -0.8528
 3        1.2741       0.8872       3        0.8980       0.6253
 4       -0.9731      -0.6776       4       -1.3470      -0.9379
 5       -0.4144      -0.2886       5        0.5389       0.3753
 6        0.1016       0.0707       6       -0.5389      -0.3753
 7        0.7582       0.5279
 8        0.1995       0.1389
 9        1.2741       0.8872
10       -1.4891      -1.0369

ORTHOGONAL WEIGHT VECTORS (UNWEIGHTED, WEIGHTED BYETA) FOR
           ** ROWS **                                ** COLUMNS **
      UNWEIGHTED     WEIGHTED           UNWEIGHTED     WEIGHTED

 1        1.3132       0.9144       1        2.7386       1.9070
 2       -2.5792      -1.7959       2       -2.7386      -1.9070
 3        2.2069       1.5367       3        2.1996       1.5316
 4       -1.6855      -1.1736       4       -2.6939      -1.8759
 5       -0.7178      -0.4998       5        1.2051       0.8391
 6        0.1759       0.1225       6       -1.2051      -0.8391
 7        1.3132       0.9144
 8        0.3455       0.2406
 9        2.2069       1.5367
10       -2.5792      -1.7959


***  CHI-SQUARE  =   15.26   WITH   10   DEGREES OF FREEDOM


*******************
** SOLUTION    2  **
*******************

 SQUARED CORRELATION RATIO =  0.3333   ;  MAXIMUM CORRELATION =  0.5773

 DELTA(TOTAL VARIANCE ACCOUNTED FOR BY  2  SOLUTION(S))  =   81.8  (PER CENT)

 NUMBER OF ITERATIONS  =  15

 RELIABILITY COEFFICIENT (ALPHA)  = -0.0000

 NEGATIVE ALPHA -- FURTHER ANALYSIS IS ABORTED
```

Example B.3 (same as example 5.1)

```
               ***********************************
               *         DUAL SCALING OF         *
                     MULTIPLE-CHOICE DATA        *
               ***********************************

    TITLE OF STUDY :  EXAMPLE B.3 (SAME AS EXAMPLE 5.1)
```

```
   ***  INPUT MATRIX  ***

    1.    1.    2.
    2.    2.    2.
    1.    1.    1.
    2.    2.    1.
    2.    1.    2.
    2.    1.    1.
    1.    1.    2.
    1.    2.    1.
    1.    1.    1.
    2.    2.    2.

   CORRESPONDING RESPONSE PATTERN MATRIX

 1. 0. 1. 0. 0. 1.
 0. 1. 0. 1. 0. 1.
 1. 0. 1. 0. 1. 0.
 0. 1. 0. 1. 1. 0.
 0. 1. 1. 0. 0. 1.
 0. 1. 1. 0. 1. 0.
 1. 0. 1. 0. 0. 1.
 1. 0. 0. 1. 1. 0.
 1. 0. 1. 0. 1. 0.
 0. 1. 0. 1. 0. 1.

** NUMBER OF RESPONSES FOR ROWS

    3.   3.   3.   3.   3.   3.   3.   3.   3.   3.

** NUMBER OF RESPONSES FOR COLUMNS

    5.   5.   6.   4.   5.   5.

   ***  MATRIX C FOR EIGENEQUATION  ***

 0.1667
-0.1667 0.1667
 0.0609-0.0609 0.1333
-0.0745 0.0745-0.1633 0.2000
 0.0333-0.0333-0.0000-0.0000 0.1667
-0.0333 0.0333-0.0000-0.0000-0.1667 0.1667

** TRACE OF C  =  1.0000



********************
** SOLUTION    1  **
********************

  SQUARED CORRELATION RATIO =  0.4849   ;  MAXIMUM CORRELATION =  0.6963

 DELTA(TOTAL VARIANCE ACCOUNTED FOR BY  1  SOLUTION(S))  =   48.5 (PER CE


 NUMBER OF ITERATIONS  =  22


 RELIABILITY COEFFICIENT (ALPHA)  =  0.4688
```

```
OPTIMAL/LOCALLY OPTIMAL WEIGHT VECTORS (UNWEIGHTED,WEIGHTED BY ETA) FOR
             ** ROWS **                               ** COLUMNS **
       UNWEIGHTED     WEIGHTED                UNWEIGHTED     WEIGHTED

 1          0.7582       0.5279          1         1.2247        0.8528
 2         -1.4891      -1.0369          2        -1.2247       -0.8528
 3          1.2741       0.8872          3         0.8980        0.6253
 4         -0.9731      -0.6776          4        -1.3470       -0.9379
 5         -0.4144      -0.2886          5         0.5389        0.3753
 6          0.1016       0.0707          6        -0.5389       -0.3753
 7          0.7582       0.5279
 8          0.1995       0.1389
 9          1.2741       0.8872
10         -1.4891      -1.0369

ORTHOGONAL WEIGHT VECTORS (UNWEIGHTED, WEIGHTED BYETA) FOR
             ** ROWS **                               ** COLUMNS **
       UNWEIGHTED     WEIGHTED                UNWEIGHTED     WEIGHTED

 1          1.3132       0.9144          1         2.7386        1.9070
 2         -2.5792      -1.7959          2        -2.7386       -1.9070
 3          2.2069       1.5367          3         2.1996        1.5316
 4         -1.6855      -1.1736          4        -2.6939       -1.8759
 5         -0.7178      -0.4998          5         1.2051        0.8391
 6          0.1759       0.1225          6        -1.2051       -0.8391
 7          1.3132       0.9144
 8          0.3455       0.2406
 9          2.2069       1.5367
10         -2.5792      -1.7959

***  CHI-SQUARE  =    15.26   WITH   10   DEGREES OF FREEDOM


********************
** SOLUTION    2  **
********************

 SQUARED CORRELATION RATIO =  0.3333   ;  MAXIMUM CORRELATION =  0.5773

DELTA(TOTAL VARIANCE ACCOUNTED FOR BY  2  SOLUTION(S))  =   81.8  (PER CENT)


NUMBER OF ITERATIONS  =  15


RELIABILITY COEFFICIENT (ALPHA)  = -0.0000


NEGATIVE ALPHA -- FURTHER ANALYSIS IS ABORTED
```

Comments Examples B.2 and B.3 are different only in the forms of the input matrix (i.e., IFORM=0 and IFORM≠0). The rank of the response-pattern matrix is four, but the run was terminated because of the negative value of Cronbach's coefficient α.

*Example B.*4 (*same as example* 6.1)

```
          **********************************
          *         DUAL SCALING OF        *
          *     PAIRED COMPARISON DATA     *
          **********************************

   TITLE OF STUDY :  EXAMPLE B.4 (SAME AS EXAMPLE 6.1)

   ***  INPUT MATRIX  ***

    1.    1.   -1.    1.   -1.    1.
    1.    1.    1.    1.    1.   -1.
   -1.   -1.   -1.    1.   -1.    1.
    1.   -1.   -1.    1.   -1.   -1.
    1.    1.   -1.    1.   -1.   -1.
    1.    1.    1.    1.   -1.   -1.
    1.    1.   -1.    1.    1.   -1.
   -1.   -1.   -1.   -1.   -1.    1.
    1.    1.    1.    1.   -1.   -1.
    1.    1.   -1.    1.   -1.   -1.

** NUMBER OF RESPONSES FOR ROWS

   12.  12.  12.  12.  12.  12.  12.  12.  12.  12.

** NUMBER OF RESPONSES FOR COLUMNS

   30.  30.  30.  30.


   ***  TRANSFORMED DATA MATRIX E  ***

   1. -1. -1.  1.
   3.  1. -3. -1.
  -3.  1.  1.  1.
  -1. -1. -1.  3.
   1. -1. -3.  3.
   3. -1. -3.  1.
   1.  1. -3.  1.
  -3. -1.  3.  1.
   3. -1. -3.  1.
   1. -1. -3.  3.


   ***  MATRIX C FOR EIGENEQUATION  ***

 0.1389
-0.0111 0.0278
-0.1333 0.0167 0.1833
 0.0056-0.0333-0.0667 0.0944


** TRACE OF C  =  0.4444


********************
** SOLUTION    1 **
********************

  SQUARED CORRELATION RATIO =  0.3131   ;  MAXIMUM CORRELATION =  0.5595
```

```
DELTA(TOTAL VARIANCE ACCOUNTED FOR BY  1  SOLUTION(S))  =   70,4  (PER CENT)


NUMBER OF ITERATIONS  =  11

OPTIMAL/LOCALLY OPTIMAL WEIGHT VECTORS (UNWEIGHTED,WEIGHTED BY ETA) FOR
          ** ROWS **                          ** COLUMNS **
     UNWEIGHTED    WEIGHTED           UNWEIGHTED     WEIGHTED

 1        0,0928      0,0519     1         0,2165       0,1212
 2        0,2006      0,1122     2        -0,0357      -0,0200
 3       -0,1290     -0,0722     3        -0,2759      -0,1544
 4        0,0566      0,0317     4         0,0951       0,0532
 5        0,2033      0,1138
 6        0,2395      0,1340
 7        0,1644      0,0920
 8       -0,2006     -0,1122
 9        0,2395      0,1340
10        0,2033      0,1138

ORTHOGONAL WEIGHT VECTORS (UNWEIGHTED, WEIGHTED BYETA) FOR
          ** ROWS **                          ** COLUMNS **
     UNWEIGHTED    WEIGHTED           UNWEIGHTED     WEIGHTED

 1        0,3215      0,1799     1         1,1859       0,6636
 2        0,6948      0,3887     2        -0,1953      -0,1093
 3       -0,4468     -0,2500     3        -1,5114      -0,8457
 4        0,1962      0,1098     4         0,5207       0,2914
 5        0,7043      0,3941
 6        0,8297      0,4642
 7        0,5694      0,3186
 8       -0,6948     -0,3887
 9        0,8297      0,4642
10        0,7043      0,3941


***  CHI-SQUARE  =   19,72   WITH   11   DEGREES OF FREEDOM


********************
** SOLUTION    2  **
********************

 SQUARED CORRELATION RATIO =  0,1074   ;  MAXIMUM CORRELATION =  0,3277

 DELTA(TOTAL VARIANCE ACCOUNTED FOR BY  2  SOLUTION(S))  =   94,6  (PER CENT)


 NUMBER OF ITERATIONS  =   5

 OPTIMAL/LOCALLY OPTIMAL WEIGHT VECTORS (UNWEIGHTED,WEIGHTED BY ETA) FOR
           ** ROWS **                          ** COLUMNS **
      UNWEIGHTED    WEIGHTED           UNWEIGHTED     WEIGHTED

  1       -0,0641     -0,0210     1         0,1754       0,0575
  2        0,2222      0,0728     2         0,1060       0,0347
  3       -0,1785     -0,0585     3         0,0201       0,0066
  4       -0,3067     -0,1005     4        -0,3015      -0,0988
  5       -0,2277     -0,0746
  6        0,0149      0,0049
  7       -0,0204     -0,0067
  8       -0,2222     -0,0728
  9        0,0149      0,0049
 10       -0,2277     -0,0746
```

```
ORTHOGONAL WEIGHT VECTORS (UNWEIGHTED, WEIGHTED BYETA) FOR
              ** ROWS **                          ** COLUMNS **
      UNWEIGHTED     WEIGHTED             UNWEIGHTED     WEIGHTED

  1      -0.2222      -0.0728       1        0.9609        0.3149
  2       0.7696       0.2522       2        0.5805        0.1902
  3      -0.6182      -0.2026       3        0.1101        0.0361
  4      -1.0626      -0.3482       4       -1.6515       -0.5412
  5      -0.7889      -0.2585
  6       0.0515       0.0169
  7      -0.0708      -0.0232
  8      -0.7696      -0.2522
  9       0.0515       0.0169
 10      -0.7889      -0.2585

***  CHI-SQUARE  =    5.96    WITH     9    DEGREES OF FREEDOM
```

Comments DUAL1 uses Nishisato's formulation for both paired comparison and rank order data. In paired comparisons, the (locally) optimal vector is collinear with the corresponding orthogonal vector. Thus we adopted orthogonal vectors in the main text. The run was terminated because the chi-square value for the second solution was smaller than the number of degrees of freedom.

Example B.5 (Johnson and Leone 1964, p.276)

```
              **********************************

              *          DUAL SCALING OF          *

              *          RANK ORDER DATA          *

              **********************************

   TITLE OF STUDY :  EXAMPLE B.5 (JOHNSON AND LEONE, 1964, P.276)

   ***  INPUT MATRIX  ***

   11.    4.    3.    6.    1.    5.    7.    8.   10.    2.   12.    9.
    8.    7.    4.    5.    3.    2.    6.    9.   12.    1.   11.   10.
    9.    8.    5.    6.    1.    7.    2.   11.   12.    3.   10.    4.
   11.    3.    2.    8.    1.    4.    5.   10.    9.    6.   12.    7.
    8.    7.    1.    9.    4.    6.    2.   11.   12.    3.   10.    5.
    9.    6.    4.   11.    3.    7.    1.    5.   10.    2.   12.    8.
** NUMBER OF RESPONSES FOR ROWS

  132. 132. 132. 132. 132. 132.

** NUMBER OF RESPONSES FOR COLUMNS

   66.  66.  66.  66.  66.  66.  66.  66.  66.  66.  66.  66.

   ***  TRANSFORMED DATA MATRIX E  ***

-9.  5.  7.  1. 11.  3. -1. -3. -7.  9.-11. -5.
-3. -1.  5.  3.  7.  9.  1. -5.-11. 11. -9. -7.
-5. -3.  3.  1. 11. -1.  9. -9.-11.  7. -7.  5.
-9.  7.  9. -3. 11.  5.  3. -7. -5.  1.-11. -1.
-3. -1. 11. -5.  5.  1.  9. -9.-11.  7. -7.  3.
-5.  1.  5. -9.  7. -1. 11.  3. -7.  9.-11. -3.
```

```
    ***  MATRIX C FOR EIGENEQUATION  ***

0,0264
0,0106 0,0099
0,0266 0,0090 0,0356
0,0073-0,0030-0,0117 0,0145
0,0372 0,0108 0,0383-0,0090 0,0558
0,0106 0,0048 0,0131 0,0021 0,0158 0,0135
0,0170-0,0011 0,0236-0,0163 0,0287 0,0011 0,0337
0,0186-0,0023-0,0253 0,0014-0,0308-0,0106-0,0174 0,0292
0,0303-0,0025-0,0388 0,0094-0,0498-0,0158-0,0337 0,0331 0,0558
0,0257 0,0025 0,0310-0,0080 0,0416 0,0140 0,0264-0,0216-0,0466 0,0438
0,0386-0,0122-0,0429 0,0140-0,0567-0,0181-0,0319 0,0285 0,0530-0,0466 0,0622
0,0064-0,0053-0,0053-0,0007-0,0076-0,0094 0,0039-0,0028 0,0057-0,0108 0,0122 0,0135


 ** TRACE OF C  =  0,3939


 ********************
 ** SOLUTION    1  **
 ********************

  SQUARED CORRELATION RATIO =  0,3004   ;  MAXIMUM CORRELATION =  0,5480

  DELTA(TOTAL VARIANCE ACCOUNTED FOR BY  1  SOLUTION(S))  =   76,2  (PER CENT)


  NUMBER OF ITERATIONS  =   7

  OPTIMAL/LOCALLY OPTIMAL WEIGHT VECTORS (UNWEIGHTED,WEIGHTED BY ETA) FOR
                 ** ROWS **                          ** COLUMNS **
           UNWEIGHTED    WEIGHTED              UNWEIGHTED    WEIGHTED

   1          0,1276       0,0699       1        -0,1163       -0,0637
   2          0,1207       0,0661       2         0,0280        0,0154
   3          0,1223       0,0670       3         0,1370        0,0751
   4          0,1247       0,0683       4        -0,0399       -0,0219
   5          0,1259       0,0690       5         0,1773        0,0972
   6          0,1170       0,0641       6         0,0547        0,0300
                                        7         0,1075        0,0589
                                        8        -0,1035       -0,0567
                                        9        -0,1768       -0,0969
                                       10         0,1491        0,0817
                                       11        -0,1904       -0,1043
                                       12        -0,0268       -0,0147


  ORTHOGONAL WEIGHT VECTORS (UNWEIGHTED, WEIGHTED BY ETA) FOR
                 ** ROWS **                          ** COLUMNS **
           UNWEIGHTED    WEIGHTED              UNWEIGHTED    WEIGHTED

   1          1,4660       0,8035       1        -0,9450       -0,5179
   2          1,3864       0,7598       2         0,2278        0,1249
   3          1,4055       0,7703       3         1,1131        0,6100
   4          1,4327       0,7852       4        -0,3245       -0,1779
   5          1,4465       0,7928       5         1,4407        0,7896
   6          1,3446       0,7369       6         0,4445        0,2436
                                        7         0,8734
                                        8        -0,8405
                                        9        -1,4361
                                       10         1,2109
                                       11        -1,5467
                                       12        -0,2177
```

```
********************
** SOLUTION   2  **
********************
```

SQUARED CORRELATION RATIO = 0,0373 ; MAXIMUM CORRELATION = 0,1931

DELTA(TOTAL VARIANCE ACCOUNTED FOR BY 2 SOLUTION(S)) = 85,7 (PER CENT

NUMBER OF ITERATIONS = 14

OPTIMAL/LOCALLY OPTIMAL WEIGHT VECTORS (UNWEIGHTED,WEIGHTED BY ETA) FOR

	** ROWS **			** COLUMNS **	
	UNWEIGHTED	WEIGHTED		UNWEIGHTED	WEIGHTED
1	0,1535	0,0296	1	-0,0741	-0,0143
2	0,1231	0,0238	2	0,1306	0,0252
3	-0,1337	-0,0258	3	0,0006	0,0001
4	0,0803	0,0155	4	0,1287	0,0248
5	-0,1430	-0,0276	5	0,0505	0,0098
6	-0,0863	-0,0167	6	0,1606	0,0310
			7	-0,2534	-0,0489
			8	0,0465	0,0090
			9	0,0641	0,0124
			10	0,0080	0,0016
			11	-0,0623	-0,0120
			12	-0,1999	-0,0386

ORTHOGONAL WEIGHT VECTORS (UNWEIGHTED, WEIGHTED BYETA) FOR

	** ROWS **			** COLUMNS **	
	UNWEIGHTED	WEIGHTED		UNWEIGHTED	WEIGHTED
1	1,7636	0,3405	1	-0,6021	-0,1163
2	1,4141	0,2730	2	1,0608	0,2048
3	-1,5357	-0,2965	3	0,0047	0,0009
4	0,9223	0,1781	4	1,0454	0,2019
5	-1,6424	-0,3171	5	0,4105	0,0793
6	-0,9917	-0,1915	6	1,3048	0,2519
			7	-2,0582	
			8	0,3778	
			9	0,5209	
			10	0,0654	
			11	-0,5059	
			12	-1,6241	

```
********************
** SOLUTION   3  **
********************
```

SQUARED CORRELATION RATIO = 0,0246 ; MAXIMUM CORRELATION = 0,1568

DELTA(TOTAL VARIANCE ACCOUNTED FOR BY 3 SOLUTION(S)) = 91,9 (PER CEN

NUMBER OF ITERATIONS = 60

OPTIMAL/LOCALLY OPTIMAL WEIGHT VECTORS (UNWEIGHTED,WEIGHTED BY ETA) FOR

	** ROWS **			** COLUMNS **	
	UNWEIGHTED	WEIGHTED		UNWEIGHTED	WEIGHTED
1	-0,0294	-0,0046	1	0,0708	0,0111
2	0,1219	0,0191	2	-0,1320	-0,0207
3	0,1359	0,0213	3	-0,0298	-0,0047
4	-0,0593	-0,0093	4	0,2351	0,0369
5	0,0491	0,0077	5	0,0040	0,0006
6	-0,2254	-0,0353	6	0,0823	0,0129

```
                                         7     -0.0812       -0.0127
                                         8     -0.2366       -0.0371
                                         9     -0.1253       -0.0196
                                        10      0.0274        0.0043
                                        11      0.1027        0.0161
                                        12      0.0827        0.0130

ORTHOGONAL WEIGHT VECTORS (UNWEIGHTED, WEIGHTED BY ETA) FOR
             ** ROWS **                         ** COLUMNS **
       UNWEIGHTED    WEIGHTED              UNWEIGHTED    WEIGHTED

  1        -0.3381     -0.0530        1      0.5750        0.0901
  2         1.4005      0.2195        2     -1.0726       -0.1681
  3         1.5617      0.2448        3     -0.2425       -0.0380
  4        -0.6808     -0.1067        4      1.9104        0.2995
  5         0.5640      0.0884        5      0.0322        0.0050
  6        -2.5892     -0.4059        6      0.6683        0.1048
                                      7     -0.6599
                                      8     -1.9220
                                      9     -1.0179
                                     10      0.2227
                                     11      0.8341
                                     12      0.6722

********************
** SOLUTION    4  **
********************

 SQUARED CORRELATION RATIO =  0.0203   ;  MAXIMUM CORRELATION =  0.1424

DELTA(TOTAL VARIANCE ACCOUNTED FOR BY  4  SOLUTION(S))  =   97.1  (PER CENT)

NUMBER OF ITERATIONS  =  15

OPTIMAL/LOCALLY OPTIMAL WEIGHT VECTORS (UNWEIGHTED,WEIGHTED BY ETA) FOR
             ** ROWS **                         ** COLUMNS **
       UNWEIGHTED    WEIGHTED              UNWEIGHTED    WEIGHTED

  1         0.0096      0.0014        1     -0.1169       -0.0166
  2        -0.1698     -0.0242        2      0.1548        0.0220
  3         0.0297      0.0042        3      0.0981        0.0140
  4         0.2116      0.0301        4     -0.0146       -0.0021
  5         0.0302      0.0043        5      0.0906        0.0129
  6        -0.1242     -0.0177        6     -0.0337       -0.0048
                                      7     -0.0397       -0.0056
                                      8     -0.1672       -0.0238
                                      9      0.1015        0.0145
                                     10     -0.2414       -0.0344
                                     11      0.0046        0.0007
                                     12      0.1639        0.0233

ORTHOGONAL WEIGHT VECTORS (UNWEIGHTED, WEIGHTED BYETA) FOR
             ** ROWS **                         ** COLUMNS **
       UNWEIGHTED    WEIGHTED              UNWEIGHTED    WEIGHTED

  1         0.1100      0.0157        1     -0.9494       -0.1352
  2        -1.9510     -0.2779        2      1.2577        0.1791
  3         0.3409      0.0486        3      0.7966        0.1135
  4         2.4306      0.3462        4     -0.1188       -0.0169
  5         0.3467      0.0494        5      0.7358        0.1048
  6        -1.4273     -0.2033        6     -0.2739       -0.0390
                                      7     -0.3222
                                      8     -1.3586
                                      9      0.8249
                                     10     -1.9614
                                     11      0.0376
                                     12      1.3317
```

```
*********************
** SOLUTION    5  **
*********************

 SQUARED CORRELATION RATIO =  0.0090   ;  MAXIMUM CORRELATION =  0.0948

DELTA(TOTAL VARIANCE ACCOUNTED FOR BY  5  SOLUTION(S))  =   99.4 (PER CENT


NUMBER OF ITERATIONS  =   9

OPTIMAL/LOCALLY OPTIMAL WEIGHT VECTORS (UNWEIGHTED,WEIGHTED BY ETA) FOR
          ** ROWS **                            ** COLUMNS **
     UNWEIGHTED      WEIGHTED            UNWEIGHTED      WEIGHTED

1     -0.0902        -0.0086        1      0.1173         0.0111
2      0.0753         0.0071        2      0.0217         0.0021
3     -0.1921        -0.0182        3      0.2435         0.0231
4      0.0445         0.0042        4     -0.1367        -0.0130
5      0.1926         0.0183        5     -0.2171        -0.0206
6     -0.0331        -0.0031        6      0.1675         0.0159
                                    7     -0.0097        -0.0009
                                    8     -0.0834        -0.0079
                                    9     -0.0309        -0.0029
                                   10     -0.0374        -0.0035
                                   11      0.0297         0.0028
                                   12     -0.0646        -0.0061

ORTHOGONAL WEIGHT VECTORS (UNWEIGHTED, WEIGHTED BY ETA) FOR
          ** ROWS **                            ** COLUMNS **
     UNWEIGHTED      WEIGHTED            UNWEIGHTED      WEIGHTED

1     -1.0368        -0.0982        1      0.9530         0.0903
2      0.8654         0.0820        2      0.1767         0.0167
3     -2.2073        -0.2092        3      1.9778         0.1874
4      0.5118         0.0485        4     -1.1106        -0.1052
5      2.2129         0.2097        5     -1.7634        -0.1671
6     -0.3805        -0.0361        6      1.3610         0.1290
                                    7     -0.0789
                                    8     -0.6774
                                    9     -0.2507
                                   10     -0.3038
                                   11      0.2412
                                   12     -0.5247
```

Comments 'Twelve variants of color film coated with slightly different emulsions were presented to a panel of judges to assess their preferences' (Johnson and Leone 1964, p.276). Johnson and Leone calculated Kendall's coefficient of concordance ($W=0.762$) and found a real consensus of opinion regarding the 12 samples of film ($F=16.0$, $df_1=11$, $df_2=53$, significant at the 0.01 level). Our optimum solution indicates that the optimal values for the six judges are very similar, showing the high degree of consensus. Furthermore, this consensus is shown to account for 76% of the total variance. The correlation ratio drops from 0.3004 in the first solution to 0.0373 in the second solution, suggesting the dominance of the first dimension. It looks as though the second and remaining solutions simply pick up very minor individual differences and differences among the samples of film.

Example B.6 (same as example 8.2)

```
**********************************
*         DUAL SCALING OF        *
*   SUCCESSIVE CATEGORIES DATA   *
**********************************

TITLE OF STUDY :  EXAMPLE B.6 (SAME AS EXAMPLE 8.2)

***  INPUT MATRIX  ***

 3.    2.    1.
 2.    2.    2.
 2.    2.    1.
 3.    2.    1.
 3.    1.    1.
 3.    3.    1.
 2.    2.    2.
 3.    3.    2.
 2.    2.    1.
 3.    1.    1.

***  SUCCESSIVE CATEGORIES DATA  ***

-1.  -1.  -1.   1.   1.   1.
-1.   1.  -1.   1.  -1.   1.
-1.   1.  -1.   1.   1.   1.
-1.  -1.  -1.   1.   1.   1.
-1.  -1.   1.   1.   1.   1.
-1.  -1.  -1.  -1.   1.   1.
-1.   1.  -1.   1.  -1.   1.
-1.  -1.  -1.  -1.  -1.   1.
-1.   1.  -1.   1.   1.   1.
-1.  -1.   1.   1.   1.   1.

***  DESIGN MATRIX  ***

 1.   0.  -1.   0.   0.
 0.   1.  -1.   0.   0.
 1.   0.   0.  -1.   0.
 0.   1.   0.  -1.   0.
 1.   0.   0.   0.  -1.
 0.   1.   0.   0.  -1.

** NUMBER OF RESPONSES FOR ROWS

 12.  12.  12.  12.  12.  12.  12.  12.  12.  12.

** NUMBER OF RESPONSES FOR COLUMNS

 30.  30.  20.  20.  20.
```

```
***  TRANSFORMED DATA MATRIX E  ***

-1.  1.  2.  0. -2.
-3.  3.  0.  0.  0.
-1.  3.  0.  0. -2.
-1.  1.  2.  0. -2.
 1.  1.  2. -2. -2.
-1. -1.  2.  2. -2.
-3.  3.  0.  0.  0.
-3. -1.  2.  2.  0.
-1.  3.  0.  0. -2.
 1.  1.  2. -2. -2.


***  MATRIX C FOR EIGENEQUATION  ***

 0.0944
-0.0556 0.1167
-0.0272 0.0136 0.1000
-0.0408-0.0272 0.0000 0.0667
 0.0204-0.0612-0.0833 0.0167 0.1167

** TRACE OF C  =  0.4944


********************
** SOLUTION    1  **
********************

 SQUARED CORRELATION RATIO =  0.2438    ;  MAXIMUM CORRELATION =  0.4938

DELTA(TOTAL VARIANCE ACCOUNTED FOR BY  1  SOLUTION(S))  =   49.3  (PER CE


NUMBER OF ITERATIONS  =  16

OPTIMAL/LOCALLY OPTIMAL WEIGHT VECTORS (UNWEIGHTED,WEIGHTED BY ETA) FOR
              ** ROWS **                           ** COLUMNS **
        UNWEIGHTED      WEIGHTED               UNWEIGHTED       WEIGHTED

 1        1.1972         0.5912        1       -0.6938          -0.3426
 2        0.8719         0.4305        2        1.0283           0.5078
 3        1.1519         0.5688        3        1.1626           0.5741
 4        1.1972         0.5912        4       -0.1410          -0.0696
 5        1.0106         0.4990        5       -1.5233          -0.7522
 6        0.8025         0.3963
 7        0.8719         0.4305
 8        0.5225         0.2580
 9        1.1519         0.5688
10        1.0106         0.4990
```

Comments Since the second and remaining solutions are not likely to satisfy the a priori order of the category boundaries, DUAL1 provides only the optimal vectors for subjects and parameters (i.e., category boundaries and stimulus values), and the run is automatically terminated.

References

Abrham, J., and Arri, P.S. 1973. Approximation of separable functions in convex programming. *Canadian Journal of Operational Research and Information Processing*, *11*, 245–52

Anderson, D.H., Center, B., and Schifflers, E. 1975. Dual analysis of frequency data. Paper presented at the Spring Meeting of the Psychometric Society, Iowa City

Ayer, M., Brunk, H.D., Ewing, G.M., Reid, W.T., and Silverman, E. 1955. An empirical distribution function for sampling with incomplete information. *Annals of Mathematical Statistics*, *26*, 641–7

Baker, F.B. 1960. UNIVAC scientific computer program for scaling of psychological inventories by the method of reciprocal averages CPA 22. *Behavioral Science*, *5*, 268–9

— n.d. Quantifying qualitative variables by the method of reciprocal averages. Occasional paper No. 7, Laboratory of Experimental Design, Department of Educational Psychology, University of Wisconsin

Baker, F.B., and Martin, T.J. 1969. FORTAP: A FORTRAN test analysis package. *Educational and Psychological Measurement*, *29*, 159–64

Barlow, R.E., Bartholomew, D.J., Bremner, J.M., and Brunk, H.D. 1972. *Statistical Inference under Order Restrictions: The Theory and Application of Isotonic Regression*. New York: Wiley

Barlow, R.E., and Brunk, H.D. 1972. The isotonic regression problem and its dual. *Journal of the American Statistical Association*, *67*, 140–7.

Bartholomew, D.J. 1959. A test of homogeneity for ordered alternatives. *Biometrika*, *46*, 36–48

— A test of homogeneity of means under restricted alternatives. *Journal of the Royal Statistical Society*, Series B, *23*, 239–81.

Bartlett, M.S. 1947. Multivariate analysis. *Journal of the Royal Statistical Society, Supplement*, *9*, 176–90

— 1951. The goodness of fit of a single hypothetical discriminant function in the case of several groups. *Annals of Eugenics*, *16*, 199–214

— 1954. A note on the multiplying factors for various χ^2 approximations. *Journal of the Royal Statistical Society*, Series B, *16*, 296–8

Barton, D.E., and Mallows, C.L. 1961. The randomization bases of the problem of the amalgamation of weighted means. *Journal of the Royal Statistical Society*, Series B, *23*, 423–33

Bechtel, G.G., Tucker, L.R., and Chang, W. 1971. A scalar product model for the multidimensional scaling of choice. *Psychometrika*, *36*, 369–87

Benzécri, J.P. 1969. Statistical analysis as a tool to make patterns emerge from data. In S. Watanabe (ed.), *Methodologies of Pattern Recognition*, 35–74. New York: Academic Press

Benzécri, J.P., et al. 1973a. *L'Analyse des données*: *I*. *La Taxonomie*. Paris: Dunod

— 1973b. *L'Analyse des données*: *II*. *L'Analyse des correspondances*. Paris: Dunod

Bishop, Y.M.M. 1969. Full contigency tables, logits and split contingency tables. *Biometrics*, *25*, 383–400

Bishop, Y.M.M., Fienberg, S.E., and Holland, P.W. 1975. *Discrete Multivariate Analysis*: *Theory and Practice*. Cambridge, Mass: MIT Press

Block, H.D., and Marschak, J. 1960. Random orderings and stochastic theories of responses. In I. Olkin, S. Ghurye, W. Hoeffding, W. Madow and H. Mann (eds.), *Contributions to probability and statistics*. Stanford: Stanford University Press

Bock, R.D. 1956. The selection of judges for preference testing. *Psychometrika*, *21*, 349–66

— 1960. *Methods and Applications of Optimal Scaling*. The University of North Carolina Psychometric Laboratory Research Memorandum No. 25

— 1963. Programming univariate and multivariate analysis of variance. *Technometrics*, *5*, 95–117

— 1965. Analysis of qualitative data (unpublished)

— 1968. Estimating multinomial response relations. In R.C. Bose (ed.), *Contributions to Statistics and Probability*: *Essays in Memory of Sumarendra Nath Roy*. Chapel Hill: University of North Carolina Press

— 1975. *Multivariate Statistical Methods in Behavioral Sciences*. New York: McGraw-Hill

Bock, R.D., and Jones, L.V. 1968. *The Measurement and Prediction of Judgment and Choice*. San Francisco: Holden-Day

Bouroche, J.M., Saporta, G., and Tenenhaus, M. 1975. Generalized canonical analysis of qualitative data. Paper presented at the U.S.-Japan Seminar on Theory, Methods and Applications of Multidimensional Scaling and Related Techniques

Bradley, R.A. 1976. Science, statistics and paired comparisons. *Biometrics*, *32*, 213–32

Bradley, R.A., Katti, S.K., and Coons, I.J. 1962. Optimal scaling for ordered categories. *Psychometrika*, *27*, 355–74

Bradley, R.A., and Terry, M.E. 1952. Rank analysis of incomplete block designs. I. The method of paired comparisons. *Biometrika*, *39*, 324–45

Bradu, D., and Gabriel, K.R. 1978. The biplot as a diagnostic tool for models of two-way tables. *Technometrics*, *20*, 47–68

Burt, C. 1950. The factorial analysis of qualitative data. *British Journal of Psychology*, Statistical Section, *3*, 166–85

— 1953. Scale analysis and factor analysis. Comments on Dr. Guttman's paper. *British Journal of Statistical Psychology*, *6*, 5–23

Carroll, J.D. 1972. Individual differences and multidimensional scaling. In R.N. Shepard, A.K. Romney, and S.B. Nerlove (eds.), *Multidimensional Scaling: Theory and Applications in the Behavioral Sciences*, Volume I. New York: Seminar Press

Carroll, J.D., and Chang, J.J. 1964. Nonparametric multidimensional analysis of paired comparisons data. Paper presented at the joint meeting of the Psychometric and Psychonomic Societies, Niagara Falls

— 1968. How to use MDPREF, a computer program for multidimensional analysis of preference data. Unpublished report, Bell Laboratories, Murray Hill, NJ

Chan, D. 1978. Treatment of missing data by optimal scaling. Unpublished MA thesis, University of Toronto

Clemans, W.V. 1965. *An Analytical and Empirical Examination of Some Properties of Ipsative Measures*. Psychometric Monographs (Psychometric Society), No. 14

Cochran, W.G., and Cox, G.M. 1957. *Experimental Designs*. New York: Wiley

Cronbach, L.J. 1951. Coefficient alpha and the internal structure of tests. *Psychometrika*, *16*, 297–334

David, H.A. 1963. *The Method of Paired Comparisons*. New York: Hafner

de Leeuw, J. 1968, *Canonical Analysis of Relational Data*. Department of Data Theory, University of Leiden, Report RN 007-68

— 1973. *Canonical Analysis of Categorical Data*. Psychological Institute, University of Leiden, The Netherlands

— 1976. HOMALS. Paper presented at the Symposium on Optimal Scaling, the Spring Meeting of the Psychometric Society, Bell Laboratories, Murray Hill, NJ

de Leeuw, J., Young, F.W., and Takane, Y. 1976. Additive structure in qualitative data: an alternating least squares method with optimal scaling features. *Psychometrika*, *41*, 471–504

Dempster, A.P. 1969. *Elements of Continuous Multivariate Analysis*. Reading, Mass.: Addison-Wesley

Dyke, G.V., and Patterson, H.D. 1952. Analysis of factorial arrangements when the data are proportions. *Biometrics*, *8*, 1–12

Eckart, C., and Young, G. 1936. The approximation of one matrix by another of lower rank. *Psychometrika*, *1*, 211–18

Edgerton, H.A., and Kolbe, L.E. 1936. The method of minimum variation for the combination of criteria. *Psychometrika*, *1*, 183–7

Escofier-Cordier, B. 1969. L'analyse factorielle des correspondances. *Bureau Universitaire de Recherche operationelle, Cahiers, Série Recherche* (université de Paris), *13*, 25–9

Faddeev, D.K., and Faddeeva, V.N. 1963. *Computational Methods of Linear Algebra*. San Francisco: W.H. Freeman and Company

Fechner, G.T. 1860. *Elemente der Psychophysik*. Leipzig: Breitkopf and Hartel

Fienberg, S.E. 1970. The analysis of multi-dimensional contingency tables. *Ecology*, *51*, 419–33

Fisher, R.A. 1935. *The Design of Experiments*. London: Oliver and Boyd

— 1940. The precision of discriminant functions. *Annals of Eugenics*, *10*, 422–9
— 1948. *Statistical Methods for Research Workers*, 10th ed. London: Oliver and Boyd
Fox, L. 1964. *An Introduction to Numerical Linear Algebra*. Oxford: Oxford University Press
Frazer, R.A., Duncan, W.J., and Collar, A.R. 1938. *Elementary Matrices*. London: Cambridge University Press
Gabriel, K.R. 1971. The biplot graphical display of matrices with applications to principal component analysis. *Biometrika*, *58*, 453–67
— 1972. Analysis of meteorological data by means of canonical decomposition and biplots. *Journal of Applied Meteorology*, *11*, 1071–7
— 1978a. Least squares approximation of matrices by additive and multiplicative models. *Journal of the Royal Statistical Society*, Series B, *40*, 186–96
— 1978b. The complex correlational biplot. In S. Shye (ed.), *Theory Construction and Data Analysis*. Tel Aviv: Jossey-Bass
Gebhardt, F. 1970. An algorithm for monotone regression with one or more independent variables. *Biometrika*, *57*, 263–71
Good, I.J. 1969. Some applications of the singular decomposition of a matrix. *Technometrics*, *11*, 823–31
Goodman, L.A. 1971. The analysis of multidimensional contingency tables: Stepwise procedures and direct estimation methods for building models for multiple classification. *Technometrics*, *13*, 33–62
Gower, J.C. 1966. Some distance properties of latent root and vector methods in multivariate analysis. *Biometrika*, *53* 325–38
Graybill, F.A. 1961. *An Introduction to Linear Statistical Models*, Volume I. New York: McGraw-Hill
— 1969. *Matrix Algebra with Statistical Applications*. Belmont, Cal.: Wadsworth
Green, B.F., Jr. 1979. Personal communication
Greenacre, M.J. 1978a. Quelques méthodes objectives de représentation graphique d'un tableau de données. Thèse de doctorat, 3e cycle, l'université Pierre et Marie Curie, Paris
— 1978b. *Some Objective Methods of Graphical Display of a Data Matrix* (the English translation of the 1978 doctoral thesis). Special Report, Department of Statistics and Operations Research, University of South Africa (November)
Greenacre, M.J., and Degos, L. 1977. Correspondence analysis of HLA gene frequency data from 125 population samples. *American Journal of Human Genetics*, *29*, 60–75
Grizzle, J.E., Starmer, C.F. and Koch, G.G. 1969. Analysis of categorical data by linear models. *Biometrics*, *25*, 489–504
Guilford, J.P. 1954. *Psychometric Methods*, 2nd ed. New York: McGraw-Hill
Guttman, L. 1941. The quantification of a class of attributes: A theory and method of scale construction. In The Committee on Social Adjustment (ed.), *The*

Prediction of Personal Adjustment, 319–48. New York: Social Science Research Council
— 1946. An approach for quantifying paired comparisons and rank order. *Annals of Mathematical Statistics, 17*, 144–63
— 1950. The principal components of scale analysis. In S.A. Stouffer, L. Guttman, E.A. Suchman, P.F. Lazarsfeld, S.A. Star, and J.A. Clausen, *Measurement and Prediction*. Princeton: Princeton University Press
— 1953. A note on Sir Cyril Burt's "Factorial analysis of qualitative data." *British Journal of Statistical Psychology, 6*, 1–4
— 1955. An additive metric from all the principal components of a perfect scale. *The British Journal of Statistical Psychology, 8*, 17–24
— 1959. Metricizing rank-ordered and unordered data for a linear factor analysis. *Sankhya: The Indian Journal of Statistics, 21*, 257–68
— 1971. Measurement as structural theory. *Psychometrika, 36*, 329–47
Haberman, S.J. 1974. *The Analysis of Frequency Data*. Chicago: University of Chicago Press
Hadley, G. 1962. *Linear Programming*. Reading, Mass.: Addison-Wesley
— 1964. *Nonlinear and Dynamic Programming*. Reading, Mass.: Addison-Wesley
Hammer, A.G. 1971. *Elementary Matrix Algebra for Psychologists and Social Scientists*. London: Pergamon Press
Hayashi, C. 1950. On the quantification of qualitative data from the mathematico-statistical point of view. *Annals of the Institute of Statistical Mathematics, 2*, 35–47
— 1952. On the prediction of phenomena from qualitative data and the quantification of qualitative data from the mathematico-statistical point of view. *Annals of the Institute of Statistical Mathematics, 3*, 69–98
— 1954. Multidimensional quantification–with the applications to analysis of social phenomena. *Annals of the Institute of Statistical Mathematics, 5*, 121–43
— 1964. Multidimensional quantification of the data obtained by the method of paired comparison. *Annals of the Institute of Statistical Mathematics, the Twentieth Anniversary Volume, 16*, 231–45
— 1967. Note on multidimensional quantification of data obtained by paired comparison. *Annals of the Institute of Statistical Mathematics, 19*, 363–5
— 1968. One dimensional quantification and multidimensional quantification. *Annals of the Japan Association for Philosophy of Science, 3*, 115–20
— 1972. Two dimensional quantification based on the measure of dissimilarity among three elements. *Annals of the Institute of Statistical Mathematics, 24*, 251–7
Hayashi, C., Higuchi, I., and Komazawa, T. 1970. *Jôhô shori to tôkei sûri*. Tokyo: Sangyo Tosho (in Japanese)
Hayashi, C., and Suzuki, T. 1975. Quantitative approach to a cross-societal research; a comparative study of Janpanese character. Part II. *Annals of the Institute of Statistical Mathematics, 27*, 1–32
Healy, M.J.R., and Goldstein, H. 1976. An approach to the scaling of categorized attributes. *Biometrika, 63*, 219–29

Hill, M.O. 1973. Reciprocal averaging: an eigenvector method of ordination. *Journal of Ecology*, *61*, 237–49

— 1974. Correspondence analysis: a neglected multivariate method. *Applied Statistics*, *23*, 340–54

Hirschfeld, H.O. 1935. A connection between correlation and contingency. *Cambridge Philosophical Society Proceedings*, *31*, 520–4

Hoaglin, D.C., and Welsch, R.E. 1978. The hat matrix in regression and ANOVA. *The American Statistician*, *32*, 17–22

Hollingshead, A.B. 1949. *Elmtown's Youth*: *The Impact of Social Classes on Adolescents*. New York: Wiley

Hope, K. 1968. *Methods of Multivariate Analysis*. London: University of London Press

Horst, P. 1935. Measuring complex attitudes. *Journal of Social Psychology*, *6*, 369–74

— 1936. Obtaining a composite measure from a number of different measures of the same attribute. *Psychometrika*, *1*, 53–60

— 1963. *Matrix Algebra for Social Scientists*. New York: Holt, Rinehart & Winston

Hotelling, H. 1933. Analysis of a complex of statistical variables into principal components. *Journal of Educational Psychology*, *24*, 417–41, 498–520

— 1936. Relations between two sets of variates. *Biometrika*, *28*, 321–77

Inukai, Y. 1972. Optimal versus partially optimal scaling of polychotomous items. Master's thesis, University of Toronto

Ishizuka, T. 1976. Weighting multicategory data to fit the common factor model. Master's thesis, University of Toronto

Iwatsubo, S. 1971. An analysis of multidimensional qualitative data structure: quantification by multiple correlation coefficient. *Bulletin of the Electrotechnical Laboratory*, *35*, 622–32 (in Japanese)

— 1974. Two classification techniques of 3-way discrete data–quantification by means of correlation ratio and three-dimensional correlation coefficient. *The Japanese Journal of Behaviormetrics*, *2*, 54–65 (in Japanese, with English abstract)

— 1975. Optimal scaling methods for multi-way qualitative data. Paper presented at the U.S.-Japan Seminar on Theory, Methods and Applications of Multidimensional Scaling and Related Techniques

— 1976. Shizen gengo no imi-kôzô to saiteki shakudohô (Semantic structure of a natural language and optimal scaling). *Mathematical Sciences*, *152*, 47–53

— 1978. An optimal scoring method for detecting clusters and interpretations from multi-way qualitative data. *Behaviormetrika*, No. 5, 1–22.

Johnson, N.L., and Leone, F.C. 1964. *Statistics and Experimental Design*: *In Engineering and the Physical Sciences*, Volume I. New York: Wiley

Johnson, P.O. 1950. The quantification of qualitative data in discriminant analysis. *Journal of the American Statistical Association*, *45*, 65–76

Johnson, R.M. 1963. On a theorem stated by Eckart and Young. *Psychometrika*, *28*, 259–63

— 1975. A simple method for pairwise monotone regression. *Psychometrika*, *40*, 163–8

Kendall, M.G. 1957. *A Course in Multivariate Analysis*. London: Charles Griffin and Company

Kendall, M.G., and Stuart, A. 1961. *The Advanced Theory of Statistics*, Volume II. London: Griffin

Kettenring, J.R. 1971. Canonical analysis of several sets of variables. *Biometrika*, *58*, 433–51

Krishnan, T. 1973. On linear combinations of binary item scores. *Psychometrika*, *38*, 291–304

Kruskal, J.B. 1964a. Multidimensional scaling by optimizing goodness of fit to a nonmetric hypothesis. *Psychometrika*, *29*, 1–28

— 1964b. Nonmetric multidimensional scaling: a numerical method. *Psychometrika*, *29*, 115–29

— 1965. Analysis of factorial experiments by estimating monotone transformations of the data. *Journal of the Royal Statistical Society*, Series B, *27*, 251–63

— 1971. Monotone regression: continuity and differentiability properties. *Psychometrika*, *36*, 57–62

Kruskal, J.B., and Carmone, F. 1969. MONANOVA, a FORTRAN IV program for monotone analysis of variance (nonmetric analysis of factorial experiments). *Behavioral Science*, *14*, 165–6 (CPA 319)

Kruskal, J.B., and Carroll, J.D. 1969. Geometrical models and badness-of-fit functions. In P.R. Krishnaiah (ed.), *Multivariate Analysis*, Volume II. New York: Academic Press

Kyogoku, J. 1967. On a method of quantifying *N*-way frequency tables. *Proceedings of the Institute of Statistical Mathematics*, *15*, 140–60 (in Japanese)

— 1971. On a method of quantifying *N*-way frequency tables. Supplementary notes. *Proceedings of the Institute of Statistical Mathematics*, *18*, 111–15

Lancaster, H.O. 1953. A reconciliation of χ^2, considered from metrical and enumerative aspects. *Sankhya*, *13*, 1–10

— 1958. The structure of bivariate distributions. *Annals of Mathematical Statistics*, *29*, 719–36

— 1963. Canonical correlations and partitions of χ^2. *Quarterly Journal of Mathematics*, *14*, 220–4

— 1966. Kolmogorov's remark on the Hotelling canonical correlations. *Biometrika*, *53*, 585–8

— 1969. *The Chi-Squared Distribution*. New York: Wiley

Laska, E., Meisner, M., Siegel, C., Takeuchi, K., and Wanderling, J. 1971. Quantifying pain scores–an analytic approach. *Committee on Problems of Drug Dependence*, *1*, 107–28

Leong, K., Nishisato, S., and Wolf, R.G. 1980. *OPSCALE, a FORTRAN Program for Analysis of Categorical Data by Dual Scaling*. Chicago: International Educational Services (in preparation)

Likert, R. 1932. A technique for the measurement of attitudes. *Archives of Psychology*, No. 140, 44–53

Lingoes, J.C. 1963. Multivariate analysis of contingencies: an IBM 7079 program for analyzing metric/nonmetric or linear/nonlinear data. *Computational Report, 2*, 1–24 (University of Michigan, Computing Center)
— 1964. Simultaneous linear regression: an IBM 7090 program for analyzing metric/nonmetric or linear/nonlinear data. *Behavioral Science, 9*, 87–8
— 1968. The mutlivariate analysis of qualitative data. *Multivariate Behavioral Research, 3*, 61–94
— 1973. *The Guttman-Lingoes Nonmetric Program Series*. Ann Arbor: Mathesis Press
— 1977. With contributions by I. Borg, J. de Leeuw, L. Guttman, W. Heiser, R.W. Lissitz, E.E. Roskam, and P.H. Schönemann. *Geometric Representations of Relational Data: Readings in Multidimensional Scaling*. Ann Arbor: Mathesis Press
Lingoes, J.C., and Raskam, E.E. 1973. A mathematical and empirical analysis of two multidimensional scaling algorithms. *Psychometrika Monograph Supplement*. Monograph No. 19, *38*, No. 4, Pt. 2
Lord, F.M. 1958. Some relations between Guttman's principal components of scale analysis and other psychometric theory. *Psychometrika, 23*, 291–6
Lubin, A. 1950. Linear and non-linear discriminating functions. *British Journal of Statistical Psychology, 3*, 90–104
Luce, R.D. 1959. *Individual Choice Behavior*. New York: Wiley
Maruyama, K. 1968. Pattern analysis with scale classification. *Japanese Psychological Research, 10*, 78–94
Maung, K. 1941. Measurement of association in contingency table with special reference to the pigmentation of hair and eye colours of Scottish school children. *Annals of Eugenics, 11*, 189–223
Maxwell, A.E. 1973. Tests of association in terms of matrix algebra. *British Journal of Mathematical and Statistical Psychology, 26*, 155–66
McDonald, R.P. 1968. A unified treatment of the weighting problem. *Psychometrika, 33*, 351–81
— 1976. A note on monotone polygons fitted to bivariate data. *Psychometrika, 41*, 543–6
McDonald, R.P., Ishizuka, T., and Nishisato, S. 1977. Weighting multicategory data to fit the common factor model. Paper presented at the Psychometric Society Annual Meeting, Chapel Hill, North Carolina
McDonald, R.P., Torii, Y., and Nishisato, S. 1979. Some results on proper eigenvalues and eigenvectors with applications to scaling. *Psychometrika, 44*, 211–27
McKeon, J.J. 1966. *Canonical Analysis: Some Relations between Canonical Correlation, Factor Analysis, Discriminant Function Analysis, and Scaling Theory*. Psychometric Monograph (Psychometric Society), No. 13
Miles, R.E. 1959. The complete amalgamation into blocks, by weighted means, of a finite set of real numbers. *Biometrika, 46*, 317–27
Mosier, C.I. 1946. Machine methods in scaling by reciprocal averages. *Proceedings, Research Forum*, 35–9. Endicath, NY: International Business Corporation

Mosteller, F. 1949. *A Theory of Scalogram Analysis, Using Noncumulative Types of Items: A New Approach to Thrustone's Method of Scaling Attitudes*. Report No. 9, Laboratory of Social Relations, Harvard University

Napior, D. 1972. Nonmetric multidimensional techniques for summated ratings. In R.N. Shepard, A.K. Romney, and S.B. Nerlove (eds.), *Multidimensional Scaling: Theory and Applications in the Behavioral Sciences*, Volume I. New York: Seminar Press

Nishisato, S. 1966. Minimum entropy clustering of test items. PHD dissertation, University of North Carolina at Chapel Hill

— 1970. Probability estimation of dichotomous response patterns by logistic fractional factorial representation. *Japanese Psychological Research*, *12*, 87–95

— 1971a. Analysis of variance through optimal scaling. *Proceedings of the First Canadian Conference on Applied Statistics*, 306–16. Montreal: Sir George Williams University Press

— 1971b. Transform factor analysis: a sketchy presentation of a general approach. *Japanese Psychological Research*, *13*, 155–66

— 1972a. Analysis of variance of categorical data through selective scaling. *Proceedings of the 20th International Congress of Psychology*, 279. Tokyo

— 1972b. *Optimal Scaling and Its Generalizations. I. Methods*. Measurement and Evaluation of Categorical Data Technical Report No. 1, Department of Measurement and Evaluation, the Ontario Institute for Studies in Education, Toronto

— 1973a. *Optimal Scaling and Its Generalizations. II. Applications*. Measurement and Evaluation of Categorical Data Technical Report No. 2, Department of Measurement and Evaluation, the Ontario Institute for Studies in Education, Toronto

— 1973b. Optimal Scaling of partially ordered categories. Paper presented at the Spring Meeting of the Psychometric Society, Chicago

— 1975. *Ôyô shinri shakudohô: Shitsuteki data no bunseki to kaishaku* (*Applied Psychological Scaling: Analysis and Interpretation of Qualitative Data*). Tokyo: Seishin Shobo Publishers, Ltd.

— 1976. *Optimal Scaling as Applied to Different Forms of Data*. Measurement and Evaluation of Categorical Data Technical Report No. 4, Department of Measurement and Evaluation, the Ontario Institute for Studies in Education, Toronto. Also presented at the Symposium on Optimal Scaling, the Spring Meeting of the Psychometric Society, Bell Laboratories, Murray Hill, NJ

— 1977. Recent developments in scaling and related area: a bibliographic review (1). *Japanese Journal of Behaviormetrics*, *4*, 74–95

— 1978a. Optimal scaling of paired comparison and rank order data: an alternative to Guttman's formulation. *Psychometrika*, *43*, 263–71

— 1978b. Dual scaling of successive categories data. Paper presented at the first joint meeting of the Psychometric Society and the Society for Mathematical Psychology, McMaster University, Hamilton

— 1978c. Errata to 'Optimal scaling of paired comparison and rank order data: an alternative to Guttman's formulation.' *Psychometrika*, *43*, 587

— 1978d. *Multidimensional Scaling: A Historical Sketch and Bibliography*. Department of Measurement, Evaluation and Computer Applications, the Ontario Institute for Studies in Education, Technical Report

— 1979a. Dual (optimal) scaling and its history *Mathematical Sciences*, No. 190, 76–83 (in Japanese)

— 1979b. *An Introduction to Dual Scaling*. Measurement and Evaluation of Categorical Data Technical Report No. 5, Department of Measurement, Evaluation and Computer Applications, the Ontario Institute for Studies in Education, Toronto

— 1979c. Dual scaling and its variants. Chap. 1 in R.E. Traub (ed.), *Analysis of Test Data* (New Directions for Testing and Measurement Quarterly Sourcebook). San Francisco: Jossey-Bass Inc.

Nishisato, S., and Arri, P.S. 1975. Nonlinear programming approach to optimal scaling of partially ordered categories. *Psychometrika*, *40*, 525–48

Nishisato, S., and Inukai, Y. 1972. Partially optimal scaling of items with ordered categories. *Japanese Psychological Research*, *14*, 109–19

Nishisato, S., and Leong, K. 1975. *OPSCAL: A FORTRAN IV Program for Analysis of Qualitative Data by Optimal Scaling*. Measurement and Evaluation of Categorical Data Technical Report No. 3., Informal Publication, Department of Measurement and Evaluation, the Ontario Institute for Studies in Education, Toronto

Nishisato, S., and Levine, R. 1975. Optimal scaling of omitted responses. Paper presented at the Spring Meeting of the Psychometric Society, Iowa City

Nishisato, S., and Sheu, W. 1979a. Dual scaling of multi-way classification data. Paper presented at the Annual Meeting of the Psychometric Society, Monterey

— 1979b. Piecewise method of reciprocal averages for dual scaling. Unpublished paper

Nishisato, S., and Torii, Y. 1971. Multifactor models of optimal scaling. Paper presented at the Spring Meeting of the Psychometric Society, St. Louis

Pearson, K. 1901a. Mathematical contributions to the theory of evolution. VII. On the correlation of characters not quantitatively measurable. *Philosophical Transactions*, Series A, *195*, 1–47

— 1901b. On lines and planes of closest fit to systems of points in space. *Philosophical Magazine and Journal of Science*, Series 6, *2*, 559–72

Poon, W.P. 1977. Transformations of data matrices in optimal scaling. Unpublished MA thesis, University of Toronto

Ramsay, J.O. 1975. Solving implicit equations in psychometric data analysis. *Psychometrika*, *40*, 337–60

Rao, C.R. 1952. *Advanced Statistical Methods in Biometric Research*. New York: Wiley

— 1965. *Linear Statistical Inference and Its Applications*. New York: Wiley

Rao, C.R., and Mitra, B.K. 1971. *Generalized Inverse of Matrices and Its Applications*. New York: Wiley

Richardson, M., and Kuder, G.F. 1933. Making a rating scale that measures. *Personnel Journal*, *12*, 36–40

Saito, T. 1973. Quantification of categorical data by using the generalized variance. *Soken Kiyo*, Nippon UNIVAC Sogo Kenkyu-sho, Inc., 61–80

Saito, T., Ogawa, S., and Nojima, E. 1972. Data analysis (1): summary review article on quantification theories. *Soken Kiyo*, Nippon UNIVAC Sogo Kenkyusho, Inc., *2*, No. 1, 23–140 (in Japanese)

Saporta, G. 1975. Liaisons entre plusieurs ensembles de variables et codage de données qualitatives. L'université Pierre et Marie Curie (Paris VI) doctoral thesis

— 1976. Discriminant analysis when all the variables are nominal: a stepwise method. Paper presented at the Symposium on Optimal Scaling, the Spring Meeting of the Psychometric Society, Bell Laboratories, Murray Hill, NJ

Schmidt, E. 1907. Zür Theorie der linearen und nichtlinearen Integralgleichungen. Erster Teil. Entwicklung willkürlicher Funktionen nach Systemen vorgeschriebener. *Mathematische Annalen*, *63*, 433–76

Shepard, R.N. 1962a. The analysis of proximities: multidimensional scaling with an unknown distance function. I. *Psychometrika*, *27*, 125–40

— 1962b. The analysis of proximities: multidimensional scaling with an unknown distance function. II. *Psychometrika*, *27*, 219–45

Shiba, S. 1965a. A method for scoring multicategory items. *Japenese Psychological Research*, *7*, 75–9

— 1965b. The generalized method for principal components of scale analysis. *Japanese Psychological Research*, *7*, 163–5

Shine, L.C. II. 1972. A note on McDonald's generalization of principal component analysis. *Psychometrika*, *37*, 9–101

Slater, P. 1960a. The analysis of personal preferences. *British Journal of Statistical Psychology*, *3*, 119–35

— 1960b. Canonical analysis of discriminance. In H.J. Eysenck (ed.), *Experiments in Psychology*, *2*, 256–70. London: Routledge and Kegan Paul

Stewart, F.M. 1973. *Introduction to Linear Algebra*. Princeton: Van Nostrand

Takane, Y., Young, F.W., and de Leeuw, J. 1977. Nonmetric individual differences multidimensional scaling: an alternating least squares method with optimal scaling features. *Psychometrika*, *42*, 7–67

Takeuchi, K., and Yanai, H. 1972. *Tahenryo kaiseki no kiso–senkei kûkan eno shaei ni yoru hôhô*. Tokyo: Tôyôkeizai Shimpôsha (in Japanese)

Teil, H. 1975. Correspondence factor analysis: an outline of its method. *Mathematical Geology*, *7*, 3–12

Thompson, W.A. Jr. 1962. The problem of negative estimates of variance components. *Annals of Mathematical Statistics*, *33*, 273–89

Thurstone, L.L. 1927. A law of comparative judgment. *Psychological Review*, *34*, 273–86

— 1931. Rank order as a psychophysical method. *Journal of Experimental Psychology*, *14*, 187–201

Thurstone, L.L., and Chave, E.J. 1929. *The Measurement of Attitude*. Chicago: University of Chicago Press

Torgerson, W.S. 1958. *Theory and Methods of Scaling*. New York: Wiley

Torii, Y. 1979. Some generalizations of optimal scaling. PHD dissertation, University of Toronto

Tucker, L.R. 1960. Intra-individual and inter-individual multidimensionality. In H. Gulliksen and S. Messick (eds.), *Psychological Scaling*. New York: Wiley

van Rijckevorsel, J., and de Leeuw, J. 1978. *An Outline to HOMALS*-1. Department of Data Theory, Faculty of Social Sciences, University of Leiden, The Netherlands

Whittaker, R.H. 1967. Gradient analysis of vegetation. *Biological Reviews*, *42*, 207–64

Wilks, S.S. 1938. Weighting systems for linear functions of correlated variables when there is no dependent variable. *Psychometrika*, *3*, 23–40

Williams, E.J. 1952. Use of scores for the analysis of association in contingency tables. *Biometrika*, *39*, 274–89

Yale, P.B. 1968. *Geometry and Symmetry*. San Francisco: Holden-Day

Yates, F. 1948. The analysis of contingency tables with groupings based on quantitative characters. *Biometrika*, *35*, 176–81

Yoshizawa, T. 1975. Models for quantification techniques in multiple contingency tables - the theoretical approach. *Japanese Journal of Behaviormetrics*, *3*, 1–11 (in Japanese, with English abstract)

Young, F.W. 1976. Optimal scaling with a variety of models. Paper presented at the Symposium on Optimal Scaling, the Spring Meeting of the Psychometric Society, Bell Laboratories, Murray Hill, NJ

Young, F.W., de Leeuw, J., and Takane, Y. 1976a. Quantifying qualitative data. The L.L. Thurstone Psychometric Laboratory Report, No. 149, University of North Carolina, July 1976 - revised July 1978. Also in H. Feger (ed.), *Similarity of Choice* (Academic Press 1979)

— 1976b. Regression with qualitative and quantitative variables: an alternating least squares method with optimal scaling features. *Psychometrika*, *41*, 505–29

Young, F.W., and Torgerson, W.S. 1967. TORSCA, a FORTRAN IV program for Shepard-Kruskal multidimensional scaling analysis. *Behavioral Science*, *12*, 498

Index